'Do you believe the earth is warming _committed environmental analyst with the unusual gift of following scientific evidence ruthlessly wherever it may lead. Taylor has done groundbreaking work on issues ranging from ocean pollution and biodiversity through renewable energy. Now he turns his relentless searchlight on climate change. His work has the ring of passion and the clarity of intellectual honesty. We can be certain his conclusions are the product of a fearless, unbiased, and intelligent intellectual journey by a remarkable mind, all the marks of genuine science. Taylor challenges us to look beyond our biases to whatever conclusions the evidence may justify. Believers in global warming such as myself may not find comfort here, but they will without question find a clear challenge to examine all the evidence objectively. At the very least, Taylor raises issues and questions that must be addressed conclusively before global warming can be genuinely regarded as "truth", inconvenient or otherwise. This book is a must-read for everyone on all sides of the climate change issue.'_

— W. Jackson Davis, professor emeritus, University of California, and author of the first draft of the Kyoto Protocol

PETER TAYLOR is a science analyst and policy advisor with over 30 years experience as a consultant to environmental NGOs, government departments and agencies, intergovernmental bodies, the European Commission, the European Parliament and the UN. His range of expertise stretches from pollution and accident risk from nuclear operations, chemical pollution of the oceans and atmosphere, wildlife ecology and conservation, to renewable energy strategies and climate change.

In addition to his advisory work, he has lectured widely in universities and institutes in Britain, Germany, Sweden, the USA and Japan, influencing the thinking and careers of several leading scientists. After graduating in Natural Sciences at Oxford University (and later returning to study Social Anthropology) he set up and directed the Oxford-based Political Ecology Research Group and pioneered the development of critical scientific review on environmental issues, both in the examination of official policy and in its use as a campaigning tool for legal reforms such as the precautionary principle (he was a leading advocate of this at UN conventions). He has sat on several government commissions and research advisory bodies. From 2000 to 2003 he was a member of the UK Government's National Advisory Group for Community Renewable Energy.

In 2000 Taylor set up a new group, Ethos (*www.ethos-uk.com*), to develop educational programmes using leading-edge computer techniques for visualizing change in the rural landscape. After an extensive review of conservation practice for the British Association of Nature Conservationists, he published *Beyond Conservation: a wildland strategy* in the spring of 2005, and helped found and organize the Wildland Network for conservationists, foresters and land managers. He is a leading advocate of *rewilding* policies in nature conservation involving minimal human intervention and the reintroduction of exterminated large mammals and sits on an advisory group for the management of National Trust and Forestry Commission land in the Lake District.

At some time he has been a member of the following professional institutes (reflecting his work and interests at different times): the Institute of Biology, the British Ecological Society, the Society for Radiological Protection, and the International Union of Radio-ecologists (at times on the editorial board of the *Journal of Radioecology*).

During his work on marine pollution and hazardous industries he both critically assessed and utilized computer models of complex marine and atmospheric pathways. He is ideally qualified to review and synthesize climate science across many disciplines, taking a broad and independent view with an unparalleled insight into the workings of science and the evolution of policy behind the scenes of public debate and thus to make recommendations that respect the essentials of social as well as environmental sustainability.

CHILL

A REASSESSMENT OF GLOBAL WARMING THEORY

DOES CLIMATE CHANGE MEAN THE WORLD IS
COOLING, AND IF SO WHAT SHOULD WE DO ABOUT IT?

PETER TAYLOR

CLAIRVIEW

Clairview Books
Hillside House, The Square
Forest Row, East Sussex
RH18 5ES

www.clairviewbooks.com

Published by Clairview 2009

A catalogue record for this book is available from the British Library

ISBN 978 1 905570 19 5

Cover by Andrew Morgan Design; photo by Paul Bodea, fotolia.com
Typeset by DP Photosetting, Neath, West Glamorgan
Printed and bound in Malta by Gutenberg Press Ltd.

Mixed Sources
Product group from well-managed
forests, and other controlled sources
www.fsc.org Cert no. TT-CoC-002424
© 1996 Forest Stewardship Council

FSC

The paper used for this book is FSC-certified and
totally chlorine-free. FSC (the Forest Stewardship
Council) is an international network to promote
responsible management of the world's forests.

Contents

Note on References

It is always a difficult choice between making science accessible for the non-specialist reader and providing supporting evidence for the argument. In this book I have adopted the standard form of scientific referencing but have aimed to explain scientific concepts in laypersons' terms. Contrary to what is portrayed in the media, significant scientific dissent exists in relation to the orthodox view of global warming and this will allow the scientifically literate readers to track the science behind the disagreements.

In the interests of this simplicity some supporting material is referenced to the website at *www.ethos-uk.com* where it can be freely read or downloaded. This contains much visual material that is clearer in colour. It is here referenced in the text as Appendices, e.g. Ethos A4.2. Wherever possible I have given sources for the diagrams and for quotes (in the endnotes after each chapter). If the references are central to my own argument they are in the list at the back of this book (pp. 380–94) and are bracketed in bold type in the text ( ); when used by quoted authors and also listed then they are bracketed in normal type as [ ]; and if not listed here but can be found in the quoted document, then they are bracketed as { }.

I welcome discussion on science and policy and can be contacted via the Ethos website where there is an email address.

Acknowledgements

The research for this book was not financed by any organization. It has been made possible by the unstinting support of a few individuals. I am particularly indebted to Matt and Owen Taylor for computer support and occasional financial aid, and to Luan Cedrowen for help with the graphics. Sonja de Winter supported several weeks of London-based library work. Sue Quatermass, Robin Stuart, Richard Waterborn and Louis Standens provided much needed encouragement, reading of manuscripts and practical support. Rick Minter, editor of *ECOS*, helped with many Internet links and maintained a forum for discussion in the pages of that journal, despite some opposition, and he and Stanley Owen advised on how radical I should be on social and philosophical perspectives in a generally repressive climate for discussion.

Although I have benefited from talking to a few specialists in various fields, and from a short visit to the Hadley Centre, where Matthew Palmer and John Kennedy found time to guide me through various data, my conclusions are my own. However, the general direction of this thesis with regard to an alternative explanation of 'global warming' is shared by many hundreds of scientists worldwide. When I started this research, I did not know this and felt very isolated. In this final year of work, however, much has changed and I am indebted to the respected science-blog of Anthony Watt for news of new papers and breakthroughs in the struggle against a closed-minded orthodoxy.

I am grateful to Sevak Gulbekian, my publisher, for the invitation to make my work more accessible and for the title, and to Eileen Lloyd for her meticulous copy-editing. I am also grateful to the sun, however impervious it may be, for turning down at this time, otherwise nobody would listen!

Overview

This book is divided into two parts: the science and the politics.

The term 'global warming' purports to be a scientifically defined and universally accepted 'fact', as if an incontrovertible signal has been identified among the variable background of the global swings in temperature. This is not so. In Part One, I work through the science base that underlies global warming and pay particular reference to the assessments performed by the UN's Intergovernmental Panel on Climate Change (IPCC), which claims a consensus of the world's experts. I show that this consensus is false and has never existed even within the UN expert groups and on some very important science issues related to the power of prediction and use of computer models. The apparent consensus is a construct of the summary material aimed at policy makers. In this process, important areas of disagreement are covered over. In Chapters 1 and 2, I show that there was particularly never a consensus on what has caused the apparently unusual signal and a substantial amount of evidence for mainly natural causation.

I go on to deal in Chapters 3 and 4 with satellite data that contradicts global warming theory but which received only the briefest of review and comment by IPCC. The data shows clearly that without the warm period from 1980 to 2005 there would be no case for unusual warming, and that during this period global cloud cover changed and allowed significantly more warming sunlight to reach the surface of the earth. In Chapter 5, I show how the oceans have absorbed and redistributed this extra heat with past cycles of warming and cooling that oscillate (see-saw from one state to another) through all the major ocean basins. In the latter decades of the twentieth century four of these cycles peaked at the same time, and at the outset of the twenty-first century these cycles reversed and the oceans have begun to cool. In Chapter 6, I deal with the special case of the Arctic 'meltdown' which has become the touchstone of warming theory and show that such Arctic meltdowns have happened before and can be explained by changes in cloud and ocean currents which tie in to long-term cycles.

In orthodox climate models natural variability is regarded as essentially random, but there is a considerable body of evidence for natural cycles. In

Chapter 7, I look at the role of the sun and its potential influence on these cycles. There is a large body of evidence that the sun may drive ocean oscillations and several theories of a mechanism. There is no consensus on the issue, and this is evident even within the IPCC documents. In Chapter 8, I discuss the most controversial of these potential driving mechanisms in a review of the science related to cosmic rays, cloud seeding and sunspot cycles. Contrary to many media reports, this controversy is alive and not settled. In Chapters 9 and 10 I formulate an alternative explanation of 'global warming' based upon these natural cycles, arguing that the contribution from human activities is small, of the order of 10–20%, and that predictions are uncertain, with some strong evidence suggesting the twenty-first century will see a prolonged period of global cooling.

In Part Two, I examine the interface between this unsettled science and those charged with developing an appropriate practical response – the policy-makers. In Chapter 11, I show how a scientific orthodoxy built upon a virtual reality model of the earth's climate system has developed, and in Chapter 12, I show how a majority has overruled dissent among other scientists and presented itself as a consensus of the world's experts. In Chapter 13, I will argue that the media and environmental campaign groups have colluded with this process of simplification and generated a dangerously naive message that now drives government action worldwide. In Chapters 14 and 15, I tackle the question of what to do in the light of shifting science, uncertain prediction and a massive prior commitment of resources to an ultimately ineffective goal of preventing climate change. The current remedies proposed are likely to prove more damaging than the ailment they seek to address. There is no real effort devoted to adaptation and the creation of human support systems that are *resilient* to whatever climate change the future will bring. I will outline sensible actions of a 'no regrets' strategy that will have a positive effect under any future. Finally, in Chapter 16, I take a philosophical perspective on the kind of thinking that led to these past errors. We are faced with an almost apocalyptic vision generated as a virtual reality computer simulation of the future. The language of campaigning draws from the psychology of warfare and military mobilization. This apocalyptic vision is used by some as a spur to repentance – the reform, in this case, of 'business as usual' or the capitalist, free market system. Critics and sceptics often fall into the camp of defenders of that free market and I will argue that we must go beyond these battle lines and narrow modes of thinking to a more balanced view of the natural reality upon which all human support systems are based.

Preface

This book is, to the best of my knowledge, the first critical look at the science and politics of climate change by a committed environmentalist. I have spent over 30 years working as a professional ecologist on some of the major policy issues of this era. My work has been largely analytical, oriented towards practical initiatives and at all times actively engaged in the creation of effective policy. In addition to my training in natural sciences, I have an abiding interest in how different cultures, including my own, perceive their environment. In my experience, the world of environmental perception always interplays with theories of causation, divinity and notions of purpose and progress. In 'western culture' these elements are separated, as if inhabiting unrelated compartments of the mind. Thus, with regard to climate change, the issue is regarded as 'scientific' as if there were some pure and objective compartment of knowledge unaffected by any of the other boxes into which some very powerful and pervasive human propensities have been so solidly put. Social scientists would beg to question such a reality, but I will address these issues only by way of a signpost to further enquiry. The main thrust of this book is to question the science on its own terms.

Critical review

As a scientist, however, I have worked as much with the tools of advocacy, critical review and legal reform as with any science. My work has been across many disciplines, and as befits a political ecologist – that is, an ecologist actively engaged in the process of developing policy – I have been concerned always with reform, aiming to make the world a better place and as much for humans and their well-being, health and safety as for the beauty of the natural world that surrounds them.

I say this because the majority of scientists in my position appear to support the current consensus that climate change, or global warming as it used to be called, is caused by human agency in the increased emissions of carbon dioxide and other greenhouse gases, and that there is an urgent need to reduce emissions and avoid potentially catastrophic ecological changes. Who am I to disagree with that apparent consensus? On the recent occasions when I have been asked to speak in public, and despite showing ample graphic material and scientific analysis, that question is always the most pronounced. This is an understandable response for most lay-people confronted with graphs, statistics and science, who will then seek some reference for the competence of the speaker. I would urge those people uneducated in science to trust their ability to think for themselves and use this material to open up discussion and further study. My own 'authority' lies fundamentally in the material I present, which is the work of other scientists, and in my past record of work assessing often dubious scientific reasoning used to support particular policies.

When I began my review of this field, I knew very little of the heavily politicized world of climate science. That work began in 2005 when there was much less publicity in Britain of the level of scientific dissent from the orthodox view of global warming. This is in stark contrast to the debate in the USA which has involved senior scientists at the very top of the institutions. Unaware of this debate, I had completely accepted the scientific premise that carbon dioxide emissions were the main driver of recent climate change. And this despite having had some contact with the science over ten years ago when I was brought in to advise the UK Government's Countryside Agency on the policy implications of atmospheric pollution and global warming. Further, in the years prior to 2003, I was involved in energy policy consultation processes on issues of sustainability. I sat for three years on the UK's national advisory board of a joint Countryside Agency and Department of Trade and Industry project for community-based renewable energy initiatives. It was as a direct result of this work that I became motivated to question the basic science of global warming. It had rapidly become clear to me that the environmental impacts of the proposed remedy for global warming were far-reaching and damaging to many of the elements of sustainability that I had worked on.

My qualifications for this task

My role in energy policy has always been to examine the impacts of policy decisions and to advise on strategies that minimize impacts. Ironically, considering the direction my arguments now take, from 1976 to 1996, I was heavily involved with the evolution of what were then dubbed 'Alternative Energy Strategies'. In 1980, my research group in Oxford carried out the first pan-European study of such strategies for the European Parliament. At that time 'alternative' was meant as an alternative to a centralized, high-technology, high-risk development strategy based upon expanding nuclear power and in particular the 'plutonium economy' of 'fast breeder' reactors.

Thus when I call myself an environmentalist I want to emphasize my lineage – it is hands-on, scientific, practical and as 'green' as they come. In this book I will go into some more detail on the body of my work because there are lessons of great relevance to this current debate. In my advisory work I specialized in the analysis and critical review of global circulation models of both the atmosphere and the oceans. Such models were first built to simulate the dispersal of radioactive pollutants – a field in which I became a specialist. I later utilized that expertise to criticize models for the release of heavy metals and organic chemicals. At times, I worked on the critical review of other scientists' models, and at times, with the aid of an expert team at my own independent research group, we ran our own simulations.

This work was not confined to the computer laboratory or consultants' office. At different times I was engaged both by government and non-governmental organizations as a legal advocate and intervener, representing parties in inquiries, commissions, parliaments and the UN's international conventions. I worked closely over a period of twelve years as chief advocate and a science advisor to Greenpeace International, and in the late 1980s and early 1990s this work led me to be consulted directly by the UN's International Maritime Organization on the reforms necessary for new legislation to better protect the ocean. This was the long process of establishing the precautionary principle in global treaties – a principle that set precautionary action as the watchword over and above scientific predictions, many of them made by erroneous computer simulation to justify apparently safe discharge and dumping practices.

I have therefore seen the process of policy formulation and its relation to science at all levels of government and on some major contentious issues. I have also seen the extent of pressure that can be brought to bear upon scientists who conduct research that runs counter to powerful vested interests, and more particularly, upon those scientists who make their views public. Many commentators in the press dismiss such pressures but in my experience they are pervasive. Fortunately, I have been able to maintain an independence of most forms of pressure, including the freedom to publish. I will explore this issue in some depth as it affects this debate.

This background now brings me to my motivation for devoting three years to this review. I have read hundreds of original science papers and spoken to many of the working scientists involved. Without the development of the Internet, I would have been severely handicapped by high costs. This work has been pursued with absolutely no funds from any interested party apart from a modest and very recent advance from my publisher to cover the three months required to simplify the more detailed work that has appeared on my website. I neither work for nor liaise with any oil or other fossil fuel interests, nor with any free-market institutes concerned about current plans for carbon taxation. However, I do have a very strong personal bias and as I have argued often enough that science can never be value-free and is always subject to personal and collective psychology, background culture and time, I believe it important for the reader to have some information on this level. It is as important as any element of academic authority.

My bias is best put with respect to the last ten years of my work – which has involved finding creative solutions to some key issues of sustainability. I care about human community on the land, though not necessarily less so about cities, *and* what is currently called biodiversity. I particularly care about the values of indigenous peoples – especially forest peoples. I also care about those communities that struggle to make a living on the periphery of our own economies, such as hill farmers. I care about the communal fabric of rural life in those countries I know well – mostly in Europe, but I have worked for the health and well-being of other such communities threatened by pollution and hazardous industrial development in other parts of the world, both developed and undeveloped. In my early years I travelled widely in Africa, later in Asia, and I care about sustainable development as it affects agriculture, forestry, water, health and cultural integrity – as well as the spectacular wildlife.

None of this means I am *against* industry, technology, industrial progress or science. In my lexicon, sustainable development means sustaining community and cultural integrity, as much as health, food and water supplies, forests or wildlife. I am *for* appropriate technology and development that respects the real meaning of sustainability and I have an abiding faith that the scientific and engineering skills exist to deal with the problems that lie ahead.

These then are my colours. I am motivated to critically review the science of climate change because the proposed cure is likely to be worse than the disease. The current policies do not meet my criteria for sustainable development. The definition of sustainability that many of us strove to get accepted at the Rio Summit in 1992 included the important elements of cultural as well as ecological integrity. I first realized the extent of this problem when reviewing the work of the UK Royal Commission on Energy and Climate and then developing landscape impact visualization tools. I worked on these issues with the Countryside Agency and the UK's National Trust to develop integrated conservation of the wild character of Britain's landscape heritage. It was obvious to me that if the UK were to meet the targets proposed by the Royal Commission (a 60% reduction of greenhouse gas emissions by 2050) there would be a huge and largely unimagined impact upon landscape, community and wildlife. And given the relatively well-protected nature of the British landscape, the impacts in less well-regulated countries would be even more severe – in parts of southern and eastern Europe particularly, but also in remote forests and highlands throughout the world.

What is at stake: the destruction of community, rural life and biodiversity

In Britain, it is well known that a 'planning bottleneck' exists with regard to the expansion of renewable energy supplies. The obstacles are continually referred to in the energy policy field as restrictions and red tape – never in respectful or positive terms that take account of the depth of feeling that largely rural communities have for their landscape, sense of place, continuity, tranquillity, safety and recreational enjoyment (including of biodiversity). And of course large numbers of city-dwellers also value these communities, landscapes and habitats. These are key elements of sustainability. If these elements are not respected, nobody has any business talking of renewable energy strategies as sustainable – their view has become narrow, technical and lowered to the common denominator of survival.

This apparent urgency, where many environmentalists argue that these values must be sacrificed in the name of survival, is supposedly based upon a now unquestionable science foundation – a prediction that in 50 or 100 years hence humanity could be facing a runaway global warming that would melt ice caps, flood cities, destroy biodiversity and bring widespread starvation and social disorder.

The basic science and the use of computer models

In 2005, I therefore began to look at the basic science. In my previous reviews I had looked briefly at the models of prediction and how helpful they would be for predicting regional impacts on relevant timescales. The answer had been that the models were so crude as to be of little use. I now began to examine the parameters of the original models. What I found was deeply shocking, even for someone well versed in the machinations of science when it gets itself closely intertwined with powerful political and financial interests. First and foremost, I was shocked at the flimsy base that supported the models. I am quite used to inadequate science in relation to ocean currents, major elements such as cloud formation or the interpretation of complex patterns in ocean sediments, but I had assumed the *basic* science of carbon dioxide and atmospheric heating would be sound. Perhaps most other scientists outside of a very small cadre of climatologists assume the same and that explains why there are so few references in the literature on this fundamental aspect. It took a huge effort to track down the basic physics – the equations used by the atmospheric model – and then only through the indirect offices of a certain dogged, scientifically literate member of the English House of Lords, Christopher Monckton, who managed to get his questioning into the newsletter of the American Physical Society.

It turns out that there is only one basic piece of work underpinning the adoption of certain equations in the physics of carbon dioxide's ability to heat the atmosphere – and that by one of the original small group of scientists that worked to make global warming an international issue as long ago as 1988, James Hansen of NASA's computer simulation laboratories. This equation *assumed* on theoretical grounds that although the capacity of carbon dioxide alone to heat the atmosphere was limited (a doubling would have a climatologically insignificant effect) there would be an additional effect on water vapour which would *amplify* carbon's role by 300%. The enhanced warming expected later in the twenty-first century could then trigger further amplifying feedbacks taking future temperatures way above those seen in the course of natural cycles.

In reality, a *lack* of consensus

It was a further shock to read, in the proceedings of the IPCC, a panel set up to review climate science and underpin the UN's Framework Convention on Climate Change (and later the Kyoto agreement to limit global emissions), that from the very first scientific meetings at least one senior climatologist had argued that this assumption with regard to carbon dioxide and water vapour was unsound. This scientist argued that the feedbacks were unproven and could readily operate in the other direction and compensate for any warming. There was thus never a *consensus* within the IPCC, rather, a majority that overruled this dissenting voice.

This lack of real consensus continued on many other key issues within the IPCC but has not been reflected in the pronouncements of the Panel's public representatives, nor in their Summary for Policymakers document. This immediately flagged up previous histories of the treatment of dissenting voices within the UN policy process. It is an issue central to the evolution of science and sound policy – in that dissent needs to be acknowledged, respected and given its voice not just at the level of scientific working groups, but at the policy level in the treatment of uncertainty. If dissent is marginalized, science travels down a slippery slope directed by the needs of policy makers for simple single-cause answers and targets, and in this, ultimately, the truth suffers.

There was another level of surprise as I began to retrieve and study the rest of the science. A very large literature existed on issues I had hardly touched upon that received very little press coverage, and yet were central to answering the doubts on this original questionable assumption of amplified warming. These issues ranged from the variability of solar irradiance (visible sunlight) over time, the electromagnetic field effects on clouds and its variance over centuries, and the nature of supposedly 'internal' cycles of variability in global temperatures – with many scientists finding correlations to 'external' solar electromagnetic cycles (I do not find the use of the terms internal and external at all useful when considering ecosystems). A large literature exists on ocean temperatures oscillating from one extreme to another in clear cycles for each major ocean basin, and on solar cycles with time-lagged land and ocean temperature responses. Finally, and most disturbing of all – instrumental data, largely derived from satellite studies, clearly contradicted the expectations of the carbon dioxide model. This data had been dubbed 'unreliable' by the very director of the space research programmes that had developed the monitoring strategies, the aforementioned James Hansen, who first claimed to have identified the fingerprints of 'anthropogenic global warming'. Hansen is now a much-travelled expert witness at government inquiries. Other science laboratories have made extensive use of this data and have made corrections for the various satellite calibration issues that seemed to deter Hansen and, as I finalize this text, his former supervisor and overall head of research operations at NASA has publicly rebuked Hansen for his bias and stated that many senior scientists doubt the ability of computer models to make reliable predictions.

In this review I rapidly began to realize that a phenomenon was at play that I was all too familiar with. Politics had intruded into science on a grand scale. It had parallels in my previous experience with the science of 'dilute and disperse' in regard to the disposal of toxic substances. There had been a paradigm of prediction and control, also reliant upon computer models and simulation, which I had watched unravel in all its manifest failures. I will review some of the lessons from that history.

Thus in the pages that follow I will outline my review of the science. I will also, in the second part of this book, offer some insights on the political process and what safeguards are required. I will also draw on my experience of energy policy and the agenda of sustainability to address the issue of what we should now be doing on a practical level in the light of what I regard as flawed science and the inability to adequately predict future climate change.

As will be evident, I am not an advocate of 'do nothing' or 'business-as-usual' for the corporate world or the consumer. We may have made an error in ascribing the main cause of climate change to carbon emissions, but this does not mean we are out of danger. Even if nothing other than natural change were to happen to the climate for the next few decades, we would still face imminent danger – not directly from natural variability, but from humanity's growing vulnerability to its cycles, most particularly with regard to food supplies. I conclude from the evidence I have seen that we may be facing a significant period of cooling comparable in severity to the Little Ice Age of AD 1400–1700.

Dangerous climate change is unavoidable

Thus a realistic look at the science suggests that there is very great danger already with us in *this* decade. There is as much urgency as ever, but the remedy involves a radical departure from current policy which is almost entirely focused upon reducing carbon emissions. In my analysis, such policies can have no meaningful effect upon future climate. The policy needs to shift to adaptation strategies in the face of *inevitable* and imminent shifts in the global climate. Indeed, my review suggests that *global cooling* is more likely than global warming over the next decade or more.

Even if my analysis is not correct and carbon dioxide *is* the villain it is made out to be, the models show that none of the current policies can *lower* levels in the atmosphere much before 2050 and most of those models assume carbon levels will *rise* over the next two decades as the global economy continues to develop.

If I am right and we face a period of cooling – more than a mere 'blip' in the trend, as current modellers now characterize the recent downturn, then this will have serious consequences. Past cycles of cooling have brought severe famine at times when the global population was very much smaller and less vulnerable to climate fluctuations. Sixty-seven countries are now dependent upon external food aid, with most of that coming from surpluses in the northern grain belt. These surpluses are very vulnerable to a cooling cycle. Further, the world population is set to double in the next decade at the same time as oil production, upon which agricultural surpluses depend, begins to decline.

The issue of food security

Ironically, one key element of future renewable energy strategy – the use of biofuels – threatens to compromise world food production even more rapidly than the climate. At present, no safeguards are in place to control and regulate international markets in bioethanol, biodiesel, woodchip, forestry wastes or other biological material that would otherwise be available for local needs. And powerful commercial interests are already engaged in clearing rural subsistence communities to make way for bio-fuel plantations.

What is so discouraging for me personally in this situation is that my former friends and allies in the environmental movement have aligned themselves, in my view very naively, with these forces of conflict and destruction. Some may call for safeguards, but only now, after decades of arguing for simplistic targets and supply-side technology rather than a strategy of demand reduction and resilience to environmental change. That is another key motivation of this book – to engage in more rational discussion and a more balanced policy.

Natural cycles not random variability

I have concluded that natural cycles are primarily responsible for the 'global warming' that has alarmed both scientists and the public. That implies a major failing of the science community upon which the public have relied for guidance. I want also to explore how this could have happened. Science has made such errors before, but science has a habit of covering its tracks! In its current defence, it argues that natural variability (there is a reluctance to refer to cycles) has the power to temporarily overwhelm the heating effect of carbon dioxide. But it follows that such variability may also have amplified the warming signal so evident between 1980 and 2000. As I will show, between 1950 and 1980 there was a natural cooling period. At the time many scientists thought an ice age could be imminent. That cooling was later ascribed to fossil fuel burning and the effect of sulphur aerosols. We now know that this science was in error. The 'global dimming', as it was dubbed, was due to a natural cycle of transparency in the atmosphere that affected unpolluted areas. Furthermore, most climate scientists acknowledge that carbon dioxide contributed little to the general warming trend that began after the Little Ice Age in the early 1800s. There has been a steady upward trend for two hundred years. And after 1950, when emissions of carbon dioxide rose dramatically, the global cycle of cooling kept temperatures down until 1980. The period 1980–2005, a mere 25 years, is now the sole 'signal' for human agency and the identification of this signal relies entirely upon the ability of computer simulations to separate natural cycles from the effect of extra greenhouse gases. For this to be reliable the computer models must 'know' what those cycles would be doing and, as we will see from expert review within the IPCC itself, this is an acknowledged major weakness of those models.

As an outsider to this community, it beggars belief that IPCC can in one sentence acknowledge such serious limitations of the models to simulate natural cycles and, in another, lay such store by the models' predictions. Furthermore, these models were validated by their apparent ability to replicate the past fluctuations thought to be caused by pollution but now known to be natural, thus showing that the mechanics of the models are wrong. A similar conclusion must be drawn from the replication of ocean heat storage, a factor now known to have been seriously overestimated.

Error at the heart of the IPCC assessment

It follows from this analysis that the early assessment of the power of carbon dioxide to warm the atmosphere was erroneous and the evidence points to that assessment being out by a factor of two or three times. Hence, a doubling of this gas does not present a serious threat. I do not address the other 'anthropogenic' (human sourced) greenhouse gases which include methane and nitrous oxide, nor the artificial chemicals such as HFCs, all of which constitute less than 20% of the computer-calculated heating effect. I am dealing with the issue of the main driving force, but there is not space to devote to such issues as forest and agricultural policy and I am convinced that these sectors need to be reformed for many reasons other than their greenhouse gas emissions.

It follows from my arguments that we do not face a significant threat from sea-level rise, other than that occasioned by natural factors. Most of the sea-level rise to date (and all other environmental effects laid at the door of 'global warming', such as the retreat of glaciers and calving ice shelves), can be accounted for by the rebound from the Little Ice Age. Indeed, the trend in sea-level rise from 1800 has been consistent, and in the last ten years, as the oceans have cooled, that trend has levelled off. It will become clear that issues of rainfall, drought, hurricanes and the spread of diseases are all consequences of the main natural cycles of warming and cooling and that adaptation is more important than fruitless attempts at mitigation.

The one area that I have not studied in depth relates to ocean acidification. Only time has prevented a deeper analysis. I doubt that it will be the problem that some biologists now sound an alarm over. The oceans and the great majority of the organisms found there are old by evolutionary standards, and if we track back tens or hundreds of millions of years the physiology of the organisms changes little, yet over this time-scale these groups have coped with very much higher atmospheric levels of carbon dioxide. This is not to argue for complacency. Modern corals in particular are not the same type. I argue for more detailed study and less special pleading and alarmism on the part of research teams.

The social and political dimensions of climate change

Finally, I make some personal reflections on the social phenomenon of 'climate change' as a *fin de cycle* all-pervasive mentality of making 'war' on global enemies − whether terror, poverty, cancer or drugs, where the language is of campaigns, targets and recruitment, with the military metaphor backed by technology, carbon-based taxation, surveillance and control. Some protagonists take things beyond metaphor, as might be expected by the embrace of a military mentality, and argue for climate crimes 'against humanity and nature' and court sentences (James Hansen), a nuclear-powered society with refugee communities in the Arctic

(James Lovelock) and even a UN enforcement army (Forum for the Future) with the power to invade and depose recalcitrant and climatically criminal regimes. One leader of the Greens in the European Parliament calls for a Carbon Army mobilized on a similar scale to the Marshall Plan that reconstructed Europe after the Second World War.

Green is fast emerging as the new Black. I despair a little at the emotional brutality of a formerly sensible environmental movement. Even wildlife conservation organizations and the culturally aware development-aid groups have joined the crusade against climate. When I look at this not from under my scientist's hat but from the stance of a social anthropologist, I ask who gains from this blatant propaganda. In any war, truth is the first victim, but to whom the spoils? The list is long and grows daily. Politicians gain advantage over more sluggish rivals and promise a future that cannot be verified, even if they sincerely believe it. Some banks are already geared up for the carbon trading economy and include some experienced ex-politicians on their boards. Renewable energy *supplies* create increased demand for turbines and aerospace expertise, precious metals and electronics. The media can sell scary climate stories and campaign groups increase subscriptions. Science institutions gain influence and funding – a considerable amount for climate science compared to the small beginnings in 1990. And above all these gains there is the avoidance of a loss, as important in the world of science as in politics, and that is the loss of face.

This is not to talk of conspiracy. That is a word commonly levelled at anyone who raises the issue of 'who gains' and it covers a dangerous naivety or blatant denial that such an issue exists. I would rather use the term *collusion* of interests. If we are to disentangle ourselves from what I see as an almighty policy 'cock up' and successfully address the changes we are facing *within* the next decade, we will need a new kind of thinking. Such new thinking is not evident in any of the myriad books on climate change, energy policy, economic development or biodiversity conservation. In my last chapter I address what only Albert Einstein seems to have understood – that the same kind of thinking that got us into this mess cannot get us out of it.

Part One

THE SCIENCE

Introduction

If there is one thing *certain* about climate, it is that it will *change*. Change is inherent in the meaning of the word climate itself. Well before linguistic evolution or human perception needed a geographical category, the ancient Greek root *klima* meant a slope or inclination, thus inferring a tendency. However, the meaning of 'climate change' has undergone a recent evolution, and something that happens naturally, cyclically, mostly quietly, sometimes dramatically, has become a global threat to humanity, implying chaos and calamity, imperilling civilized values, international justice and ecological sustainability. The term has also come to imply human responsibility for the change and, by further implication, the imperative to prevent this threat by better directed human agency.

Science has played an integral role in this rapid linguistic evolution, and indeed has become so bound up with the social processes and meaning behind the term – processes that have generated a global awareness, political action and large-scale financial investment – that a twenty-first-century phenomenon has evolved that has yet to be appreciated in its full extent and implications. These political elements feed back into the science on a scale that I doubt has been seen in the whole history of science. 'Climate change' has long ceased to be a scientific concept – it is a political movement and an ideology.

A few years ago, 'global warming' was the villain, and it could only be identified by careful statistical analysis. Then the threat shifted to any change in climate – perhaps because even in a warming world some regions might cool. The recent *global* cooling was not predicted, yet it is now included, at least linguistically, as if it had been anticipated all along. Scientists have colluded in this accretion of linguistic meaning, and the IPCC, as if in final admission of this reality, added a footnote to its 4th Report – that the term climate change, previously defined in its 3rd Report as caused by human agency, should now be regarded as applying to both natural and human agency combined.

When the current chairman of IPCC, the economist Rajendra Pachauri, states on a tour to the USA, that the science is 'settled', he sends a message to all the institutes that this is what those in authority and in a position to dispense funds and favours actually believe. Such language sets the way the wind is blowing. Yet there is a growing number of scientists crying foul, and my own estimate that carbon dioxide may be responsible for as little as 10–20% of the global warming signal is now shared by a few senior analysts. In Part One I present my evidence:

i) that the main driver of global warming has been an unprecedented combination of natural cycles operating through a system of connected ocean basins that have oscillated and peaked together;

ii) that these cycles of warming and cooling are caused by small variations in cloud cover with consequent effect upon the flux of sunlight and accumulating warmth in the ocean surface waters;

iii) the mechanism driving these cycles is now under intensive examination in major science laboratories and involves a combination of visible and UV radiation pulsing over the 11-year solar cycle, as well as a little understood magnetic or electrical mechanism that amplifies the cycle by reducing cloud cover sufficient to create a strong warming pulse in the period 1980–2000;

iv) satellite data confirm this pulse of warming sunlight, and measurements show that it can account for virtually all of the late twentieth-century warming.

The use and abuse of consensus

Most of these issues are dealt with in the technical detail of the IPCC's most recent Working Group Reports. It is clear that working scientists are not in agreement on key issues, yet this lack of agreement does not resurface in the Summary Report. The consensus upon which the Summary claims agreement is, upon closer inspection, confined to descriptions of the extent of warming whilst leaving considerable uncertainty on the causes. This uncertainty is obscured by use of phrases such as 'likely', which are then defined as probabilities of being correct. For example, 'likely' carries a probability of more than 66% and within that other 33% are hidden the disagreements. These percentages are not derived from scientific or statistical treatments, but by expert judgement, and that might just as well equate to a majority rule – it is not consensus.

In the outside world, representatives of the IPCC then claim a consensus for the view that most of the warming of the past century is caused by human activities. This would mean that all scientists who had contributed to its assessment would agree with its conclusions. As we shall see, this is not the case. Use of the term implies that there are no major areas of scientific disagreement within the body of experts. This is also not the case. The question then arises as to the means whereby IPCC achieves agreement on publication of its Summary for Policymakers. It will be clear from this review that scientists who question many of the assumptions and who report contradictory findings are not asked to agree the final drafts. Thus, by subtle forms of editorial control, dissent is marginalized and an appearance of 'settled science' is portrayed.

A true consensus report limits its conclusions and recommendations to those areas upon which all assembled experts are in agreement and then highlights any significant disagreements, at the same time outlining the policy implications of any uncertainty and the implications of differing policy options.

The disadvantages of this approach are that policy makers may defer action until consensus emerges. The advantages are that dissenting voices have often been proven correct, especially on environmental risks, and expensive policy errors may be avoided.

In the chapters that follow, I outline a lack of consensus in several key areas of science which have not emerged into the public arena. I also detail the evidence from real-world data that contradicts the IPCC view and show how that view has been constructed around the relatively narrow community of computer scientists and the virtual reality of climate models.

1

The Uncertain Signal

'Correlation is not cause'
(basic science)

The global warming signal has been communicated to the world as if it is unequivocally a sign of human interference in the climate and that the late twentieth-century warming was not only unusual, but inexplicable other than by human cause. In his Oscar-winning documentary *An Inconvenient Truth*, the Nobel Laureate and former Vice President of the USA Al Gore shows an apparently convincing graph of global temperature and carbon dioxide cycles over several ice ages, indicating how they run in parallel, thus implying that the greenhouse gas causes temperatures to rise. But in order not to obscure the message he neglects to point out that detailed study shows that carbon dioxide *lags* the temperature by about 800 years (**Monin et al., 2001**). It is a tiny gap on the scale of a graph dealing in cycles of hundreds of thousands of years, and of course the audience are not there to question or debate and he doesn't draw attention to the issue. There have always been uncertainties as to what drives global temperatures and the climate feedbacks. It is very obvious that they are cyclic in nature, yet these cycles are all but ignored in the interests of a simple message.

In this chapter we will examine the nature of the global temperature rise as a signal for human-induced climate change. For a signal to point to human interference it must stand out against the natural background variability. There is no doubt that the twentieth century experienced a warming compared to the nineteenth century and that this increased markedly in the last two decades. At the same time carbon dioxide levels rose to levels above those recorded in previous periods between ice ages and some effect would be expected from standard atmospheric physics. But that physics has never been agreed; even Al Gore's teacher and mentor, the atmospheric physicist Roger Revelle, lionized in the film, disagreed that carbon dioxide would be a problem. And contrary to much common understanding, this warming may not be unusual, with clear evidence of natural cycles having produced a peak.

The science of climate change is not settled, despite what some leading scientists have been saying. If we look closely at the relevant texts we will find that IPCC have never stated that the science was unequivocal, despite what their public representatives may have said. Instead they have hidden the true nature of the uncertainty by the choice of words such as 'very likely' and 'likely' which purport to have a basis in probability.

We will have to examine what these terms mean and how they have been used. IPCC defines 'very likely', which it uses to affirm the warming signal is unusual, as a probability of 90% but does not then clarify to policy makers that this would not satisfy the criteria for confirming a scientific hypothesis (95% is required, leaving a 1-in-20 chance of being wrong). As noted, the term 'likely' denotes much less probability of being right – at 66%, leaving a 1-in-3 chance of being wrong, and this is the level of confidence applied to the attribution of human cause for the warming. We can thus immediately see that the cause of global warming is far from settled science. We can also see that a large group of scientists covering many disciplines and areas of doubt, discussion and disagreement might not voice dissent from statements that leave such considerable leeway.

As we shall see, the signal of global warming – whether it stands out from normal variability – has always been uncertain. Between IPCC's 2001 Assessment Report and that in 2007, confidence that there was a real signal increased from 90% to 95%. However, the Panel were only agreeing the nature of the signal, not its cause, and they actually narrowed the time period over which the signal could be regarded as unusual. When it comes to considering causes (the Summary Report talks naively of natural *or* human rather than the reality of multiplicity), the 2001 assessment left a large probability of 33% that the cause was natural.

Random variability versus repeating cycles

The science hangs upon the ability of climate studies to distinguish between natural cycles of change and the human signal. As I will show, the science has never been able to do that at the level appropriate to confirm the assumption that greenhouse gases have caused most of the warming and this is because the science of past climate cycles is very uncertain. The IPCC recognize this, but instead of highlighting the importance of further study it reverts to models and their dictates that assume natural variability is essentially random. It is curious therefore to find that the 2007 report again emphasizes this uncertainty yet at the same time claims to have identified the human signal with greater confidence. This claim is not supported by the body of the science in the report and it is a major contradiction at the heart of their assessment.

I am not alone in coming to this conclusion. As recently as 2001, following IPPC's Third Assessment Report (often referred to as TAR), the US National Academy of Sciences constituted a special committee to assess the report and advise the US Congress. Though broadly agreeing that the world had warmed significantly in the twentieth century and the warming was relatively unusual, their expert panel found greater uncertainty as to the causes (**National Academy of Sciences, 2001**):

> Because of the large and still uncertain level of natural variability inherent in the climate record and the uncertainties in the time histories of the various forcing agents (and particularly aerosols), a causal linkage between the build-up of greenhouse gases in the atmosphere and the observed climate changes during the 20th century cannot be unequivocally established. The fact that the magnitude of the observed warming is large in comparison to natural variability as simulated in climate models is suggestive of such a linkage, but it does not constitute proof of one because the model simulations could be deficient in natural variability on the decadal to century time scale.

As I will show in this and the next chapter, subsequent developments in our understanding of this natural variability confirm this view. This 2001 NAS statement by an eleven-person panel included a recognized dissenting voice from the supposed consensus, Richard Lindzen, Professor of Meteorology at MIT, as well as James Hansen, an outspoken proponent. I will show in subsequent chapters how the steady suppression of such a dissenting voice in the pronouncements of other institutions has led the scientific community astray.

As we shall see, the NAS panel went on to advise its government that the UN report would not be a sound basis for US policy and it drew attention to several unresolved scientific issues relating to causal mechanisms present in the body of the UN report. However, by 2005, the NAS had joined with ten other science academies worldwide (the G8 countries of Japan, Russia, Canada, Germany, the UK, France and Italy, with the addition of Brazil, China and India) to issue a call to all governments, despite uncertainties in the science, to instigate carbon emission reductions. In the text of that declaration the academies quote IPCC-3: 'It is likely that most of the warming in recent decades can be attributed to human activities.'

Thus, the academies present no new science, admit without stating it clearly that there is still a substantial possibility that the warming is natural, and yet call for immediate action. Such simple statements disguise a great deal of disagreement as to causation, which is clear within the body of the UN report but becomes obfuscated in the Summary for Policymakers. Certainly, the late twentieth-century rise in global temperatures coincided with a steep rise in carbon dioxide emissions, and the basic atmospheric physics of greenhouse gases argues for a contribution, but the language of the IPCC's Summary Report uses concepts of single cause and estimates of certainty that have little basis in science. The IPCC admits that its quoted probabilities are no more than 'expert judgement'.

Simplifying science in communication to policy makers

When I first began reviewing the science, this amount of latitude surprised me greatly. It had clearly not been communicated in that way to policy makers, and the majority of my colleagues in the environmental science field were equally unaware. When I first began discussing my theories of natural causation, professorial friends in the field of environmental sciences would say, 'How can you possibly be right, when you disagree with all the world's experts, national academies and the UN?' Only when one looks deeper into just what those experts have agreed to is it obvious that there is plenty of support for theories of natural causation as the main cause of the warming.

However, in the years since the 3rd Assessment in 2001, IPCC has apparently revised its levels of confidence upwards despite acknowledging greater uncertainty in the understanding of natural processes! As we noted, by the 2007 Assessment confidence in the signal itself had improved to 95% (i.e. that warming had taken place). Yet on the issue of how this rise compares to previous variability, the level of confidence dropped markedly:

Average Northern Hemisphere temperatures during the second half of the 20th century were *very likely* (above 90% confidence) higher than during any other 50-year period in the last 500 years and *likely* (above 66% confidence) the highest in at least the past 1300 years. Some recent studies indicate greater variability in Northern Hemisphere temperatures than suggested in the Third Assessment Report (2001), particularly finding that cooler periods existed in the 12th to 14th, 17th, and 19th centuries.

When it comes to causation the confidence again *apparently* increases, to 90%, albeit still not within the bounds acceptable for confirming a hypothesis in science. But the caveats involve a complex set of negatives: '*very likely* that it is not due to known natural causes alone'.

These statements disguise a great deal. The increased confidence is actually limited to a defined period and does not highlight the considerable disagreement that exists on the magnitude of previous warm periods. The studies that suggest *greater* variability include severe criticism of the IPCC's 3rd Report in which natural variability in the ecological records of the past had been 'smoothed' out.

On causation, the last phrase 'not due to known natural causes' clearly leaves open the question of unknown causes, but does not state that there is considerable evidence reviewed in the body of its work that indicates some unknown causes may be at work – such as solar-cloud effects, which we review in Chapters 7 and 8.

The IPCC is careful not to talk of cycles. It places itself firmly in the camp that believes only in random variability and hence the unpredictability of natural processes. As we shall see in Chapter 2 and when we consider the oceans and poles in Chapters 5 and 6, there are many hundreds of scientists engaged in a study of oceanic and solar cycles. The difference between something being variable and cyclic is that cycles, such as a previous warm period, will repeat. This is not made clear to policy makers. There is a significant body of evidence that past cycles produced periods as warm or warmer than the current century, with some disagreement on the detail. IPCC do not clearly represent this lack of consensus.

The reliance upon computer-generated realities

IPCC makes clear that the *only* method available to distinguish the current pattern of warming from natural fluctuations in the global mean temperature is by computer simulation. In that process, a virtual planetary ecosystem, or model, is created that attempts to mimic the past pattern of temperature fluctuation. This is the fundamental basis of the IPCC approach. The great majority of climate studies are built upon these models.

The only way that such models can be validated is if they replicate the past fluctuations of temperature. But as we shall see, even this test is not reliable. The Panel conclude that the suite of models used is reasonably successful in mimicking this past variability but they do so only if they include the factors for enhanced concentrations of human-sourced greenhouse gases. If the models are run using natural factors alone, then they diverge as seen in Fig. 1, taken from the latest 4th Assessment Report of the IPCC in 2007. The Panel holds that the spike of the 1980–2000 period cannot be simulated without the input of these emissions.

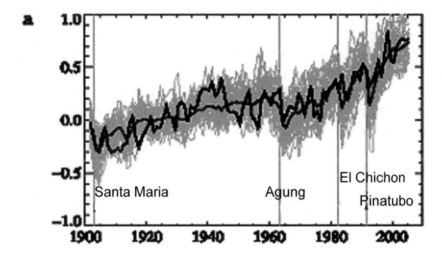

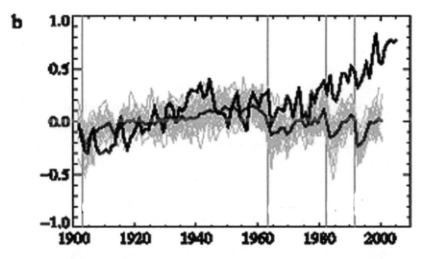

Fig. 1 Comparison between computer simulation and observation of the global surface temperatures (°C), from observations (thick-black) and computer simulations (dark grey) using (a) both human and natural factors and (b) natural factors only. The vertical grey lines mark major volcanic eruptions. (Source: IPCC-4 WGI, Chapter 3, 2007)

This is the crux of the IPCC case and the first question a critical reviewer asks is whether the model has included all of the relevant factors relating to the natural environment. Has there been anything unusual happening *naturally* that parallels the temperature rise? And how reliable are the inputs relating to greenhouse gases? It is not uncommon for models to replicate a pattern but not the actual mechanisms involved. In the analysis that follows, I shall demonstrate that this is exactly what has happened. There *are* unusual natural circumstances in the late twentieth century. Furthermore, the models have recently been shown to have falsely replicated the pattern, something admitted to but obscured in the IPCC Working Group Reports. In the scientific detail the Panel regularly admit that the modellers' grasp of natural fluctuations is very limited. This ought to mean that the flat line from 1950 onwards in Fig. 1 is not reliable, yet it is upon the difference between these two lines that IPCC rests its whole case.

The mathematical simulation of natural variability is unsound

In Fig. 1, the temperatures are expressed as *anomalies*. This is done by finding the global mean temperature for a particular period – in this case 1901–50, and expressing each year in relation to that period, generally within about 0.5°C above or below the line. In the diagram major volcanic eruptions affecting global temperatures are shown by vertical lines. The grey shading varying around the dark grey line represents the 'variance' in the computer predictions. In actuality it is the operation of chance in the computations that simulates natural variability and it is standard practice to run a simulation using the same starting point many times because of the chance factors operating within the mathematics. Each 'run' of the programme generates a slightly different result. In this model no amount of runs regenerated the observed pattern of temperature rise unless the factor of increased human emissions of greenhouse gases was included. This is the difference between graph a and graph b.

It is upon this work that the entire edifice of 'climate change' and 'global warming' rests. I would say that 99% of all climate studies rely upon this basic model and do not question the reliability of its initial premises. The vast majority of further computer studies simply build upon it. Further-more, because it is unquestioned in its own field and largely impenetrable to other disciplines of climate research, such as oceanography, sediment studies and solar-terrestrial physics, many scientists in those fields refer to 'anthropogenic' global warming as if it had been established. They will often introduce their papers assuming this is the case when they have no competence to judge either way.

However, that edifice is now beginning to crumble. Not only is it clear that natural factors were not well known enough to be modelled, as the US National Academy of Sciences suspected, but recent work has shown that the models falsely replicated key elements of the past pattern. In particular, the 'global dimming' period of falling temperatures between 1945 and 1978 was assumed to be caused by sulphur particles from fossil fuel emissions. The models incorporated assumptions about the power of sulphate aerosol to create the dimming, resulting in three decades of cooling despite the increases of carbon dioxide. The models also built in erroneous assumptions about upper ocean heat storage derived from a monitoring system now known to be flawed. Models replicated the ocean warming that had been reported, but the later work showed the reality had been 200% less.

Thus the models had been validated because they had replicated the past pattern using assumptions about carbon dioxide and sulphate aerosol. The latter were supposed to have counteracted the rising greenhouse gas effect. As we will see in more detail in Chapter 2, that model was wrong. The cooling was largely natural and this means that the mathematical assumptions for carbon dioxide's effect in particular have not therefore been validated. We will see that those mathematics have been under intense criticism as they involve 'gain' factors for which there is no direct evidence.

Any natural scientist familiar with the operation of computer models and the process of simulation knows that models are fraught with such difficulty. The successful mimicry of the past pattern does not guarantee that the real-world mechanisms have been effectively modelled. As we shall see in the more detailed analysis that is to follow, the evidence is convincing that this model, which was developed and used in earlier IPCC reports, does not replicate those processes.

We shall look in more detail at the very recent science that has caused the revisions in understanding. The received wisdom was not challenged until 2005 when major satellite monitoring data was reassessed. The inescapable conclusion from this reassessment is that the decline in temperatures was part of a natural cycle. We shall look at the dynamics of this cycle in more detail as it involves oscillations within ocean basins that periodically warm the atmosphere, even over land, for periods of 30 years, followed by cooling periods of the same length.

Looking within the texts of the IPCC Working Group reports, there are many occasions when they refer to major areas of uncertainty with regard to natural cycles and processes, yet they do not highlight these uncertainties with regard to this all-important model. Clearly, one cannot build a reliable model of natural causes when those causes are poorly understood. And this also illustrates the limitation of using modelling to underpin major investment decisions – it is relatively easy to revise the models, but not the decisions.

There are two areas where the models fail to incorporate key natural features of climate: (a) the periodic cycles of warming and cooling in different ocean basins and their 'teleconnections' (how what happens in one basin affects what happens in adjacent basins), and (b) solar cycles, in particular the long-term periodic fluctuations of both visible light which warms the oceans and the magnetic flux which is suspected of causing changes in cloud cover. In the case of the ocean cycles, the mathematics of the various interactions and irregular periodicity makes incorporation into models very difficult; the most recent attempts show potential global cooling for the next decade. In the case of solar cycles, there is no consensus on past solar variability (estimates would give between 12% and 70% for its contribution to warming from 1800 to 1950) and no consensus on the interaction between solar magnetic cycles and clouds. However, these are very real possibilities with undetermined likelihoods (probabilities). The test of the standard model is whether it predicts what happens next, and we will see that the evidence points to the need to revise the models and incorporate both oceanic and solar cycles. Some attempts are under way and I report on what is now a breaking area of climate science.

Finally, with regard to Fig. 1, we should note the short timescale from 1900 to 2000 in which a steady upward 'trend' is apparent. This is an artefact both of the selectively short timescale and the use of global 'means' and the annual change or 'anomaly'. The 'anomaly' of 0.5 degrees over a 50- or 100-year time period looks startling. But the reality is that half a degree is only 3% of the global 'mean' of about 14°C which varies naturally by about 10% in an irregular pattern over many centuries in what is regarded as a relatively stable pattern between less stable ice age fluctuations. A graph that showed the last five or ten thousand years in *absolute* terms rather than the relative anomaly would not be at all impressive. We will see that even over this stable period there are cycles of warmth and cold that are not indicated in the approach taken in Fig. 1.

Constructing global temperatures

One reason for the focus upon shorter timescales is the difference between the instrumental record (from about 1850) and the various methods of estimating global temperature prior to the instrument record. Computer modellers prefer data that can be treated statistically. Prior to the instrument era, 'proxies' for temperature were used that have much greater levels of uncertainty and that require different forms of statistical treatment. In the proxy record, patterns are more apparent, but exact temperatures are not reliably calculated. In fact, even the instrument era is not without controversy. Calculation of the global mean from instrumental records requires an extensive database and all manner of techniques to make up for areas of the globe with poor coverage. I do not propose to critique the accuracy of this record though it is subject to some debate. I am concerned more with the preferential treatment of this record compared to 'proxy' records. The non-instrumental inference of global temperatures is derived from a variety of means such as: the ratio of oxygen isotopes in the ice crystals in sequential records of deposition, as on the ice caps of Greenland and Antarctica; in cave stalagmites; from sediments laid down in river and ocean current systems; in deep boreholes which reveal an imprint of varying surface temperatures; and in tree-ring studies.

Whilst an attempt to provide a global mean is understandable from the perspective of creating a usable annual index of global change, it places an undue value upon the last 150 years of the instrumental record. The longer-term proxy data is constituted from regional sources and it is not a simple matter to create a global picture. Thus any previous pattern revealed by, for example, the Greenland ice-cores, cannot be readily extrapolated to global levels. Thus, an undue focus is placed on the instrumental record on account of its greater level of certainty and amenability to statistical treatment.

The 'hockey stick' controversy

In this context, the 3rd Assessment Report in 2001 generated considerable controversy when the Panel laid great emphasis on a figure now infamously known by climatologists as the 'hockey stick' graph. By what emerged as very questionable statistical treatment, Michael Mann of Pennsylvania State University led a team that smoothed out all the past cycles and was left with the last 150 years of the statistically robust instrumental record as the steep 'handle' to the smooth shaft, thus making the recent warm period appear highly unusual (**Mann et al., 1999**). In between this 2001 Assessment and the 4th in 2007, this approach was heavily criticized (**McIntyre & McKitrick, 2003; 2005a, 2005b**). IPCC now acknowledge the reality of a weight of evidence showing greater variability in the past and admit to major uncertainty with regard to natural cycles, in particular the Medieval Warm Period around 1000 years ago, which some argue was as warm as the late twentieth century, and the Little Ice Age of 400 years ago.

One reason for the discrepancy in knowledge of natural cycles compared to recent instrumental records is the huge disparity of resources invested in monitoring temperature and building models compared to the basic science of natural variability. The latter has plodded on in mostly academic institutions throughout the world with painstaking and unglamorous fieldwork. The longer-term natural cycles can only be studied in the disciplines of palaeoecology by use of mundane environmental indicators contained in the sediment patterns, fossils shells and assemblages, stalagmites, tree-rings and ice-cores which are much less precise than instrumental records. The fieldwork is tedious with laboratory measurements coupled to complex statistical treatment. The literature is, however, extensive and conclusive with regard to the cyclic nature of past patterns.

It has become evident during the course of my review that this considerable imbalance and bias in the climate science has affected judgements. Study of the deeper past inevitably stimulates enquiry and methodology relating to cycles and uncertain multiple causes, whereas reliance upon computer models operates in the other direction, producing a desire to simplify and fix parameters, settle the science and get on with constructing the future. It also tends to view variability as essentially random, and there is a distinct tendency among many climatologists to studiously avoid the use of the term 'cycle'. The problem for the simulators is that if you do not know where you are in an irregular cycle you cannot incorporate it into the simulation. Mathematical algorithms readily mimic random variability, but natural cycles are not regular and predictable enough to be accommodated in models, so they are simply left out.

Natural cycles

Palaeoclimatologists are now in general agreement that global temperatures are in recovery from a down period in such long-term cycles and hence would have been going up in the natural course of events. This is most clearly elaborated by Professor Syun Ichi Akasofu, a leading geophysicist and, until recently, head of the International Arctic Research Center at the University of Fairbanks in Alaska, in a document available from IARC entitled *Recovery from the Little Ice Age* (**Akasofu, 2009**). Akasofu and his colleagues are well placed to study cycles in the Arctic climate system and we shall review evidence for cyclic warm and cold periods later in some detail.

Although prior to the instrumental era adequate proxies of the global temperature are a problem, there are clear indications that in previous warm cycles temperatures were higher than present. Those who believe we are seeing a human imprint have argued that it is the unexpected *rate* of temperature change that indicates man-made or anthropogenic global warming, but the problem is that those few indices that reflect global change, such as data from deep ocean sediments, tend not to reflect shorter-term changes. In contrast the regional data – for example, from the Greenland ice cap – show that major regional change has happened very quickly over timescales of less than a decade and that rapid change in certain key locations, such as the North Atlantic, can be quickly propagated across the whole northern hemisphere. Further, there is evidence of major cycles even *within* each 100,000-year ice age, as shown in Fig. 2, and that these continue through the interglacial period in a less dramatic form.

Thus, to be certain that 'global warming' (by which I now mean the late twentieth-century rise in the global instrumental record) is not mainly due to natural factors operating at the same time as the rise in carbon dioxide emissions requires that these natural factors be adequately known. Yet it is clear from IPCC Working Group Reports that a sufficient level of scientific confidence does not exist and there is no consensus on the matter. In this respect, the IPCC Working Group Reports contradict the Summary Report.

When I began to look in more detail at what was known with regard to natural changes it rapidly became clear that other factors of direct relevance to the climate system *had* changed considerably over the global warming time period, and more particularly since the beginning of the century (we will look in detail in Chapter 4 on clouds, 5 and 6 on ocean cycles and 7 and

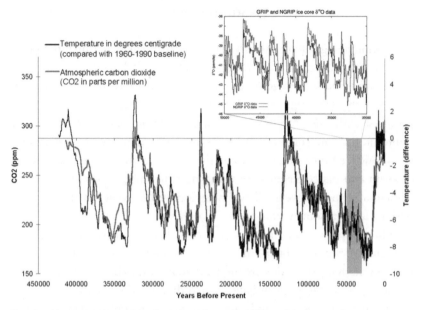

Fig. 2   *Temperature and carbon dioxide varying in cycles during the ice ages, as recorded (a) in Antarctica for the last 400,000 years, (b) the Greenland ice cap between 30,000 and 50,000 years ago. (Source: NOAA Paleoclimatology World Data Center)*

8 on solar science). I was disturbed to find that no attempt had been made to incorporate these factors, some of which were the result of scientific research reported only after the first models had been built.

The basis for prediction

These questionable models that have apparently succeeded in 'predicting the past' are used by IPCC to predict the future (see Ethos A1.1), with the results varying according to differing assumptions about the amount of carbon dioxide that will be released. The middle range forecasts are for carbon dioxide levels to double around mid-century and for temperatures to be forced up by 2–4°C above 'normal'. Some models incorporate more extreme 'feedbacks' whereby increased warmth leads to higher greenhouse gases released from vegetation or sediments, and these can produce rises as high as 6–10°C.

It is generally accepted that a human-induced rise to 1°C above the expected natural range would not be unduly dangerous and that anything above 2°C would be, and much policy debate has centred on how to keep future carbon dioxide levels down such that 2°C will not be exceeded. I find this proposition dangerously simplistic. The 'danger' limit is largely based on the fact that past records of climate, both in this interglacial period of 10,000 years and in the previous interglacials, show a two degree limit above the current mean in their fluctuations. If the planet has not been any hotter in the protracted 'era' of glaciations which goes back hundreds of thousands of years, then – so the reasoning goes – we had better not stray outside of that regime.

As I will argue, we are *already* dangerously vulnerable to the natural climate, but not because of anything unusual that the climate may do, rather because we as human society have changed drastically, multiplying our population and resource demands with every generation and becoming ever dependent on narrower margins of production, whether it be food, water or construction materials. We have colonized places that any palaeoclimatologist would have advised against, such as low-lying coastal areas in hurricane regions and floodplains in monsoon zones; we have decimated forests that protect against mudslides and which store water and release it slowly, and we have crowded vast numbers into vulnerable housing projects – whether they be energy demanding high-rise apartment blocks or huge insanitary shanty towns.

It is a curious and disturbing experience as an ecologist to watch huge investments being made now to solve a problem in 50 years time when it is clearly obvious that problem exists here now and we need very large investments in adaptation to deal with them. Investment in adaptation is minuscule in comparison to attempts at mitigation (by reducing emissions).

I will make a more detailed critique of these computer predictions. It is clear to me that they overstate the future impact of carbon dioxide and underestimate the power of natural cycles. If I am wrong, then even within their own terms these models and policies based upon them distract attention from the fact we are *already* committed to an increased danger involving amplification of the impacts that we are already experiencing and this will happen with certainty over the *next two or three decades*, whatever the success of the emission control scenarios.

The signal and the noise

In the IPCC graph used to ascribe the cause of global warming to greenhouse gas emissions (Fig. 1) the sudden post-1950 rise looks significant.

When it is shown as a major rise on a graph of 150 years, as in the Hadley Centre presentation in Fig. 3, one of the most common representations of global data, it still looks impressive. Thus, we can see the apparent reality

Northern Hemisphere

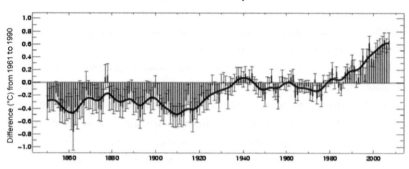

Southern Hemisphere

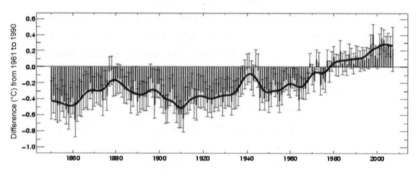

Global Average

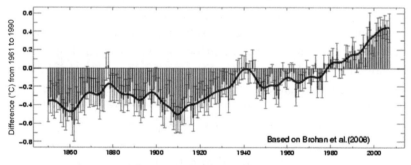

Fig. 3 *Average near-surface temperatures 1850–2007. Temperatures are expressed as annual anomalies. The grey bars indicate the uncertainty of the data points; the thick line is an 11-year running average. (Source: Hadley Centre, UK Met Office)*

of global warming – a steep rise when it 'shouldn't have', according to the model. If this graph were extended backwards to include what we know about northern hemisphere temperature variations over the past 10,000 years not only would the 50-year signal disappear, our attention would be drawn to cycles of peaks and troughs running at roughly 1500- and 400-year intervals. But because this prior period is only accessible through the proxy record with much greater uncertainty in calculating a global mean, an exact comparison cannot be made.

In addition to the surface record, which shows this steep rise, it is also worth looking at the temperature record higher up in the atmosphere. It is known to closely follow the surface temperature in pattern but with much less of a pronounced trend. Atmospheric temperature has been measured since 1979 using either instruments on weather balloons or microwave sounding units (MSU) from satellites. Some specialists argue that air temperature measurements in the lower troposphere (about 3000 m) using these techniques give a more accurate picture of global change than surface installations. But the satellite methods also have their detractors, who argue that trends are difficult to establish as satellite orbits change and instruments wear out more rapidly. An example of satellite derived data is given in Fig. 4.

Here the monthly data of the period of satellite observations is presented and more clearly shows what appears to be an irregular cyclic pattern with a recent fall back towards the long-term mean. For the first 20 years of this period there was no significant trend in lower troposphere

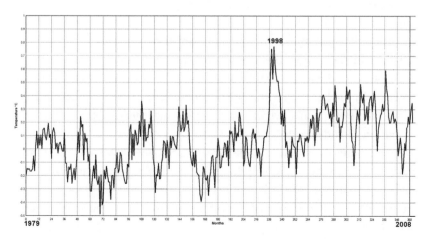

Fig. 4   Monthly mean (anomaly) of lower troposphere global temperature, 1979–2008. (Source: University of Alabama at Huntsville, USA)

temperatures until the major El Niño in 1998, which marks the peak in this graph. Some specialists regard the following period as strongly influenced by that event. Volcanic eruptions disrupt any cyclic patterns and there are two in this time's series, El Chichon in 1982 and Pinatubo in 1991, both suppressing temperatures for nearly three years by as much as 0.25°C and both occurring at times when temperatures might have been elevated by El Niños as in 1998.

This atmospheric data shows what may be a more immediate response of the planetary system to changing natural conditions. At the surface, temperatures are more influenced by the stored heat of the oceans. We shall see in Chapter 5 when we consider the role of the oceans that recent studies have shown that the rise in land temperatures is driven by transfer of heat from the oceans (rather than by trapping of heat over land by greenhouse gases). However, we can see from Fig. 3 that the southern hemisphere surface temperatures have peaked and may now be in decline, and that these cause the overall global average to form a plateau.

Longer-term regional data reveals cycles

There are greater fluctuations than we see in the twentieth century if the record is extended beyond the period for which we have reasonable global measuring stations. However, we don't have to go to proxies entirely in order to see a cyclic phenomenon at play. These cycles are sometimes obvious on a regional level where data goes back sufficiently. In Fig. 5 we can see that temperatures taken from the instrumental record in the North American continent show clear evidence of a warm period between 1750 and 1800 that just misses the previous global data graphs.

This data shows how the signal is damped across the whole northern hemisphere and more variable over a single continent, such as North America. In the latter case, the late twentieth-century rise is only 20–25% above the 1940s peaks, which are generally regarded as little influenced by carbon dioxide levels.

A recent recalculation of data in the USA now places 1934 as the warmest year in the US record. In data sets such as Hadley in Fig. 3, the cut-off at 1850 fails to show any previous warm period. Thus the peak around 1940 in the middle of the Hadley set would not be suspected as part of a cycle.

The importance of the longer-term cycles, as indicated here, will become evident when we consider the role of the oceans in global

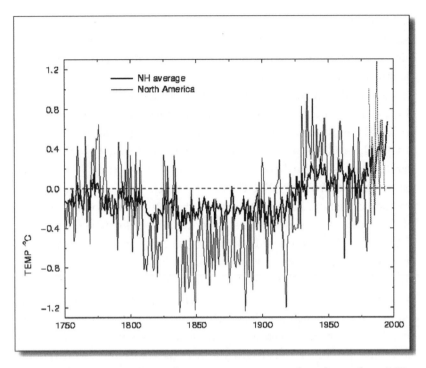

*Fig. 5   Long-term temperature fluctuations: average northern hemisphere (NH –
darker line) and North America (more variable lighter line), 1750–2000,
expressed relative to the 1902–1980 average. (Source: NOAA)*

warming, but proponents of anthropogenic greenhouse gas as the main
driver focus upon the 'general trend' over the century rather than on
such cycles. However, trend data can also work the other way, and if
trends were plotted for the last ten years they would be negative (see
Ethos A1:2). The problem with trend-thinking is that it ignores and
effectively disguises the cyclic phenomena which are evident in all of
these graphs. Such cycles have assumed much greater significance in
the last year of debate, partly because temperatures have fallen despite
expectations (for example, both Hadley and NASA observers predicted
a record year in January 2007 when in fact this year showed a very
marked fall),[1] and partly because the recent fall can be ascribed to the
influence of a Pacific Ocean cycle of approximately 30 years duration.

There are also longer-term cycles. Many analysts regard the steady trend
from the beginning of the nineteenth century as a long climb out of a 400-
year trough in global temperatures that is part of a low-frequency cycle
evident in the northern hemisphere and in parts of the southern. The last

such global low was marked in western Europe by freezing winters (roughly from 1650 to 1850) as well as cloudy, cool summers that affected crop production and brought widespread famine and social unrest (**Lamb, 1995**).

This periodic fluctuation appears as part of another low-frequency cycle of about 1500 years duration that is discernible in a range of past environmental indicators such as tree-rings and sediment patterns. The wealth of scientific evidence for this cycle is well summarized in Fred Singer and Dennis Avery's recent book *Unstoppable Global Warming – every 1500 years* (**Singer & Avery, 2007**). They collate much of the data that supports a natural causation. The problem is that these authors go on to support a laissez-faire and business-as-usual approach to development.

A great deal of the scientific literature on these cycles contains correlative data with proxies of the sun's activity, in particular the strength of the solar wind. Fig. 6 shows the fluctuations of the solar wind as recorded by the proxies of isotopes[2] in annual layers of ice in both Greenland and Antarctica. As we shall see, this cyclic pattern is mirrored by the proxy data for temperature, such as sediment patterns and ice-rafting in the North Atlantic. Not all scientists agree that this isotope record provides an accurate picture but certainly a large body of evidence supports the existence of powerful cycles, even if there is no consensus on exact tem-

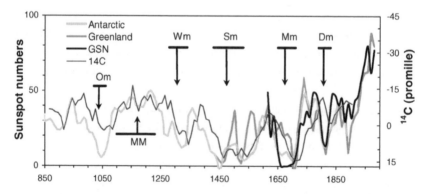

Fig. 6   *Fluctuations in the isotopes of carbon and beryllium and sunspot numbers as indicators of changing solar activity between AD 850 and 2000. The sunspot numbers (GSN) are known from 1610 onwards.*

Om – Oort Minimum; MM – Medieval Maximum; Sm – Spörer Minimum, Mm – Maunder Minimum; Dm – Dalton Minimum; Wm – Wolf Minimum

(Source: Usoskin, 2002)

peratures (these proxies are also used as temperature proxies because of their correlation during the instrumental record).

There is a good correlation between the Medieval Maximum of solar wind activity and a warm period recorded in northern and western Europe during which the Vikings settled and grew crops in Greenland, and a very good correlation between the troughs of 1450 and 1650 shown in the isotope record and a dip in temperatures throughout the globe estimated at 1°C in the northern hemisphere and 0.5°C in much of the southern.

The isotope and sunspot number record shows a striking rise of solar activity from 1810 to 1990. This science was so badly reported at the time I began this review that I had no idea there was a large body of solar-terrestrial physics that documented the rise and its very unusual nature. We shall give more detailed attention to this subject in Chapter 7; here we are mainly concerned to establish the presence of such strong cycles. It is still a missing factor in many official analyses and presentations – something I find quite astounding, considering the body of science that underpins its relevance.

Cycles, variability and ascribing likely causes

As we noted, references in IPCC's 4th Summary Report cover over an enormous controversy in the science literature following their previous publication of the infamous 'hockey stick' graph that smoothed out all the proxy data associated with the Medieval Warm Period between 900 and 1200 years ago and a cool period, the Little Ice Age, between 1250 and 1850. This assessment was curtailed at 1000 years BP. If it had been taken further back it would have had to examine the Roman Warm Period and, prior to that, the Holocene Climate Optimum of 8000 BP, reckoned to have been up to 2 degrees above the 1900–2000 mean for the northern hemisphere.

In the first of these major warm periods the Viking settlements in Greenland were able to grow crops and raise cattle. In a previous warm period about 6000 years ago, recent evidence from fossil beaches in northern Greenland and Canada shows that the Arctic must have been very largely ice-free in the summer (see Chapter 6). Part explanation of these longer-term cycles lies with the tilt and orbital variations of the earth in relation to the sun – the Holocene Climate Optimum was certainly influenced by these orbital factors, which do not imply any change in the source output of the sun. Another factor relates to potential long-term oscillation of the oceanic system at about 1500 years, which may be entirely 'internal' to the dynamic of the earth. Shorter ocean cycles in various ocean basins and harmonics of their relationship may explain much of the remainder.

Set against such natural variability, the late twentieth century does not look unusual and could readily be assumed a natural cycle recovering from a low in about 1810 (we will examine the work of several respected palaeoclimatologists who share this conclusion). The tightly worded paragraphs of the IPCC on this past history above must be seen in that regard. Their statement of high confidence applies only to the northern hemisphere and to the last 500 years, and this confidence drops when the timescale is extended to 1300 years. There is no highlighted commentary on longer timescales when the cyclic nature of warm periods would become apparent.

If the extent of natural *variability* has increased as the Panel acknowledge, and they admit to not knowing the causes at work, how can they then have *more* confidence that the late twentieth-century increase is not part of that natural variability? In fact, the IPCC's increased confidence stems entirely from an ever greater number of computer simulations, rather than any greater understanding of natural cycles. All of these simulations apparently come to the conclusion that when natural factors (i.e. random variability) are put into the models they cannot generate the warming that has been seen, whereas once carbon dioxide is added to the mix they do. When these simulations are examined, it is clear that they do not incorporate a model of ocean cycles, neither do they possess mechanisms to link the solar wind and cloud formation. In fact it is widely acknowledged that the relation of warming oceans, water vapour and clouds has never been reliably computed, yet the response of natural water vapour to the increase of carbon dioxide is, as we shall see, central to computer simulation and prediction.

Carbon dioxide science is not well established

If the arguments relating to the power of natural cycles are correct then the power of carbon dioxide to influence the climate has been seriously overestimated. It is generally assumed that this power is simple to calculate, but this is far from the case. There are two chief reasons for this: one that is frequently referred to and the subject of much 'controversy' (and lack of consensus within the IPCC) is the relation of carbon dioxide to water vapour and cloud; the other has only arisen more recently and concerns what might be called the 'saturation' effect, where each additional unit of carbon dioxide brings diminishing power to warm the atmosphere.

The role of water vapour and cloud

In relation to water vapour, carbon dioxide is one of a number of naturally occurring greenhouse gases, without the panoply of which global average temperatures would fall to $-15°$Celsius. The earth would be a frozen planet. However, carbon dioxide is not the *main* greenhouse gas, though it is the main anthropogenic addition (and the reason we shall focus upon it in this review). The main natural greenhouse gas is invisible water vapour and the greenhouse 'effect' that results is also mediated by condensed water vapour in the visible form of clouds.

As all textbooks report, the natural or pre-industrial level of carbon dioxide is 280 parts per million (that is roughly one molecule in every 3000 molecules of air). This has risen to about 380 parts per million under the influence of fossil fuel burning and land clearance. It is generally agreed that natural levels of carbon dioxide have not risen appreciably. I am not going to examine this assessment, although concerns have been raised as to the accuracy of the baseline data, with some assessments showing considerably higher levels in the nineteenth century (**Beck, 2007**). It is less generally known that all modellers are in agreement that on its own account even a doubling of carbon dioxide concentrations would have a minimal effect on the heat balance (see later argument). All the models *assume* an amplifying factor relating to carbon's interplay with water vapour. This amplifying factor is estimated in IPCC models at 300% – but it is *entirely* theoretical. There is no evidence for it outside of the models.

This unproven amplification is based on the assumption that any warming caused by carbon dioxide also increases the capacity of the atmosphere to hold more water vapour, which is a potent greenhouse gas and hence a positive feedback is created. The Summary Reports and all references to this basic science do not refer to it as controversial – yet controversy surrounded the assessment from the very outset and within the IPCC body of experts. Richard Lindzen, professor of meteorology at MIT, questioned this assumption (**Lindzen et al., 1982; 2001**). He has been a long-term member of both the IPCC and the US National Academy of Sciences special panel on climate change (convened to advise on the reliability of the IPCC's 3rd Assessment Report in 2001), and has consistently argued that additional water vapour could readily turn to cloud and hence reverse the supposed 'feedback' effect (**Lindzen, 1991**).

To emphasize the importance of this issue: if the computed mid-range future projected warming expected from a doubling of carbon dioxide levels (to 560 parts per million) were an admittedly dangerous 2.5 degrees Celsius, this expectation would have to be reduced to 0.8 degrees and become of no serious consequence. The planet's ecosystems have regularly handled such variations in the ecologically recent past (when species and ecosystems were much the same as today). It was for this reason that Al Gore's famous mentor and professor of atmospheric physics, Roger Revelle, who monitored the carbon dioxide increase in the 1950s and to whom Gore paid due homage in his documentary, never agreed with his pupil that doubling would be a problem. (Gore did not mention this disagreement and we look at this saga later when we examine the politics of this science.)

We should note at this stage that the 'doubling point', which depends on the rate of economic growth and the mix of fuel used to generate that growth, as well as the continued capacity of the oceans to absorb carbon dioxide, is variously estimated to arise between 2050 and 2100.

The law of diminishing returns: carbon dioxide and saturation
The second issue is more obscure but nevertheless real, and in my view will become much more controversial. It relates to what is termed the 'saturation' point for carbon dioxide's effects – a point at which adding more carbon to the atmosphere has virtually no additional warming power. Strictly speaking, it is not a saturation point – rather, the power to warm diminishes rapidly in relation to increasing concentration such that each additional unit has a tenth of the power of the previous addition. This logarithmic relationship should already be built in to the computer models for atmospheric heating, but those computer models are not so readily accessible to outside specialists, nor transparent to non-specialists. The atmospheric model is complex – it involves differing concentrations of many gases at different altitudes, all with interactions in relation to heat exchange at infrared wavelengths. It is likely that there are many choices of variables; in fact although the models all predict warming, the wide range of the projections obscures just what variables are used.

Furthermore, when the output of the atmospheric models upon which all projections of future global warming is examined, all models show a hotspot developing in the mid-atmosphere (the troposphere at between 5 to 10 km) yet such a hotspot has not been recorded in the instrument data. This has been cause for several leading atmospheric physicists, including those on the IPCC Working Groups, to question the reliability of the model (for example, John Christy, professor of atmospheric physics at the University of Alabama at Huntsville, custodian of one of four major global atmospheric data sets – see Douglas & Christy, in press).

Increased controversy has emerged with the more recent work of the Hungarian physicist Ferenc Miskolczi, an atmospheric physicist formerly at NASA, who places estimates of the warming power much lower than the IPCC and moreover develops an argument that predicts a *fall* in water vapour content in the troposphere as a consequence of increased carbon dioxide – something Miskolczi claims has actually been observed. This theory argues that compensatory mechanisms reduce the carbon effect virtually to zero (**Miskolczi, 2007**).

And most recently, in the July issue of the newsletter of the prestigious American Physical Society, Christopher Monckton outlines in a long critical essay the way in which the IPCC has chosen its physical variables and relies upon this unproven water-vapour feedback in its equations. This is the point made at the very outset by Richard Lindzen, who, as we will document, has been consistently sidelined by the committee process. But what is unusual is that the APS have considered the matter important enough to devote a forum piece to the issue (**Monckton, 2008**).

The role of the APS in this controversy is of great interest to me, as I recall how, in 1976, this prestigious society similarly broke ranks with the scientific establishment (mostly of engineers) who had covered up all knowledge of the 'meltdown' potential of American nuclear reactors. The APS revelations and the subsequent publication of the Rasmussen Report (and the Three Mile Island incident in 1979) effectively ended the expansion of nuclear reactors in the USA. I have not been able to follow the discussion that this opinion piece has furthered, and reference here to highlight this issue of the controversial 'gain' factor in the carbon dioxide equations.

In summary, therefore, the foundation of carbon dioxide science is not without controversy and the assumption of a consensus on the issue is unsound. Unless, that is, one is content with a politically constructed consensus where known dissenters are simply excluded from the discussion.

Divisions in climate science reflect mentality

The world of climate scientists appears to me to be divided into two sorts – like the hemispheres of the planet which behave differently with regard to climate, and the hemispheres of the brain which think differently. There are those who see natural variability as simply a statistical quality of the data – these are mainly the computer modellers, mathematicians and physicists. They may have little or no actual field experience of living systems and posses only limited ecological understanding of either the present complexity or past history. I can find many situations were there is an assumption made that such variations are simply random or the operation of *chaotic* elements, such that these small random variations are amplified (as in the problems of predicting the track of weather systems).

And then there is another camp that studies *cycles* and *periods*, dealing with oscillations and wave phenomena, with the potential for *harmonics* when two or more waves come into phase and peak together or cancel each other out when phases interfere. This camp is mostly drawn from hands-on oceanographers and works with complex mathematical wave-form analysis to see beyond the noise and discern cyclic patterns. These researchers are also alert to *timing* mechanisms that may lie outside of the earth's system, such as the solar cycle, and of the operation of time lags and teleconnections (long-distance connections where what happens in one ocean basin, such as the Pacific, influences what happens in another, such as the Atlantic).

There are many oceanographers who have extended their work to include an examination of solar cycles and the peaks and troughs of temperature, sometimes well aligned and sometimes with time lags. There is now a growing body of solar-terrestrial science (which I will review) that argues for a causal relationship to the late twentieth-century temperature rise because the correlations are so strong. But this body has grown only since the first papers in 1991, some time after the IPCC and many more players had nailed their colours to the carbon dioxide mast.

Constructing a false consensus

One thing is clear, that by the end of the last century the massive scientific effort that had been devoted to climate studies had failed to provide sufficient evidence to establish beyond scientific doubt that the observed warming was *not* a natural phenomenon. Yet when one considers the language of the media, many science magazines, environmental lobby groups and, of course, national governments and their agencies, the question of scientific uncertainty has been replaced with terms such as 'settled science' as if the question had been put beyond doubt *by the findings of the UN panel itself*. Much store was set by the 'consensus' of 1500 or more experts and reviewers from the science community.

We will see clearly in the following chapters that the 'consensus' that is reported within the UN panel is very limited, especially on issues of causation. Statements of consensus must be examined closely. There is little value, for example, in a statement that simply states 'emissions of greenhouse gases from human activities are changing the climate' when no indication is given as to the quantitative effect – by 10% or 80%? Almost all climatologists would agree that some effect is highly likely, if difficult to prove beyond scientific doubt, but the crucial issue relates to the power of the driving force. Representatives of the IPCC have stated that 'most' of the warming of the twentieth century is anthropogenic, but this statement is not supported by a consensus of scientists within its own Working Groups. A reading of the Summary Report is ambiguous in this regard, stating only that the 'net effect of human emissions has been of warming'.

From the outset scientific members of the Panel's Working Groups, senior within their disciplines, have not been in agreement on the most important issue of what drives the changes we have seen. It follows, therefore, that a similar lack of consensus has existed with regard to projections of future warming. There is a large discrepancy between what is debated and discussed in the Working Groups of the IPCC – each of which provide technical reports, and the summaries of those technical reports; and then even more so, the construction of the Summary for Policymakers that is the main and often only document that policy makers read.

Knowing how science, committees and the UN operate, I do not believe these discrepancies are simply matters of incompetence. I believe there is a strong directorial effort to find a language that provides media and policy makers with the simplistic statements they need to mobilize opinion, and that many of the scientists sincerely believe it is their responsibility to do so. They would argue that presenting alternative hypotheses would not only be confusing but would undermine faith in the science and provide little basis for effective action.

There is a major problem in this approach in that, however sincere, it sets the committee up for conflict over new data and analysis that might 'rock the boat'. This is the issue of what I will term 'prior commitment'. In this respect, solar-cloud connections were mooted in 1991 as a possible explanation for an observed correlation of northern hemisphere temperatures and strength of the 11-year solar cycle. The first definitive papers arose between 1997 and 1999. In its 2001 Assessment, the Panel felt obliged to refer to the new theories as 'controversial' and did not allow them to significantly disturb the central tenets relating to carbon dioxide. However, having taken such a stance, they are necessarily more inclined to maintain it in the face of further research. We will examine the solar-cloud story in some detail as it played out in the IPCC's 2007 Assessment.

There are other areas of research that were just breaking into print as the 2007 Assessment was being finalized and which would have had serious consequences for the level of consensus. It is always the case that such major assessments as that of the IPCC, which take several years to compile, are out of date as soon as they are published. One such area related to a key element of the predictions of future warming which incorporate a significant contribution from the past heating of the oceans. For example, in the 2007 Summary Report, the Panel state:

> Observations since 1961 show that the average temperature of the global ocean has increased to depths of at least 3000 m and that the ocean has been absorbing more than 80% of the heat added to the climate system. Such warming causes seawater to expand, contributing to sea level rise.

Yet more recent analyses by expert oceanographers in 2006–07 cut by a half the estimate of past increases in heat content, and the most recent data showed a global cooling of the oceans. As expected, sea-level rise has tailed off. These controversial oceanographic data were not referenced by the Panel, yet they included other very recent (and disputed) analyses, even to the year of publication in 2007, when these new papers supported their position, for example on a lesser role for the past fluctuation of solar radiation.

The term *controversial* is used regularly within science journals and committees to describe new data and analyses, but it is necessarily imprecise and subjective, reflecting the operation of lines of authority and majority within the institutions of science. The history of all scientific disciplines readily demonstrates that scientific truth has generally only progressed through a process of controversy that necessarily involved a minority analysis being at first subject to opposition and resistance by the majority. The institutional record does not often reflect that such controversy is not necessarily polite and restricted to scientific discourse – *ad hominem* attacks and displays of emotion are commonplace and a reason why many scientists are reluctant to enter controversial territory.

That does not mean that all such controversy denotes the appearance of a new truth, but it does mean that seldom has the appearance and eventual acceptance of a new truth progressed without such opposition. In Part Two, I will give some examples of this process in other areas of environmental science which will prime the reader to the issues of *prior commitment*. It is a classic response of the scientific establishment, most particularly in those areas involving analysis of public safety and environmental risk, that where new data challenges such prior commitment the scientists concerned are marginalized or excluded as controversial, and a false consensus is constructed in their absence.

Towards a natural explanation for global warming

I will outline here a predominantly natural explanation for global warming and the following chapters will provide the evidence. It is a position that is entirely tenable within the accepted bounds of science, with every tenet supported by peer-reviewed work in the specialist literature and within the levels of confidence outlined by the IPCC. It is only the campaigners and those now caught in the machinery of government commitment who cannot allow such a proposition to gain credence and influence policy.

On this latter point, campaigners are quick to pick up on new science that appears to support the scary future (see analysis of the 'tipping points' in Chapter 6), quoting papers that have appeared since IPPC-4 (but which are entirely based upon extrapolation of trends) and arguing for greater urgency in emission controls, but do not highlight the new papers on the errors of the models and assumptions about global dimming and ocean heat content (reviewed in Chapters 3 and 5), nor on new studies which show land temperatures dependent upon ocean heat content (**Compo & Sardeshmukh, 2008**), the levelling off of that warming (**Willis, 2008**), criticisms of the lack of cycles in the models (**Koutsoyiannis et al., 2008**), or recent model runs using enhanced solar factors (**Scafetta & West, 2007**). I will deal with this issue in Chapter 6.

Naturally, much of the work I quote is deemed 'controversial' by the

IPPC and all who have allied themselves to the apparent 'consensus', but I will show that the evidence for this alternative analysis is strong enough to have been discussed by the IPCC's Working Groups and to represent a significant *lack of consensus* among its own experts, something that has been deliberately obscured by the UN machinery and colluded with by most of the world's scientific institutions.

This evidence is strong enough to justify, at the very least, the creation of an alternative model – a 'what if' analysis relating to solar-cloud theory and ocean cycles that would parallel the standard model. Though such theory may not allow exact mathematical formulation, approximations would be as justified as the current use of clearly flawed computer models providing spurious levels of accuracy. Allowing for such a high level of uncertainty, the issues of prediction, mitigation and adaptation should then be readdressed. The alternative argument is this:

i)   that the main driver of recent global warming has been an unprecedented combination of natural cycles of warming (and cooling), primarily in ocean basins and operating through a system of teleconnections, that have oscillated and peaked together;

ii) this peak has risen on top of a longer-term cycle of solar heating known as a Solar Grand Maximum – an increase of 230% in the strength of the solar wind (a plasma flux of electrons and protons) over the last century (from 1900 to 1990) – together with a small but not insignificant increase in visible short-wave radiation and a larger and potentially very significant increase in UV radiation;

iii) this solar heating involves a combination of increases in short-wave and UV radiation that occur in pulses of 11-year cycles and a little-understood solar-terrestrial mechanism that has reduced cloud cover sufficient to create a strong warming pulse over the period 1980–2000;

iv) such cycles are a feature of the current 10,000-year interglacial and a combination of long-term ocean circulation and shorter-term ocean basin cycles, particularly in the Pacific, that remix and re-circulate the accumulated heat of each cycle – with much of it dissipating on a much shorter timescale than current models assume.

This argument or hypothesis can be directly tested by *observation*. In particular:

i)   observations which support the concept of teleconnection between Pacific and Atlantic Ocean cycles, Arctic Ocean temperatures and ice-melt (and there are good data to show previous Arctic melting cycles);

ii) proxy evidence for the previous solar magnetic 'highs' and 'lows' and coincidence with proxies for temperature variations;

iii) satellite data demonstrating a decrease in low cloud cover from 1980 to 2000 and a consequent increased flux of short-wave radiation to the ocean surface (of sufficient power to account for virtually all of the late twentieth-century warming);

iv) evidence for a phase change in 2001, with increased cloud cover, cloud albedo (reflectance of light to space) and consequent falling global temperatures;

v) breaking research on the mechanisms that connect the sun's behaviour to the earth's climate system.

It is important to note that I have introduced the term 'main driver' rather than an 'either-or' with respect to 'is it natural or is it man-made'? Surprisingly, this either-or type of thinking, creeps into even the highest levels of scientific discourse. In truth, as virtually all working scientists would readily admit, the causation for global warming involves multiple factors.

Notes

1. In February 2007, writing in *ECOS*, the quarterly journal of the British Association of Nature Conservationists, I drew attention to the unusual advance prediction of the Hadley Centre, drawn from their increased confidence in their computer simulations. Not only did they expect 2007 to be a record year, they expected an El Niño in the Pacific. I pointed out that Theodore Landscheidt, a much-maligned critic of the IPCC and someone working with solar cycles, had predicted in the refereed literature in 2003 that the globe would first register a cooling in 2007 and that 2002 would be the last major El Niño year for some decades.

2. An isotope is a variant of an element, for example, carbon, which is chemically identical but physically lighter or heavier. In this case they are produced by the impact of cosmic radiation on gases in the atmosphere. They can be incorporated into tree-rings, ice or sedimentary material.

2

Natural Causes

Thus far, we have determined that significant uncertainty exists with regard to the causes of the late twentieth-century warming and that the perception that human influences dominate lies entirely upon computer models. We have also discussed that, as a general principle, a computer model is not 'validated' simply because it replicates a pattern. It is well known within the modelling community that all manner of tweaking occurs in order to get a model to replicate the past. The fact that it has done so does not mean that it actually simulates natural processes, and we have noted that major flaws have become apparent in those models. We now turn to a more detailed examination of the natural factors at play.

In the warming case based upon these models only the general curve for 1900–2000 is replicated together with cooling blips from the major volcanic eruptions. The actual centennial instrument record includes finer and periodic fluctuations. These cycles include:

- the El Niño Southern Oscillation (ENSO), which is an irregular 4-to-5-year cycle which at its peak in 1998 drove world temperatures to a record high;
- the Pacific Decadal Oscillation (PDO), which has a 30-year cycle only recently fully elaborated and which in 2007 depressed temperatures world-wide as it entered a 30-year reversal of the previous 'positive' phase;
- and the Arctic Oscillation (AO), which has a 60–70 year cycle linked also to a North Atlantic Oscillation (NAO) of similar period that strongly affected northern hemisphere temperatures during its previous peak in the 1940s.

A controversy is now unfolding as it becomes evident that these linked cycles peaked towards the end of the century. They have clearly driven global temperatures to their late twentieth and early twenty-first century high, and as they turn are now responsible for a distinct cooling. The modellers and defenders of the orthodoxy are responding with admissions that the models were inadequate and global temperatures may now fall, but that greenhouse warming will return with a vengeance at a later date. At least one revision of the models has now incorporated an ocean oscillation, and is predicting that a cooling will indeed ensue – until at least 2015 (we will examine this when we look at ocean cycles later).

The defenders of the orthodox now admit that natural factors are dominating and producing cooling. Their defence is that the current cooling is random natural variability rather than cyclic change, but this conveniently ignores evidence that these same factors could have created virtually all of the warming in the first place! The main reason for this disjointed logic is the compartmentalized nature of climate research and modelling. This ought not to be the case – after all, the climate system requires systems thinking and a broad grasp of planetary ecology. But only in the last few years have the links between oceanographic patterns of temperature fluctuation and cloud cover been appreciated. And even so, few oceanographers then appraise themselves of the pioneering work of solar-terrestrial scientists on the fluctuations of the magnetic field, or experts on clouds and the satellite measurements of albedo (the amount of light and hence warming energy that is reflected from the earth's surface, including clouds and aerosol particles). Often, researchers in these fields will add an obligatory obeisance to recent anthropogenic global warming as if proven and accepted but without any knowledge of atmospheric physics or any awareness of the major uncertainty with regard to causation.

When a broader view of the planetary system is taken, it becomes clear that the long-term behaviour of clouds is the controlling factor in climate. As little as a 1% reduction in reflective low-level cloud is enough to create a global warming signal equivalent to that computed for carbon dioxide. Yet it is freely admitted by climate modellers that the models do not have a reliable cloud component. In fact, the observed cloud changes that we will document below are simply taken as 'feedback' from the carbon dioxide greenhouse effect – even though the models do not contain reliable feedback mechanisms and have not predicted the clear decline in low-level cloud that has been observed (and which we shall review in detail in the following two chapters).

The key factor of variable cloud cover

We shall see from the satellite data that cloud cover changed by 4% overall between 1983 and 2000. Furthermore, in some data sets, cloud cover can be seen to fluctuate with the 11-year peaks and troughs of the solar cycle. These changes have not been widely discussed. Yet they present an opportunity to test the hypotheses with regard to natural cycles. Cloud cover is related to ocean temperature and atmospheric circulation patterns and it varies with the oscillations we have mentioned above – indeed, it is likely to be part of the causal chain for the periodic changes in temperature and atmospheric pressure characteristic of each cycle of ocean warming and cooling. Further, the hypothesis that solar cycles affect cloud cover directly by acting upon the seeding process (reviewed in Chapter 8), if correct, would provide a mechanism for connecting ocean cycles with solar cycles – and there is strong correlative evidence that they are so linked.

The data that I present in the following chapters has not been widely discussed in the public realm, yet it is available from the International Satellite Cloud Climatology Project and has been reviewed in the specialist literature. It has even featured in detailed but closed discussion in the 2007 IPCC Technical Working Groups. In my view, it is conclusive evidence that natural factors dominated the global warming period but IPCC does not draw this conclusion. The cloud data and additional radiation measurements at the earth's surface made by satellite shows clearly that a measured increase in short-wave (SW) radiation, or visible sunlight, reached the oceans and is quite enough to drive the whole of the late twentieth-century warming. This is the physical, observable evidence that supports the natural explanation and it provides no direct support for the carbon dioxide theory.

The long-term trends in the cloud and radiation record were fully analysed only very recently and published in the journal *Science* in 2005. This was within the time-frame for IPCC-4, and the Panel report these reviews, together with some doubts and questions relating to reliability of satellite data, but fail to represent in their Summary Report the discussion and a significant lack of consensus. The two factors of ocean and cloud variability should have caused serious questions as to whether the IPCC models truly reflect natural processes. The IPCC graph shown in 2007 (Fig. 1) only extends to the year 2000. In the ten years from 1998 to 2008, global temperatures have remained flat, and one of their sharpest falls was recorded in 2007–08.

In defence, modellers claim that all simulations show variability, and the current dip is within the (outer) bounds of that variability. It is therefore merely a 'blip in the general trend'. The problem with this defence is that the blip is proving as large as the previous downward blips caused by major volcanic eruptions, which they have successfully simulated. In this argument, the modellers are making an assumption that all past variability is essentially random in nature. As we shall see, there are grounds for believing that solar factors determine this variability and it is not random and not all modellers are disregarding the solar factor (**Scaffeta & West, 2007**).

There is accumulating evidence that the 11-year cycle of fluctuation in the solar wind causes a 3% variation in low-level reflective cloud cover. If correct, this would lead to a pulse of warming sunlight (radiation of short wavelength) and we can test this assumption by perusal of satellite data on the amount of SW radiation reaching the earth's surface. The data show clearly that such pulses exist. When this evidence is coupled to evidence of a long-term rise in solar output from 1900 to 1995, which by inference would lessen cloud cover, then it is clear that these factors are powerful enough to account for the global warming signal in the twentieth century. The final confirmation would be a downturn in global temperatures as solar output declined from its peak of 1990–95, cloud increased and more warming sunlight was reflected. This is what has been observed, with a time lag in peak temperatures at 1998, a reversal of cloud trends in 2001 and a sharp fall in surface temperatures in 2007.

In the detail, we will look at the flux of *reflected* SW radiation at the top of the atmosphere as measured by satellites. We will see that satellite data confirm the hypothesis that significant changes in cloud cover have occurred. The strength of this effect is expressed in the same units used to measure the carbon dioxide effect – watts per square metre, and these observed fluxes are seen to dwarf the computed effect of the greenhouse gas increase. This brings into serious question the reliability of conclusions based upon the IPCC graphs in Fig. 1.

This evidence for cloud change and short-wave radiation as the dominant factor needs to be seen in parallel with the evidence that the orthodox model is flawed with regard to its ability to model the past pattern in relation to aerosols. Global temperatures first began their marked twentieth-century rise between 1920 and 1940, well before carbon dioxide began its own steep rise, and as carbon dioxide emissions rose in the post-war boom global temperatures actually dropped and did not return to 1940 levels until 1980. The interpretation of this disjunction is important with regard to the models' abilities to replicate the past.

The 1945–80 cooling period and false attribution

As we noted earlier, modellers claimed success in tracking this past 'global dimming' by replicating the action of sulphate aerosol – a consequence of both volcanic activity and fossil fuel burning. There was a major volcanic eruption in 1962 (Agung in Indonesia), and sulphur emissions in the northern hemisphere rose rapidly in the 1960s. However, they have placed great store by an apparent ability to simulate the cooling period as a consequence of sulphate aerosols resulting from increased fossil fuel burning. As we shall see, this assumption and the methodology were flawed.

We need to spend a little time exploring this cooling period. It coincides with two important elements in the global warming debate: the first is the almost linear rise in carbon dioxide levels over that period; and second, the pattern of peaks and troughs in temperature coincides with the pattern of the varying length of the solar cycle, reported in 1991 by the Danish meteorologists Eigel Friis-Christensen and Knut Lassen. We shall look in detail at this controversial solar factor, but first we need to look at the 'global dimming' phenomenon as it is central to the claims that computer simulation can provide accurate predictive power.

The 1940–80 dip in temperature was assumed to be due to the dimming of sunlight falling on the earth's surface. Industrial pollution of the atmosphere with sulphate aerosols mostly from coal burning was widely believed to be the culprit and the dimming was held to have stopped due to the operation of the Clean Air Act in the USA as well as European Union efforts to combat acid rain and reduce sulphur emissions. These pollution control measures then apparently led to a clearer atmosphere and a resumption of 'global warming'. It is a key aspect of the modellers claim to be able to reproduce the past patterns of temperature fluctuation by including this component in their model of the atmosphere.

The global dimming thesis was often quoted in respected peer-reviewed journals, but with no reference to the data that would support the conclusion that global-scale industrial emissions had followed the necessary pattern. As with many modelling studies, the input data were not readily available in the literature and the process of incorporation into models was not transparent.

Scientific opinion has, however, shifted in the last few years. It is now clear that sulphate pollution from either industrial emissions or other sources such as volcanoes could not have been responsible for the dimming because it was too localized. The most recent reviews of satellite data show that changes in natural aerosols and cloud patterns are implicated and that attributing the source of 'global dimming' to industrial aerosols was led by an artefact of measurement protocols that were biased to land and certain polluted regions of the northern hemisphere.

When global data are analysed the effect is seen to occur in areas where anthropogenic pollution is not significant. Furthermore, recently published indices of the 'dust veil' from volcanoes provide no supporting evidence for their contribution to this dimming. Volcanic dust has to be injected into the upper atmosphere to have lasting effect. There were major eruptions: in 1962, Agung; 1982, El Chichon; and Pinatubo in 1991. Each of these injected sufficient dust to depress global temperatures but the effect lasted for only 2–3 years. Mount St Helens in 1980 did not penetrate the upper levels and had no noticeable effect. Instead, the latest research shows natural factors are implicated that relate to the *transparency* of the atmosphere to sunlight. The key research is summarized in the seminal paper 'From dimming to brightening: decadal changes in solar radiation at the Earth's surface', by Martin Wild, of the Institute of Atmospheric and Climate Science in Zurich, and a team of contributors from the US National Oceanographic and Atmospheric Agency (NOAA) and Russia's Geophysical Observatory (**Wild et al., 2005**):

> Variations in solar radiation incident at the Earth's surface profoundly affect the human and terrestrial environment ... a decline in solar radiation at land surfaces has become apparent in many observational records up to 1990 ... newly available surface observations from 1990 to present, primarily from the northern hemisphere, show that the dimming did not persist into the 1990s ... instead a widespread brightening has been observed since the late 1980s – this reversal is reconcilable with changes in cloudiness and atmospheric transmission and may substantially affect surface climate, the hydrological cycle, glaciers and ecosystems.

This is a clear statement that changes in short-wave radiation at the earth's surface (visible light) need to be studied and their implications realized. The authors do not overtly criticize global warming models but comment that these models do not adequately incorporate cloud changes and solar radiation trends. In fact, Wild became a member of Working Group I of the IPCC's 2007 4th Assessment, yet these comments are not represented in the Summary Report, and we shall look in more detail at how the

Working Group dealt with his findings. I should emphasize at this point that Wild does not necessarily support any of the explanations based upon solar-cloud theory, which he regards as speculative (personal communication).

His 2005 study reports a general decrease of sunlight over land surfaces of the order of 6–9 watts per square metre from 1960 to 1990. However, 'dimming of solar radiation fades after 1985 over Europe and a reversal to brightening is found'.

The authors comment:

> ... the transition from decreasing to increasing solar radiation is in line with a similar shift in transparency of the cloud-free atmosphere which shows a general tendency to decrease up to the early 1980s and a gradual recovery thereafter.

They speculate that this may be due to a decline in the economy of eastern Europe in the late 1980s, but then report that they found the same pattern of reversal around the rest of the world – even in the North and South Pacific and Antarctica. The shift from decrease to increase is also found in Russia, Japan and in 85 rural stations in China. Data on continued dimming are restricted to India, and some central African data shows no tendency for recovery from an earlier dimming in the region.

> The changes in both satellite derived and surface measured insolation data are in line with changes in global cloudiness ... which show an increase until late 1980s and a decrease thereafter, on the order of 5% from the late 1980s to 2002.

which corresponds to

> an increase of 6 watts per square metre in absorbed solar radiation by the globe.

The authors then go on to speculate and draw parallels: first that the increase in transparency in the early 1990s 'reflects the recovery from Pinatubo aerosol loadings'; and in addition, 'air quality regulations and the decline of the eastern European economy may have affected the large-scale aerosol concentration'. And they note that the overall increase for 'clear sky measurements' is comparable to that for all sky conditions – thus 'transparency of the cloud-free atmosphere also contributed to the increase in insolation'.

However, these speculations in relation to the eastern European economy are not consistent with the data and references, which clearly show that any dimming due to man-made aerosols was a localized northern hemisphere phenomenon, and that another more powerful and possibly cyclic factor was at work that affected both the transparency of clear sky *and* the degree of cloud cover. It is evident that transparency increased across the globe and not just in pollution affected areas – the exception is in parts of India and Africa where smog and forest fires are persistent.

The authors make another unsupported statement:

> . . . whereas the decline in solar energy could have counterbalanced the increase in down-welling long-wave radiation from the enhanced greenhouse effect before 1980, this masking of the greenhouse effect and related impacts may no longer have been effective thereafter, enabling the greenhouse signals to become more evident during the 1990s.

No masking the greenhouse effect

No quantitative data is shown to support this statement. The authors draw no emphasis as to the *scale* of the radiation inputs they have monitored compared to the assumed greenhouse gas effect. The latter is computed by the IPCC at 2.5 watts/sq m as the cumulative effect of the build-up since pre-industrial times, with 0.8 watt/sq m of that sum added between 1980 and 2000 from carbon dioxide alone. This compares with 6 watts/sq m observed by the team for the 1980–2000 period, a factor of 7.5 greater than the additional carbon dioxide effect, leaving its computed greenhouse effect at about 11% of the observed late twentieth-century 'radiative forcing' – the term used to describe the driving mechanism within the modelled global atmosphere. Furthermore, no data is presented for sulphur emissions and the timing of their reductions, which in the case of the decline of the eastern European and Russian economy was most pronounced *after* the noted reversal of insolation trends in the mid-1980s; the reversal also occurs in China where pollution levels have been reported as increasing rather than falling. The major economies of the USA and China certainly instigated flue-gas desulphurization much earlier, in the 1970s, with West Germany following in 1980, but other states did not catch up until the mid-to-late 1980s. Perusal of global sulphur emission data shows little change over three decades – the clean-up programmes of the USA, western Europe and the demise of the Russian and eastern European economies appear to be counterbalanced by increased global production. A recent review of global emissions shows a small increase from 1985 to

1990 from 65 million tonnes to 71 Mt (See Ethos A2:1 and **Lefohn et al.,
1999**).

Wild's research team actually *conclude* that previous 'global' dimming
data were exaggerated by a bias to northern hemisphere land measurements (e.g. Germany 1960–80) and that oceanic patterns of cloud cover
and transparency are more important. The IPCC eventually accepted this
conclusion in 2007 but failed to point out the significance for past modelling studies (**IPCC Working Group I Report, 2007**):

> A reduction in downward solar radiation ('dimming') of about 1.3% per
> decade or about 7 W/m^2 was observed from 1961 to 1990 at land
> stations around the world ... However, the stations where these analyses
> took place are quite limited in domain and dominated by large urban
> areas, and the dimming is much less at rural sites or even missing
> altogether over remote areas, except for identifiable effects of volcanic
> eruptions, such as Mt. Pinatubo in 1991. At the majority of 421 analysed
> sites, the decline in surface solar radiation ended around 1990 and a
> recovery of about 6 $W m^2$ occurred afterwards (**Wild et al., 2004; 2005**).

And the Working Group Reports also reference the work of Wild, Wielicki
and Pinker:

> The increase in surface solar radiation ('brightening') agrees with
> satellite and surface observations of reduced cloud cover (**Wang et al.,
> 2002b; Wielicki et al., 2002a; Rossow and Dueñas, 2004; Norris,
> 2005b; Pinker et al., 2005**), although there is evidence that some of
> these changes are spurious. In addition, the satellite-observed increase
> in surface radiation noted by Pinker occurred primarily over ocean,
> whereas the increase observed by Wild was restricted to land stations.

And the conclusion (my emphasis):

> *Possible causes of the 1990s reversal are reduced cloudiness and increased
> cloud-free atmospheric transparency* due to the reduction of anthro-
> pogenic aerosol concentrations and recovery from the effects of the
> 1991 eruption of Mt. Pinatubo.

The latter part of this sentence is internally contradicted by other conclusions in the Panel's Report regarding 'global dimming', but what is
remarkable is that this cloud data – discussed in a different chapter and
specialist group – is not regarded as controversial or contradictory.
Nevertheless, the issue does not emerge as significant in the Panel's

Summary for Policymakers. Furthermore, the comments on Wild and Pinker are not reflective of the degree of agreement in those papers: the increased flux or 'brightening' was observed over *both* land and ocean; it was the 'dimming' by pollutants and the recovery from it noted by Wild that was largely a land-based phenomenon. Thus, although IPCC summarize and reference this material and the trends, they do not translate the issue from Working Group to Summary Report. They thus downplay its significance – upon which there is at the very least *no consensus*, with several of these authors stating in their papers the importance of these data for climatology and assessments of global warming potential. In fact the level of agreement on cloud changes and increased SW radiation flux at the surface is much stronger than any other agreement.

Thus, anthropogenic aerosols are not responsible for the 'dimming' period from 1945 to 1980, nor are pollution controls the source of the brightening from 1990 onwards. It is possible that the major eruption of Agung in 1960 contributed via natural aerosol to the trough of global temperatures and that both El Chichon and Pinatubo dented the rises in later decades, but the evidence points to another driving force connected to clouds, atmospheric transparency (natural aerosols) and the periodic flux of SW radiation to the surface.

Whereas it can be argued that clearer skies in the years following the major eruptions in 1982 and 1991 have enhanced the warming, there are no eruptions capable of having caused the current downturn, the forces of which are clearly powerful enough to counter both the greenhouse gas build-up and the apparently clearer skies.

Wild and his colleagues additionally comment upon recent reported changes in albedo (reflectance of SW radiation from mainly cloud surfaces):

> ... the dramatic increase in surface albedo for 2002–2003 (reported by Big Bear Solar Observatory) lies outside the period of available measurements and is controversial.

In this short comment, the researchers are referring to an abrupt change in the pattern of cloudiness. This is seldom commented upon in the global warming debate and IPCC do not highlight it, yet it would obviously presage a period of cooling. In fact, an anomalous high in albedo in 2002–03 is part of a major change of trend that occurred in 1999–2001. Until that date, the satellite data shows a decadal declining trend in cloud cover and this reversed around 2000.

This abrupt change is still subject to intense discussion. 'Controversy', as we have observed, is a strong term in science circles for data that may go against the grain or clash with other measurements, but this data is nevertheless important and has clear implications for alternative explanations of recent climate change.

These findings should have led to some comment upon the fact that *natural* aerosols contribute significantly to aerosol loading and transparency, particularly over the oceans. The authors make no mention at all of solar theory alongside greenhouse gas theory (GHG). They also fail to point out that GHG theory and modelling has no component that will adequately explain SW flux variations, particularly the noted reversals of trend. And whilst the effect expected from carbon dioxide theory can be computed (it would be markedly a 'down-welling' flux of long-wave radiation and amount to about 1 watt per square metre over the 1950–90 period), it can hardly be 'masked' by a decline of 8 watts per square metre which is then *followed* by a rise of 6 watts. It would be more accurate to say that the computed effects of carbon dioxide are *dwarfed* by the natural cycles of short-wave insolation (and as we shall also see, the long-wave response from clouds). Thus, it would be true to say that any carbon dioxide effect does not stand out above the natural variability and hence the conclusions of the models with regard to a human signature are not supported by the satellite data.

This paper is a classic example of a narrow focus on the part of specialist atmospheric scientists. They appear out of touch with solar theory and make speculations on pollutant emission trends without reference to hard data; and they deliberately ignore other interpretations, however controversial, put forward in refereed science journals. If these scientists *are* aware of wider work, then is it a matter of institutional expediency that they do not address that work? It also reflects badly upon the peer-review process in a journal that is well aware of the 'controversy'. The lack of acknowledgement that such controversy exists contributes further to perceptions that there is a broader consensus outside of the IPCC.

The central role of clouds in determining natural variability

The most important element in this recent analysis of the global warming signal relates to the percentage changes in cloud cover, which increased between 1960 and 1980 when global temperatures fell, decreased after

1980 when temperatures began to rise, and then increased again during a time when warming stalled and the oceans began to cool.

The effect of the changes as reported by this team is very significant. They document a 5% decline of low-level cloud associated with a rise of 6 watts/sq metre in the mean annual flux from 1990 to 2000. *This is a very large figure* in relation to the IPCC estimate for the additional contribution of carbon dioxide over the same period.[1] The IPCC figure for the overall long-term rise is 2.5 W/sq m for the whole range of greenhouse gases over a 150-year period, 1.6 of which is allocated to carbon dioxide.

This team is quick to comment upon the potential masking of the much smaller carbon dioxide influence, but completely fails to comment upon the overall pattern of warming and the implications of its research both for the cooling 'blip' *and* the post-1980 warming. Thus, the 'masking' of the carbon dioxide effect by *natural* cloud and aerosol variability is also followed by an *enhancement* of the warming – what we might more accurately term an *overwhelming* rather than an overshadowing of greenhouse gases by the natural thinning of cloud cover and increase in transparency of the cloud-free atmosphere.

Trends, pulses and periodicity

We also have to bear in mind that these overall means, averages and trends over two decades only tell a part of the story – equally important is the evidence that they are *pulsed* by the 11-year solar cycle. There is a considerable body of evidence correlating solar 11-year and 22-year cycles to other indicators of climate such as sediment, stalagmites and tree-rings, which an inter-disciplinary team would be aware of. The importance of pulsed patterns should be recognized and it is not. There is a distinct tendency on the part of physicists in climatology to focus upon *trends* to the exclusion of *pattern*, periodicity and the operation of cycles.

This team makes no reference to patterns of solar activity, perhaps because that activity has been reported as containing no trend over the period of its study. However, there have been two 11-year cycles from a peak in 1980 to a peak in 2000, with some significant differences. Overall activity in the visible spectrum may have shown no overall trend, but the amplitude of the solar cycle rose to an all-time high in 1990 and fell by 30% in 2000.

These researchers could not have been unaware of the controversial developments in other fields of climate research, in particular the published papers on potential links between solar activity, cosmic ray flux and cloud formation. In the study period, the amplitude of the relevant cosmic ray pulse varied significantly (as we shall see in Chapter 8). And the team should have been as much aware of this and its importance to solar theory as of the issues of global dimming and anthropogenic sulphate pollution (the previous science of which they had effectively shown to be erroneous).

Thus there is strong evidence of cloud changes over timescales of decades that would account for the bulk of global warming, as well as global dimming. This matter has only recently been included in the IPCC Review but the Panel summarize the findings and then fail to make any conclusions on the basis of one or two papers critical of satellite-derived trends, and one relatively limited study that contradicted the trend (we return to this treatment in the following chapters).

Trends in cloud cover and the global albedo

Further research relating to cloud cover was covered in the same 2005 issue of *Science*. Bruce Wielicki's team at NASA's Langley Research Center published the results of global albedo measurements made from satellites. The measurement of albedo is one of the fundamental monitoring exercises with regard to climate change. It is a measure of the amount of energy reflected from earth and hence directly impacts upon warming and cooling patterns. Any trends or patterns here are of great relevance to 'global warming' theory and in particular relate to changes in cloud cover, by far the largest factor in the planet's albedo. Wielicki's work is very important in that it is one of the few recent reports that collate and analyse the past 20 years of satellite data (**Wielicki et al., 2005**).

The researchers emphasize the importance of albedo in that a change of 0.01 (1%) in the index represents 3.4 watts/sq m energy balance change – equivalent to the computed effect of a doubling of carbon dioxide levels. Albedo is influenced by the earth's cloud fractional coverage, cloud thickness, aerosol amount, forest-, snow- and ice-cover. The average reflectance due to clouds, for example, is about 25% of the total irradiance at the top of the atmosphere (see Fig. 7).

In Fig. 7, the input of 100 units represents about 1365 watts/sq metre at the top of the atmosphere, averaged over the globe (it is higher at the Equator, lower at the poles). We should note that Hadley's image is a static model already in a dynamic equilibrium. This is what simulations use as

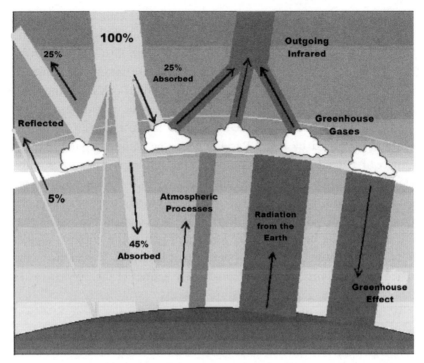

Fig. 7   The radiation balance of the earth. (After the Hadley Centre, Met Office, Exeter)

their starting point, but this is far from reality as it ignores cycles that effectively create a permanent lack of equilibrium. This is an inherent limitation of all models that assume a disturbance of this fictional equilibrium in their prediction programmes as well as any future recovery to an equilibrium point. With regard to albedo, therefore, and the single most important controlling factor of warming and cooling episodes, there is always a pattern of change and none of the current models simulates either short- or long-term cycles. Nor would they attempt to as long as the knowledge of those cycles is inadequate, because the modellers would not be able to accurately locate themselves within the cycle.

In Wielicki's research, study of the effects of volcanic eruptions on the albedo of the earth confirmed the accuracy of monitoring equipment. Following the eruption of Mount Pinatubo monitored in June 1991, stratospheric aerosols increased global albedo by 0.007 and an additional 2.5 W/sq m was reflected over the following two years. This event, which was accompanied by detailed measurements of ocean surface temperatures, depressed the global mean by 0.25°C for a period of two years.

This simple study gives us one form of broad estimate of what to expect in terms of temperature changes per unit of radiative forcing (RF). An eventual doubling of carbon dioxide at an IPCC-assumed eventual RF of 3.4 watts/sq m leads to a 0.34°C rise for the doubling, of which 52% or 1.8 watts/sq m has already accrued (IPCC's estimate of the carbon dioxide proportion of all anthropogenic greenhouse gases). This gross estimate is also of the same order proposed by Richard Lindzen and supposed by Roger Revelle. It can be counter-argued that the impact of Pinatubo was short-lived and the full impact of carbon dioxide has yet to feed through because oceans store the heat; but as we shall see, ocean heat storage has also been grossly overstated. On the timescales of decades, we shall learn that the ocean surface waters cannot integrate the heat into lower levels and that ocean currents rapidly move surface waters to areas of permanent heat loss. Variations in the speed of these currents are the most likely source of the 30-year cycles that we see in the global temperature patterns.

We can also calculate that the 6 watts per square metre excess observed in the decades from 1980 to 2000 would produce, by this relationship, a rise of 0.6°C, which is close to what was recorded. If the maximum temperature anomaly for that decade was 0.7 as NASA data suggest, then only 14% of that could be laid directly at the door of carbon dioxide.

Wielicki's paper also drew attention to the report by Pallé in *Science* (2004, see below) that global albedo had suddenly increased between 2001 and 2003, and this claim by Pallé seems to have prompted Wielicki to publish his recent work because it contradicts those findings. He presents graphs of albedo from the CERES radiation budget instrument operating from NASA's Terra spacecraft. It shows a continued steady *decline* in albedo measured in terms of reflected SW radiation from January 2000 to January 2004. These observations registered a small decrease over this period of about 1 watt/sq metre. This contrasted with the Pallé results, which were calculated by a different method known as 'earthshine' − whereby reflected light from the earth to the moon is measured − and which showed a change of trend in 2001 with a sharper rise from January 2003. This rise was of the order of 4 watts/sq metre or an 0.01 *increase* in the albedo index.

Wielicki argued that such a large increase in albedo ought to cool the earth to a greater degree even than the Pinatubo eruption, and that as this has not been reported it substantiates the CERES instrumental readings where the continued decline of albedo is as expected associated with increasing global temperatures (although in 2005 the researchers were not to know that temperatures had reached a plateau):

> If such changes had occurred ... a global cooling of twice the Pinatubo eruption would be expected even over short time periods — such a cooling has not been observed.

> A second possibility would be a large decrease in global ocean heat storage ... observations of mean annual global heat storage from 1992–2002 show an increase from 2000–2002 of 0.7 Watts/sq metre (sampling noise is 0.4) and this is consistent with the CERES results.

The authors note that cloud changes that do not affect ocean heat storage could be responsible for the earthshine data if thermal infrared radiation to space was blocked at the same time as SW reflected, and state that such cloud feedback in the climate system cannot yet be evaluated and more research is required. The contradictory nature of these two methods of measurements accounted in large measure for the IPCC not drawing attention to the issue in its Summary for Policymakers — yet there was clearly a major issue, with no consensus on the science.

In fact, at the time this analysis was under way, there *was* data showing an unexplained loss of ocean heat content just after this period — from 2003 to 2005. However, this data was not published until November of 2006. Researchers at the Scripps Institution of Oceanography reported a large change (loss) in heat content of the oceans for the years 2003–05, immediately after the time period to 2002 for which Wielicki had data. Before this period, ocean heat content had apparently been increasing. This later data showing heat loss coincided with the earthshine measurements of increased albedo over the same time period.

This new data would support the earthshine data over the CERES data and the cooling response that ought to follow. (As this goes to press, Pallé reports that there has been a reconciliation of the conflicting data sets and the data clearly shows a step-change towards more cloud in 2001, but no further overall increase (**Pallé et al., 2009**). However, the oceanographic data was withdrawn (April 2007) with apologies for instrument error. It has since been redone without the offending instrument set and a much smaller cooling is evident. We shall return to this issue in Chapter 5.

In their comments on ocean heating, these authors do not consider the complexity of the oceanographic response to increased SW radiation and cloud cover. The earthshine data may be picking up cloud changes in regions where they have different impacts on temperature – for example, increased cloud in polar and sub-polar regions during winter months would insulate polar waters from heat loss to space as well as adding to reflectance.

Before we look further at the importance of ocean heat storage, we should note that Wielicki's work is one of a number of papers that refer to a period of changing short-wave (SW) flux throughout the 1990s affecting the oceans' heat storage. The authors comment that these changes are 'not reconciled' with climate models that assume changes in LW flux are responsible for global warming. This is yet another example of low-key statements that are not overt criticisms of the current global warming hypothesis when in fact they are of critical importance and give rise, once more, to the ability of partisan commentators to argue that no real scientific dissent has been published or, as with the IPCC, that a consensus exists on causation.

The possibility of an imminent global cooling

It is often stated in a very loose way by commentators that the correlation of solar cycle length and temperature noted by the Danish team, which correlated to the cooling noted from 1945 to 1978, broke down after 1996, when global temperatures continued to rise. In the downturn cycle length was long and in the upturn it was shorter (associated with higher solar output), whereas when the solar cycle lengthened once more temperatures continued to rise. According to the Danish theory the longer solar cycle of 1986–96 should have heralded a cooling (between 1997 and 2004, Nigel Marsh and Henrik Svensmark added to the work of Knut Lassen and Eigil Friis-Christensen).

In fact there are time lags evident in these relationships, and cooling did not become evident until after much of this controversy played out. Global temperatures peaked between 1998 and 2005, eight years after the cycle peak of 1990, which held relatively high and then fell back, as can be seen from detailed study of the actual temperature profiles. That peak was driven by the El Niño Southern Oscillation (ENSO) and further maintained by a smaller El Niño in 2002. If a running average is used (as in the Hadley graph in Fig. 3), then it appears that temperatures continue to rise after 1998 until the mid-2007 data. Such a running average was used by many of the critics of the Danish theory. The downward turn in the running average was only evident when full data for 2007 was included.

If the trend in yearly averages were computed for the nine years following the 1998 peak there would be a clear reversal of trend or 'cooling' (see Ethos A1.2). Many would argue it is too early to call this a trend, particularly as the 1998 year was affected so strongly by a very large ENSO event. They would argue that the average in 2005 was also high as in one US compilation (NASA's Goddard Institute of Space Studies data set) it comes out higher than 1998 both globally and in the northern hemisphere. NASA lays some emphasis on the 2005 record, which was reached without the effects of a major ENSO event, as evidence of strong global warming due to carbon dioxide.

In the current debate, the recent lower temperatures are being referred to by orthodox theorists as a 'dip' in the overall trend, and by others as a reversal of that trend. Interestingly, the one commentator whose work has unusually spanned the three fields of satellite data, oceanographic research and solar cycles and who published his expectations in 2003, *predicted* that 2007 would be the first year to show a clear change in trend (**Landscheidt, 2003**).

We can readily see that oceanographic factors play an important role in the overall pattern and that these factors, in addition to the solar-cloud data and an alternative proposed hypothesis, are left out of key scientific papers that suffer from a narrow disciplinary focus. We are also left with compelling data and published peer-reviewed analysis in 2005 that cloud cover has changed over decadal timescales and affected the flux of energy to the earth's surface. The importance of this cannot be over-stated: hitherto the climatology community and in particular computer modellers assumed that cloud cover was relatively constant in the long term and in the short term fluctuated only randomly without clear trends. This community has, however, been acutely aware that the Danish research had correlated cloud cover with cosmic ray flux over the 11-year solar cycle and hence of the implications if this proved indicative of a real relationship. The issue of cloud cover thus becomes of paramount importance for climate studies – with the evidence beginning to point away from the carbon dioxide model as the main driver to an enhanced role of the sun and natural cycles.

Note

1. The effect is expressed by the concept of radiative forcing, or RF, developed by the IPCC Working Groups as an index of some external 'forcing' of the global temperature in terms of a unit of heat per square metre of surface for different

sources. Many papers use an RF expressed in watts per square metre as if all such measures expressed in this unit are comparable, for example as with short-wave radiation from the sun – but IPCC warns that they are not. There is an IPCC special publication on their concept of radiative forcing, but I have not found it helpful. It is not at all transparent how the measure has been arrived at for carbon dioxide and other greenhouse gases, nor how this measure, which relates to the atmosphere, can be related to other measures. I follow the lead of other authors who have simply compared the published measures for the differing sources.

3

Satellite Data

Thus far we have questioned the basic assumptions of the computer models, concluding they contain unproven feedback assumptions related to fundamental elements of atmospheric physics concerning the warming potential of carbon dioxide. Further, it is widely acknowledged in the scientific literature that these models are not able to adequately replicate natural periodic fluctuations inherent in the oceanic system. The anthropogenic global warming thesis rests upon the ability of the virtual-earth models to identify what ought to be happening naturally in order to identify something that is happening unnaturally. Considering that the global warming signal prior to 1950 is now regarded as predominantly natural – in effect, a rebound from the Little Ice Age of AD 1400–1800 – and between 1950 and 1978 there was a major dip in the general trend, which is also now regarded as predominantly natural in origin, and further, that in the period 2000–08 there was no net warming, then everything rests upon the period 1980–2000 when global temperatures rose by about 0.6°C above the longer-term mean.

This period coincides with an era of enhanced satellite coverage of the earth and monitoring of cloud cover and radiation flux, both short wave and long wave, at the earth's surface, the top of the atmosphere and within the atmosphere; in this chapter we examine this data in more detail. We will look to confirm the expected flux of SW radiation to the surface of the earth that would be expected from a thinning of clouds and earthshine data. We shall also look for evidence of cycles.

There are acknowledged problems with identifying trends, even over two or three decades, because satellites 'drift' in their orbits and instruments are subject to various stresses and biases over time. Added to this problem, each satellite has a limited lifespan and when replaced presents problems of inter-calibration of data sets. Nevertheless, after corrections for bias, these data sources have been used extensively in the published literature to build up a picture of radiation flux and cloud cover since 1983; their limitations notwithstanding, they constitute an important body of evidence in relation to the global warming thesis and the testing of assumptions made by the climate models.

Testing model assumptions in the real world

Whatever the cause of the observed warming and the fluctuations of the past, the computer models can be tested by observation of the radiation fluxes in the atmosphere. Short-wave radiation or visible light is the main source of incoming energy, and the greenhouse effect merely traps this energy as heat in the atmosphere or radiates the energy back to the surface. The atmosphere itself stores very little of this heat. It is the oceans that absorb the SW radiation and the resultant heat is circulated around the globe by ocean currents, extraction of heat by wind and evaporation of water vapour, condensation into cloud and precipitation as rain or snow.

The general systems models of the earth assume an equilibrium state such that any increase in greenhouse gases will perturb this equilibrium, trapping more heat and raising the global temperature. However, this assumed equilibrium state is never found in the natural state of the planet because there are cycles of varying output of the sun's energy, as well as delayed responses of the world's oceans which retain and recycle the heat and in particular determine a fluctuating cloud cover. At the outset, it was assumed that these natural cycles were of minor importance compared to the build-up of greenhouse gases, but as we have seen that assumption is likely to be wrong. There are now emerging criticisms of the very process of modelling based on assumed equilibrium states rather than actual initial starting conditions (**Koutsoyiannis et al., 2008**). We will look more closely at the importance of this issue when we consider oceanic cycles.

We can test the carbon dioxide model by looking at the radiation flux data and searching for the expected signal of greenhouse gas warming. The models predict that the atmosphere will warm markedly in the mid-troposphere at about 5 km altitude, it will cool at the top of the atmosphere (the stratosphere), there will be a surface air temperature increase most marked over land and at night, and there will be an enhanced downward flux of infrared radiation to the surface during clear sky conditions. The water vapour feedback model assumes that water vapour will increase in the lower and mid-troposphere but not contribute to an increase in reflective clouds which would act as a negative feedback.

As we shall see from the data outlined below, the satellite data do not entirely confirm the model expectations:

i) the mid-troposphere 'hotspot' has not appeared;

ii) night-time temperature increases which were first reported have since been shown not to differ from daytime trends (also admitted in IPCC-4);

iii) the enhanced downward flux of infrared radiation cannot be discerned against natural fluctuations from cloud;

iv) water vapour has increased in the lower troposphere, but in line with expectations from a warming ocean (IPCC-4);

v) there has been a marked decrease between 1980 and 2000 in low-level reflective cloud, and an increase in high level *cirrus* cloud, neither of which is predicted by the models and both of which have global warming potential.

The data sets

There are detailed satellite data on solar short-wave and long-wave flux at the top of the atmosphere, in the middle and at the bottom, as well as surface measurements of solar short-wave flux. There are also nearly three decades of satellite-derived cloud monitoring data (reviewed in Chapter 4). In addition, satellites have carried microwave sounding units (MSU) which measure the temperature of the middle and lower troposphere.

Data for atmospheric temperature confirms only that the upper atmosphere has cooled, but there is no evidence of the warm band in the mid-troposphere predicted by the models. Some researchers have recently sought to correct the data for the mid-troposphere, claiming a potential instrument error due to high-level wind influence and that when corrected there is a warming signal, but there is no general agreement on the issue. The upper atmospheric cooling is the main phenomenon in line with global warming theory, though there are other potential factors at play in relation to cloud cover and aerosols.

However, the radiation flux data provides the most telling indictment of the models, and most of the analysis on the decadal trends and patterns has been published in the period after the IPCC's 3rd Assessment in 2001 and before the 4th in 2007. These analyses show:

• there was a gradual increase in the flux of short-wave radiation (visible light) to the bottom of the atmosphere over the decades from 1980 to 2000;

• there was a gradual decrease in low-level reflective cloud, and hence a decrease in global albedo from 1980 to 2000; the trend then appears to have reversed in the period 2000-2005;

- there are indications of pulses, phase-change and periodicity in the data.

The first of these trends is extremely significant. The SW flux (the amount of sunlight reaching the surface of the earth and hence potentially heating up the planet) has increased substantially and this radiative effect is much larger, at between four and six times, than the change ascribed to increasing greenhouse gases over these decades by the computer models. It should be emphasized that this increase in the SW flux is *not* predicted by the computer models and there is no computational link relating this change to the increase in greenhouse gases. It cannot thus be safely inferred as a simple feedback response to warming by greenhouse gases. I review the detailed papers below.

Do satellites detect trends in surface radiation?

In 2005, along with other key papers in *Science*, R.T. Pinker and colleagues at the Department of Meteorology, University of Maryland, as well as other researchers at the Goddard Space Flight Center and NOAA's Climate Monitoring and Diagnostics Laboratory, published an important analysis of the satellite data regarding the detection of surface radiation trends (**Pinker et al., 2005**).

Their paper drew together all of the data for the long-term monitoring of solar radiation (visible light) at the earth's surface and found an overall increase from 1983 to 2001 at the rate of 0.16 watts per square metre per year. This trend constituted a decrease until about 1990 followed by a sustained increase, and the authors point out that this is equal to or greater than the centennial forcing calculated for carbon dioxide and hence that the finding has great significance for climate studies.

It is important to note that several authors of these recent solar flux papers make this point but do not then state the implications! The implications are profound, but in the time-honoured manner of scientific reporting they do not make clear such implications for the entrenched orthodox view that greenhouse gases have driven the warming and limit themselves to reporting the data. Thus an important paper does not register under the category of anything that would challenge or contradict the global warming model (noting that in recent debates one group of researchers had claimed that there were no published scientific papers that contradicted the models).

Thus, we have a set of data that shows an increased flux of *short-wave* radiation to the surface that is greater than the supposed warming effect of carbon dioxide. It follows therefore that carbon dioxide cannot then be the *main* driver of the changes observed at the surface – such as surface air temperatures. GHG theory relies upon a model of atmospheric heating with greenhouse gases such as carbon dioxide, methane and water vapour absorbing LW radiation emitted from the warming earth (and, more importantly, the oceans), hence warming the atmosphere itself, but also re-radiating heat as LW radiation back to the earth's surface. In order to confirm GHG theory the satellite data would have to show an increase in the down-welling of LW radiation to the surface. As we shall see, the natural flux of LW radiation has so much variability that no clear greenhouse gas signal is evident in the data.

Pinker and colleagues dedicate a large part of their analysis to the phenomenon of 'global dimming' that Wild also dealt with, where between 1960 and 1990 land-based measurements showed a decline in the amount of sunlight reaching the earth's surface. Pinker confirms the dimming – about 10% in Europe and about 0.2% per year from 1958 to 1992 over land globally. In the former USSR this decrease was higher at 20%:

> Speculation about the causes of global dimming include cloud changes, increasing amounts of human-made aerosols, and reduced atmospheric transparency after volcanic eruptions.

However, they report again that the satellite record does not confirm the 'dimming' hypothesis based upon pollution and shows recovery in an increasing trend from 1983 to 2001. The trend is not uniform – with a very slight decrease from 1983 to 1992 and an increase thereafter.

Arctic data is referred to in detail – showing a fall in surface radiation between 1983 and 1992 for the 60–90°N zone 'consistent with an increase in cloudiness during spring and summer when most of the radiation is received' and then a rise thereafter. This would imply that cloud cover changed post-1992 and allowed more radiation through – something of great relevance to the recent warming in the Arctic.

A key set of data comes from Samoa – a pollution free zone in the Pacific, which shows the same inter-decadal trends and the shift post-1994:

It is claimed (by the authors) that the observed changes in radiation budget are caused by changes in mean tropical cloudiness, which is detected in the satellite observations but fails to be predicted by several current climate models.

This is an important statement and obviously critical of the climate models, but perhaps not transparent enough to be regarded as a dissenting voice or a problem for the consensus. The authors then look at the long-term dimming trends for land and ocean, concluding:

...it is possible that land-based observations are not representative of Earth as a whole, and therefore brightening at a global scale is possible.

...a global scale decrease in cloudiness was found which is consistent with an increase in surface solar radiation ... because clouds are the major modulation of the solar radiation that reaches the surface.

...since the 1960s the melt date in Alaska has advanced by 8 days as a result of decrease in winter snowfall, followed by a warmer spring, which are believed to be caused by variations in regional circulation patterns.

Referring to data from the earthshine monitoring centre at Big Bear Solar Observatory, the authors note:

A steady decrease in Earth's reflectance from 1984 to 2000 was shown, with a strong drop during the 1990s [and then] a reversal of the decline from 2001–2003.

Thus, Pinker confirms the picture that emerged from Wild and Wielicki's papers we reviewed earlier and shows some confidence in the albedo observations at Big Bear Observatory,

The trends in albedo and cloud

The earthshine data are the result of pioneering work by E. Pallé at the Big Bear Solar Observatory in Chicago in a collaboration with Martin Wild at the Zurich-based Institute for Atmospheric and Climate Science and others. In a paper published in *Geophysical Research Letters* in 2005 he summarized the albedo data showing a clear trend from a variety of different data sources (**Pallé et al., 2005** and see Ethos A3.1) concluding that:

There is a consistent picture among all data sets by which the Earth's albedo has decreased over the 1985–2000 interval. The amplitude of this decrease ranges from 2–3 W/m^2 to 6–7 W/m^2 but any value inside these ranges is highly climatologically significant and implies major changes in the Earth's radiation budget.

The authors then discuss the recent period from 2001 to 2005 where there is some discrepancy between the different methods of calculating albedo and where the earthshine and other data show a reversal of the trend but the NASA CERES satellite data continue to show a falling albedo. They conclude:

> In any case, regardless of discrepancies over the very recent years, all the observational estimates of the Earth reflectance presented in this paper are broadly consistent and suggest changes in the Earth's short-wave forcing, both at the surface and at the top of the atmosphere, that will have a large impact on the planet's radiation budget. Clearly a more in-depth inter-comparison and integration of the several data sets presented here is needed, together with an improved model representation of the changes in the shortwave branch of the Earth's radiation budget.

This conclusion – that these changes 'will have a large impact on the planet's radiation budget' – made by specialists in the field contrasts strongly with the position taken by the IPCC, which we shall review below.

Climatologically significant changes in solar flux, cloud and temperature that are not predicted by climate models

All of these authors – Wild, Pinker, Pallé, and others – make clear that these changes are climatologically significant and are not predicted by current climate models. These authors make criticism of those models and effectively state that these changes in cloud-modulated solar flux are of an order that is very significant. What they do not say is that when compared to the relatively much smaller radiative forcing signal com-puted for carbon dioxide this enhanced flux of SW radiation can account for all of the recent warming. All of the papers point to the modulation of the sun's rays by clouds as the main source of warming.

In a reply to some criticism that his data did not show the required degree of globality, Pallé comments in 2006 (*Geophysical Research Letters*) that his and all other data have confirmed that in the period 1983–2000 the albedo of the planet changed by 2% – accounting for an increased flux of solar radiation over that period of about 6 watts/sq metre.

Furthermore, since 2000, the trend reversed.

> The updated time series of these data now show a decrease in sunlight reaching the ground in 2000–2004, following the ISCCP cloud data.

And in relation to the top of the atmosphere (a measure of reflectance by low-lying clouds):

> ...data show a maintained decrease in TOA reflectance 1985–2000, and an increase over the period 2000–2004 of about 2 W/m^2.

Thus, Pallé's paper is the latest in a series of analyses that show:

i) a significant increase in short-wave radiation reaching the earth's surface over a period of two decades – exactly the period of 'global warming' – that is of the order of 6 watts/sq metre, which dwarfs the increased carbon-dioxide effect over the same period estimated at 0.8 watts/sq metre (and the total of all anthropogenic greenhouse gases at 2.5 watts/sq metre);

ii) a change in the trend from 2000 to 2004 (and strongest beginning in 2003) which shows less sunlight reaching the surface – at about 2 watts/sq metre.

These papers are thus of immense significance. Even if there had been no recent shifts in temperature patterns, these data would require a rethink on the mechanics of global warming, and as these data were published in time to be reviewed by the 4th Assessment of climate science by the IPCC (February 2007), it has been instructive to review the Panel's treatment (see Ethos A3.2 for the full text). They confirm the 4% decline in cloud cover in the ISCCP data sets and refer to several sources that are consistent with this data. However, having reported on this important data, IPCC then report doubts relating to the instrumentation records, mostly by one author, Norris, whose work is not discussed in any of these seminal papers (but Norris *is* a member of the IPCC Working Group) and the Panel conclude as follows (my emphasis in italics):

> In summary, while there is some consistency between ISCCP, ERBS, SAGE II and surface observations of a reduction in high cloud cover during the 1990s relative to the 1980s, there are substantial uncer-

tainties in decadal trends in all data sets and at present *there is no clear consensus on changes in total cloudiness over decadal time scales.*

This conclusion is strongly at odds with the main arguments and data presented in the main text of the Working Group Report and with these papers published in the peer-reviewed literature. The summarizing IPCC author switches between comments on high-level cloud cover and total cloud cover, whilst the main interest is in the highly reflective *low* cloud data for which a strong set of data and scientific opinion exists – albeit with some of the disagreement noted but arising from only one source, Norris, who reviews limited synoptic reports from ships. The Panel make statements about conflicting data for the tropics as if this were an adequate treatment of the overall pattern and enough reason to dismiss the significance of the other work. It would be quite feasible for tropical cloud cover to remain constant and other areas to show significant changes. The tropics are generally low in cloud cover compared to higher latitudes.

Having raised doubts and effectively dismissed the data trends, the IPCC then fail to signal the importance of the whole issue and in particular the acute *lack of consensus* within their Working Group. Yet the significance of this research cannot be over-stated: these radiative trends, which consistency across the data sets suggests are real, are the result of cloud formation changes which are not predicted by GHG climate models and can account for the very rapid increase in global temperatures, in particular the rise in sea surface temperatures. This influx of SW radiation cannot be absorbed by land surfaces, but penetrates the oceans to 100 m where all of its energy is absorbed. Oceanic surface waters constitute 70% of the globe and hence account for the greater part of 'global warming'.

In my view, it is not an appropriate scientific response for a Working Group leader to adjudicate in this way on such important data, trends and conclusions made by specialists in their fields, particularly given the importance of the issue with respect to the adequacy of global climate computer models and predictions upon which major policy decisions and large investments are based. At the very least, this lack of consensus should be reported back to the main panel and presented in the Summary document.

This is another example of the construction of a false image of consensus. In a matter of this importance alternative explanations should be sought and recommendations made such that research organizations are advised to construct alternative models and to report conclusions regarding the consequences of each model proving correct. To do otherwise is to lay the Panel open to the charge of institutional and political manipulation of the science as a consequence of their prior commitment.

It is ironic that on an issue as central as cloud cover trends the IPCC should refer to a *lack of consensus* among experts in this field in the Working Group Reports, and at the same time the Panel's Summary Report and conclusions are heralded as a consensus of the world's climate experts – which the chairman of IPCC then confidently announces to the world press (as he did in September 2008 during a pre-election tour of the USA)!

The specialists cited above fail to refer to oceanic influences on surface temperature patterns. Total cloud cover may be less relevant to the earth's radiation budget than the spatial patterning and timing, because ocean currents move the accumulated heat in surface water from the tropics pole-ward to regions of permanent heat loss. Thus, an increase in cloud cover in the tropics would have a cooling effect by blocking SW radiation coming in, whereas an increase at the poles would produce a warming by blocking heat escaping. Further, large effects would be possible if cloud simply shifted in its spatial pattern – for example, by degrees of latitude south from the northern cloud belts that insulate ocean waters from the dark northern sky. A spatial movement whilst overall percentage cover did not change could nevertheless engender strong cooling in the ocean heat stores. We will look at this in Chapter 5.

Pallé summarizes the earthshine and albedo data – see Fig. 8 – and adds a note to show the comparative estimate for carbon dioxide gas forcing in units of watts/square metre (the vertical grey bars g1 and g2 to the left in the figure). The downward trend in albedo means a rising temperature would be expected as less sunlight is reflected by cloud. The peak in 2003 is now known to be a data error. The computed effect due to carbon dioxide is seen to be small compared to the cloud albedo changes and, further, the figure for the *incremental* change in greenhouse gases over this 20-year period would be even smaller at half the length of the line shown. Pallé has recently resolved a series of data and calibration problems and notes that after these corrections there is still a 'step-change' between 1998 and 2001 and no further increases in albedo (**Pallé et al., 2009**).

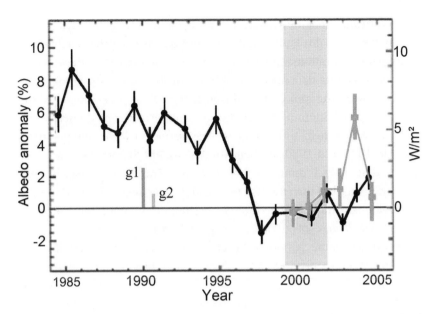

Fig. 8   *Globally averaged albedo. The black line is derived from ISCCP cloud amount. The observed earthshine albedo anomalies are in grey as measured from 1999. The shaded area from 1999 through mid-2001 was used for calibration and is the reference against which anomalies are defined. (After* **Pallé et al., 2005**)

Cloud cover changes are more significant than greenhouse gas emissions

All of these reviews of satellite data are reporting climatologically significant results in relation to clouds and short-wave radiation fluxes. Some researchers go out of their way to tie these in with global warming orthodoxy, but initial discrepancies such as 'continued warming' in the face of a recent increase in cloud cover no longer have force in view of the marked cooling during 2007–08 as well as new data showing that ocean heat content has not risen since 2002.

None of the high-level reviews cited above, often involving several institutes worldwide, and published in the leading science journals, consider that calibration problems in the satellite data are sufficient grounds for discounting the key finding that SW radiation flux increased in the global warming period and was due to decreasing cloud cover. In my opinion, the IPCC is therefore culpable in covering over this important data and, at the very least, a lack of consensus. They do not provide any detailed treatment of the satellite data itself, merely noting the reviews and conclusions of Wild, Pinker and Pallé and questioning the reliability of instrument calibration and whether any trends are real. Yet, these recent reviews of the flux data make important contributions to the whole issue of global warming and cast doubt on the previous assessments of the causes of global dimming, which the Panel hitherto accepted uncritically, and which was held to account for the period of lower temperatures between 1945 and 1980. In their 4th Assessment Report the Panel now accept that 'global dimming' conclusions were biased towards land measurements, but then they fail to investigate the full implications for the global climate models that built in assumptions about the power of anthropogenic aerosols to 'mask' the global warming signal.

This crucial time period of lower temperatures is not explained by greenhouse gas theory and, in the absence of anthropogenic pollutants as an explanation and with no major volcanic eruptions, natural solar and oceanic factors are the only candidates. This means that the entire suite of climate models prior to this date (around 2005) which claimed to be validated by hind-casting the pattern of the past century were flawed. And whatever the detail of cloud changes, particularly the very recent patterns since 2000, this does not detract from the central critique of modelling – that the whole of recent global warming could be explained by changes in cloud cover. In this latter respect, recent work by Roy Spencer and William Braswell of the Earth System Science Centre at the University of Alabama (Huntsville), underscores the criticism that climate models have assumed cloud variability is a response to ocean warming, when cloud patterns have their own non-feedback dynamic which then affects ocean temperatures (**Spencer and Braswell, 2008**).

NASA's computation of the global radiation flux

Some of the data that has led to the above reviews is available on line from NASA's Goddard Institute of Space Studies (GISS), which also provides a model for calculating global averages at the top, middle and bottom (surface) of the atmosphere. This model integrates the data input from a series of satellite monitoring programmes. In 2006 NASA produced the graphs I have assembled in Ethos A3.4 which are worth looking at in detail. Some of the trends may not be reliable because of instrument calibration problems and NASA discuss this (which I will return to). As the reliability of this data is questioned by NASA itself (voiced by the head of GISS, James Hansen, who curiously prefers ground-based measurement to satellite data sets), I quote from their commentary (May 2006) in Ethos A3.5.

Despite NASA's reservations as to the size of the change, there is a spike in the LW flux in late 2001, a time when other global data sets begin to show significant shifts in cloud patterns, in heat loss from the oceans, and in global albedo. NASA do not regard this as reliable, but they do conclude (my emphasis in italics):

> *The overall slow decrease of upwelling SW flux from the mid-1980s until the end of the 1990s and subsequent increase from 2000 onwards appear to be caused, primarily, by changes in global cloud cover* (although there is a small increase of cloud optical thickness after 2000) and is confirmed by the ERBS measurements [my note: this is another independent main source of instrumental data].

And

> *The cooling at the end of the record* is probably exaggerated because of the inhomogeneity of the atmospheric temperature dataset used for the calculations.

> The overall slight rise (relative heating) of global total net flux at TOA between the 1980s and 1990s is confirmed in the tropics by the ERBS measurements and exceeds the estimated climate forcing changes (greenhouse gases and aerosols) for this period. *The most obvious explanation is the associated changes in cloudiness during this period.*

Thus, although NASA have reservations about several aspects of the apparent trends in the satellite data, there are some trends that are confirmed by other data sets (ERBS and CERES), in particular the albedo changes at the top of the atmosphere and the overall cloud thinning throughout the 1983–2000 period. The shifts in the post-2001 period in up-welling LW that are regarded as spurious nevertheless coincide with other data from the earthshine measurements detailed above as well as cloud analyses that show a global shift around 2001 (which we shall look at in a later section), and although the magnitude may be in question, the correlation is of some importance. The issue was not resolved at the time of IPPC-4, but by the beginning of 2009 the discrepancy between earthshine and conflicting CERES was resolved and confirmed a significant step change in albedo in the period 1999–2001. This now combines with revised ocean heat content data which shows no further heat build-up since 2002 (see Chapter 5).

Most analyses of cloud data focus upon the overall amount of cloud and global average changes; there are no analyses to my knowledge of the spatial distribution. Cloud patterns are relatively stable (though likely subject to long-term cycles not yet evident in the 25-year satellite record), with certain northern regions of the subtropical and high-latitude regions maintaining extensive cover. A shift in the spatial distribution such that warm waters were exposed to northern clear skies during winter could produce a net heat loss without any percentage change in overall cloud cover.

From a reading of the analysis by Pinker and colleagues, we can begin to construct a global picture that explains the trends in short-wave and long-wave flux:

- Between 1983 and 2001 there was an increase in solar SW radiation at the surface, most markedly after 1995, of the order of 0.16 watts/sq m/ year. Regional analyses show a shift in mid-1990 from a small decline to a rise (e.g. recorded in Barrow, Alaska, and in American Samoa).
- This flux change occurred for both cloudy and clear sky conditions, indicating both a shift in cloud patterns or thickness and in the overall transparency of the atmosphere, with the latter most likely due to changes in natural aerosol loadings.
- The trend over oceans was greater at 0.24 watts/sq m with the overall trend for land-based observations showing no significant rise.
- This data was regarded by Pinker as consistent with ISCCP data showing a global scale decrease in cloudiness.
- The magnitude of this trend over 19 annual means, at 4.5 watts/sq m, is very significant for climate compared to the computed total over 20 years for additional greenhouse gases at 0.8 watts/sq m.

Although NASA regard interpretation on trends as problematic due to inter-calibration issues, these data are consistent with the results of the albedo measurements of earthshine that show a steady decline of reflectance until 2001 (see Fig. 8), with the strongest drop during the 1990s. Thus it seems a reliable conclusion that with decreasing cloud and increasing transparency significantly more SW radiation has reached the earth's surface, particularly over the oceans, during the warming period 1983–2000.

The International Satellite Cloud Climatology Project (ISCCP) has published similar data based on global radiative flux calculations based on a global model of the atmosphere developed by NASA's Goddard Institute of Space Studies. The results are consistent with the GISS interpretation but without the sudden changes in 2001. They state *(my emphasis in bold and italics)*:

> Notable features are: (1) a decrease of the net SW at the surface and TOA, as well as in the atmosphere produced by the Mt. Pinatubo volcanic aerosols in 1991–92; (2) *an overall increase of the net SW at TOA and the surface, but not IN the atmosphere, from the 1980s to 1990s* **associated with a decrease in low-latitude cloud cover**; (3) three (possibly four) decreases in net LW at the surface and increases in the atmosphere, but not at TOA; and (4) a small decrease of net LW at TOA and *in* the atmosphere and a larger increase of net LW at the surface occurring in the late 1990s.
>
> (*http://isccp.giss.gov/projects/browse_fc.html*)

The IPCC, however, regard the data series as not reliable enough to make any firm conclusions and none of this data is deemed sufficiently important to bring the greenhouse gas models into question, despite the fact that this huge global investment in radiation monitoring provides no obvious support for greenhouse gas theory and confirms that cloud changes are responsible for the variation in the fluxes of warming radiation at the surface of the earth. There is no clear signal of increased longwave radiation within the atmosphere as would be expected from greenhouse gas theory. The cooling of the upper atmosphere is attributed to cloud effects rather than the blocking of long-wave radiation that is also expected from global warming theory, and the LW patterns at the surface are also attributed to cloud changes.

There clearly *are* some issues of calibration that arise, because the working life of a satellite is relatively short and calibration between one set of data and another is not a simple matter. But it should also be borne in mind that the head of NASA's Goddard Institute of Space Studies, James Hansen, has a long-standing commitment to the carbon dioxide hypothesis, being one of the leading figures in its original exposition as well as having a high public profile in the arguments for government action on emissions and a leading role in the IPCC. At the time of writing the questions regarding calibration and instrument error have not been resolved (the NASA trend data were first analysed in June 2006). It is perhaps for these reasons that IPCC did not highlight these data. However, calibration issues are not enough reason to spurn these sources, which contain many indications of global patterns. Furthermore, there have been instances in the past where satellite data have not been believed because they run counter to expectations, as when the ozone hole first registered on US systems. And, naturally, when data run counter to such a strong prior commitment within both NASA and the IPCC there will be questions relating to the reliability of the instruments.

We shall see that questioning the reliability of instruments is a rather regular feature of the science when those instruments record patterns that run counter to the predictions of the computer models. Thus far we have seen the global albedo data questioned, and NASA's own lack of faith in the LW flux data (the SW flux and its implications for cloud thinning seems to have survived the doubting but are not highlighted in press briefings). Later we shall see that the absence of mid-tropospheric heating is also questioned, as is the sudden loss of heat from the oceans.

The sudden flux increase of LW radiation in 2001 has no obvious explanation and NASA assumes the whole signal is spurious, despite the fact that the continuing flux indicates a step-change had taken place. An explanation for such a step-change would be a shift in the *spatial distribution* of cloud, particularly in the regions of planetary heat loss in the northern seas. When we look at cloud data in more detail we will note this step-change and a clear percentage shift in 2001–02.

These shifts and trends in both SW and LW radiation budgets are much greater than the changes computed for carbon dioxide and other greenhouse gases, and it is these changes that are clearly driving the climate shifts noted since 1980. The change in trends around 2000 more readily explain the flattening of global temperatures since the ENSO-induced peak in 1998. The fall in 1999 and small peak in 2002 are also consistent with sea surface temperature oscillations caused by ENSO. A record ENSO expected by both NOAA and the Hadley Centre observers in 2007 did not materialize and there is now a clear post-1998 downward trend in sea surface temperatures. On land, temperatures in the northern hemisphere have remained high, with 2005 equalling 1998; but there is a plateau rather than a sustained increase, with 2007 marking a downturn in the 11-year running average and now likely to persist through 2008.

This flux data suggests a trend of increasing transparency of the atmosphere to SW radiation. This may represent an interplay of natural aerosol components other than volcanic (oceans release large quantities of aerosol), cloud percentage changes and spatial distribution, and oceanic heat storage (and loss). Major volcanic emissions notably depress temperatures for a period of two years only. There were two in 1982 and 1991 that did so, but the normal pattern returned after three years in each case.

Cycles and pulses in the flux data

The overall patterns of flux change shown in the NASA GISS data show correlations with solar maxima (1990 and 2000) and minima (1986, 1996, 2006) though not always exactly in phase (see Ethos A3.4), and there were two major disruptions to the pattern with Pinatubo erupting at the time of a solar maximum, and ENSO peaking just after a solar minimum. However, it is clear *from the scale of the flux changes* that major natural cyclic factors are at play. Moreover, the amplitude of these flux changes obscures any expected signal from greenhouse gases. This underscores the point that none of the radiation measurements provides any evidence for anthropogenic greenhouse gas theory. If GHGs were the main driving force of the global warming then it should be evident in the data, but no such obvious anthropogenic signal is revealed simply because the 'variability' of the fluxes is so great. The anthropogenic signal, such as it is, can only be derived from longer-term computer-simulated 'attribution' studies that attempt to delineate the recent change from past natural variability. As we have seen, this approach is flawed because the models cannot adequately reproduce that natural variability. And it is this variability that exists in clear pulses and patterns probably relating to the solar cycle that accounts for the warming pattern between 1980 and 2000, and for the 'global dimming' that occurred between 1945 and 1978.

We will see further, when we look in more detail at oceanographic work and solar cycles, that IPCC have underplayed the current knowledge of longer-term cyclic changes, much of which points to a solar 'amplifier' – that is, a process that occurs alongside the small irradiance changes and which must account for the surface temperatures correlating with the solar cycle. This is the rationale for the Danish group's search for the 'missing link' in climate studies. It is clear from the satellite data that the missing link is changes in cloud patterns. In fact all of the NASA and ISCCP satellite data point to the existence of a cloud-based amplifying mechanism that correlates to the solar cycle.

As noted, these radiation flux papers and NASA data were reviewed by

IPCC in their Working Group technical reports. In addition to the IPCC Reviews quoted above, another section of the Working Group (Chapter 3 of IPCC-4 Working Group I) passed judgement on whether the above changes noted by Wielicki, Pallé, Pinker and others are 'real'. This material is presented in Ethos A3.6 with conclusions relating to the correlation of ocean heat storage and their interpretation of the radiation fluxes. However, their conclusions were seriously out of date with regard to ocean heat storage discussions. Willis, whose 2004 work is referenced, was part of the team under John Lyman who in November 2006 reported the major loss of ocean heat between 2003 and 2005, and also party to the retraction of that data in April 2007, together with discussions of the downward revision of previous ocean heat storage estimates. The latter were published early enough in 2007 to be referenced by Willis and Lyman but not to be taken on board by IPCC. I deal with the ocean heat storage issue in some detail in Chapter 5.

IPCC should have been aware of these important developments and included reference to them even at such a late stage. In other areas of debate they have included various references to 2006 and even 2007 in their final Report in May 2007 and have commented in another chapter of their Working Group Report where they are aware of Pallé's arguments in papers from 2004 to 2006, in which he explains the anomalous high of 2003, but also shows that there has been a phase change in cloud cover from 2001. There is also corroborating data from NASA GISS data sets showing less SW radiation reaching the surface after 2000. We shall see that when some late papers have supported the IPCC's former position, they have made it into the final assessment, for example the 2007 paper by Raimund Muscheler that discounts long-term solar energy fluctuations (see Chapters 7 and 8).

In their comments on ocean heat storage IPCC appear completely unaware of the 2006/2007 debate on the ocean heat content data sets – a debate that is now concluding that ocean heat storage, upon which the anthropogenic signal of climate change very much depends, has been significantly overestimated for the centennial rise *and* had ceased by 2002. The Panel appear to make extraordinary efforts to track papers that would contradict an emerging alternative explanation of warming related to the SW flux and albedo, whilst failing to report analyses that contradict some basic global warming assumptions.

The IPCC conclude their judgements upon the work of peers published in the specialist journals and for which they are not necessarily in a position to pass such judgement. Although some of the specialists such as Wild and Wielicki are members of the Working Group and hence will have an opportunity to provide drafts and make comments, it is not at all clear whether their judgements contribute to this summary conclusion. Some specialists with more confidence in the reliability of trend data, such as Pallé and Pinker, are not on the Working Group list.

One section of IPCC's Working Group I concludes:

> In summary, although there is independent evidence for decadal changes in TOA radiative fluxes over the last two decades, the evidence is equivocal. Changes in the planetary and tropical TOA radiative fluxes are consistent with independent global ocean heat-storage data, and are expected to be dominated by changes in cloud radiative forcing. To the extent that they are real, they may simply reflect natural low-frequency variability of the climate system.

There is a body of recent oceanographic work that elaborates upon 'natural low-frequency variability', showing it to be well correlated with solar cycles, and IPCC make no comment upon the fact that computer models do not replicate this natural low-frequency variability any more than they replicate higher frequency variability such as ENSO. Yet, upon this highly unsatisfactory state of affairs, IPCC place great faith in the 'attribution studies' from computer models that apparently demonstrate global warming between 1980 and 2000 which cannot be explained by natural cycles.

These Working Group discussions and conclusions of the group leaders then get translated into the simplistic language of the Summary for Policymakers. Uncertain and contradictory data and opinions translate not into a 'lack of consensus' but into 'poorly constrained' areas of science. Bold statements are then made from which it is not at all clear that an issue over radiative flux data even exists. By labelling material either controversial, questioned, contradictory or poorly constrained, IPCC is able to sidestep material that would substantially contradict its own prior commitment.

In my view there is a substantial body of observational data which do not confirm the IPCC models and point instead to cloud cover changes as the key factor in the global warming record. In the next chapter we will look more closely at the pattern of these changes.

4

Cloud Cover

As we discovered in the previous chapter, there has been a recent trend of increased solar radiation to the earth's surface maintained over the crucial decades from 1980 to 2000 and the warming power of this increase dwarfs the computer-calculated power of carbon dioxide by at least a factor of four. The claim to have identified a 'global warming' signal – that is, a signal due to human activity – is not supported by the satellite data on radiation flux. In all of the papers we have examined, the cause of this increased flux of SW radiation to the earth's surface is identified as changes in cloud cover or cloud optical thickness and also an increased transparency under clear sky conditions. In this chapter we will look in more detail at global cloud patterns.

The computer models upon which IPCC places such reliance do not predict changes in cloud cover and modellers admit to no reliable mathematical treatment of the cloud feedback consequent upon 'global warming'. The satellite data show no clear signal of warming due to changes in the long-wave (infrared) radiation flux and this is only to be expected given the variability of that flux, which is very much determined by the degree of cloud cover.

In my view, the satellite data, with whatever reservations about reliability and trends, provide more evidence for a solar-cloud effect than support for the greenhouse gas model and any detection of an anthropogenic signal.[1] The data indicate a series of warming pulses and this heat will be absorbed by the surface waters of the ocean where it is then subjected to circulating currents and gradual heat loss either by emitted LW radiation back to space or by evaporation. The oceans are subject to cycles in current patterns, atmospheric pressure and temperature, and the heat returned to the atmosphere will be subject to these cycles. Several key cycles are *inter-decadal* and span more than the two to three decades of satellite observation. These cycles also clearly involve time lags, sudden phase changes and destructive as well as constructive interference as different cycles in different ocean basins interact.

Cloud cover varies with the natural cycles of the oceans

The IPCC in its conclusions places great faith in general circulation models that cannot yet adequately model key elements of this planetary system such as these ocean oscillations, and as a consequence the models cannot explore the interplay of ocean currents and ocean heat content with these other cyclic factors. Yet, cloud cover is known to vary significantly with the cycles.

One of the chief critics of such modelling, William Kininmonth, has a career spanning 40 years as a professional meteorologist, having headed Australia's National Climate Centre and co-ordinated the UN task force on El Niño. He was also closely associated with the development of the Framework Convention on Climate Change. Since the release of the IPCC's 2001 3rd Assessment Report, he has been one of its most ardent critics, arguing that IPCC warnings had extended beyond sound theory and evidence.

His main criticism relates to the role of predictive modelling and how it underlies all of the IPCC assessment. In his book *Climate Change: a natural hazard* he argues that the predictive models exaggerate the influence of carbon in the atmosphere, and by their reliance upon linear computation they cannot model a complex, non-linear response system such as climate where the water vapour cycle is at the heart of things (**Kininmonth, 2005**). The models cannot replicate the type and behaviour of clouds involved in the mass transfer of heat between equators and poles, in particular the differential role of clouds in equatorial regions of heat gain where the heat from short-wave radiation is trapped in the upper 100 m of ocean and not balanced by long-wave heat loss at night, as is usually the case on land, and at higher latitudes during winter months of permanent heat loss where cloud cover insulates the oceans. We shall return to these issues when we look more closely at ocean currents, upper ocean heat content, and the oscillations of temperature and atmospheric pressure that occur in each ocean basin. This aspect of differential heat gain and heat loss with latitude is central to a natural explanation of global warming.

It is clear from the satellite studies that large changes have occurred over the past two decades and there is mounting evidence that the temperature cycles of the ocean are also timed by the solar cycle itself (which I review in Chapter 5). If this is the case, then a natural mechanism of timing must exist and the IPCC shows continued resistance to admitting to this. The orthodox position is that ocean oscillations are simply part of the natural *internal* variability of climate. It is clear to many oceano-graphers and palaeoecologists that this natural variability is cyclic and as we shall see in the next chapter there are many studies that link these cycles to the electromagnetic status of the sun. This will return us to the issue of solar-cloud theory, but first let us take a look at the available data on cloud cover.

It is clear from our previous analysis that at least two and possibly three factors related to cloud cover are important:

i)   The overall *trend* of global cloud cover. This is measured as percen-tage cloud cover change, and we are particularly interested in the optically thick, low-level cloud types that have the capacity to reflect the short-wave solar radiation that is the only significant surface *source* of heat for the planet. Here, all the available evidence shows a global decline in reflective cloud between 1980 and 2000, and an increase from 2000 onwards; this is corroborated by the SW radiation flux data on the amount of sunlight reaching the planet's surface, which we reviewed earlier. We don't have reliable data for the rest of the century.

ii) The *spatial* distribution of cloud cover. This factor has received less attention, but in view of our later analysis of global ocean current patterns, heat sinks and areas of heat loss, it deserves a careful review (we can attempt only a very basic analysis).

iii) The relative abundance of high-level, mid-level and low-level cloud. The abundance of high-level cloud also has global warming implica-tions because this cloud, though it is less reflective, acts to trap out-going LW radiation.

The data sets of the International Cloud Climatology Project

We can establish the overall global trends from ISCCP's most recent data sets which are separately available for nine cloud types and heights and these are aggregated into low-level, mid-level and high-level.

In the analysis that follows we are primarily concerned with trends in optically reflective low-level clouds, and hence stratus is the key type; thin high-level cloud is also important for insulating against heat loss, thus cirrus and cirrostratus become important.

The global data is provided in sets according to visible and infrared

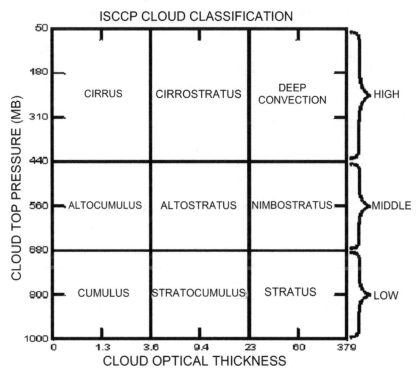

Fig. 9   *Cloud types (as classified by the International Satellite Cloud Climatology Project)*

surveying technology. There are differences between the two methods, with the visible wavelengths detecting some peaks and troughs and the IR slightly differing amplitudes of longer-term trends. I have presented both data sets here, but because of limitations of space, various combinations have been placed in the Appendices on the Ethos website. This is not an exhaustive analysis. I am not aware of any detailed work on patterns of evolving cloud changes and I think this should be done as a matter of urgency by a body independent of the IPCC. The data for global total cloud trends is shown in Fig. 10 for both types of measurement.

There is a clearly marked trend of reducing low-level cloud and the IR data sets show a correlation of increased cloud with the solar cycle troughs of 1986 and 1996, whereas this correlation appears to break down after 2001. In the VIS-IR data, Pinatubo's effects are seen in 1991, and a fainter correlation to solar cycles. The decline continues beyond 2000–01, whereas in the mid-level data there is an abrupt increase in 2000–01 after relative stability from 1983 to 2000. In the high-level IR data there is a

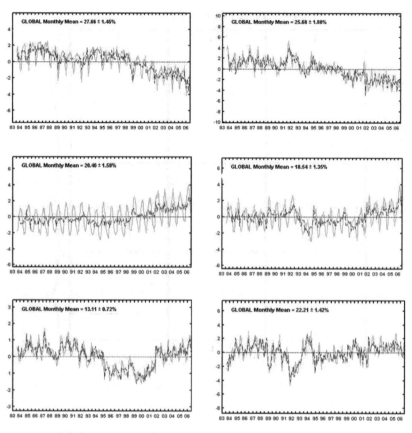

Fig. 10   *Global average cloud cover changes: low-level cloud (top row), mid-level cloud (middle row) and high-level cloud (bottom row) 1983–2006. Expressed as % deviation from mean over the period (taken from: ISCCP). Note: the left column shows IR (infrared) data and the right VIS-IR (visible).*

marked trough during the solar peak of 2000 (interestingly the pattern of peaks and troughs reverses with the solar cycle reversal of polarity over the 22-year Hale cycle) and the VIS-IR data pick up a Pinatubo depression, but there is no marked long-term trend. All data show a sharp change in 2002 with rises for mid-level and high cloud and falls for low-level.

This ISCCP data shows that low-level cloud continues to decrease after the break-point in 2000, whereas mid-level and high-level cloud record increases. It would appear that the increase in mid- and high-level cloud is capable of outweighing the decrease in low-level cloud in terms of the radiation budget at the surface as well as in the albedo measurements of reflected light from earth.

Global trends in cloud type

I have looked at various sets of cloud data, some of which is presented in the Appendices. When global patterns are analysed according to cloud type a shift is apparent in convective high cloud and cirrostratus, and this is likely to account for the increased albedo because the optically thicker, low-level cloud types have continued to decline. The only cloud types registering a marked decline are low-level cumulus and stratocumulus, which declined 2% from 1999 and 4% since 1983. These types are of low and medium optical thickness and constitute about 24% of cover; the more reflective stratus shows little overall trend and is in any case only 2% of cover, so cumulus and stratocumulus types probably account for the documented surface radiation flux changes we saw earlier. A more detailed analysis of regional change would be instructive – here we can attempt only a broad picture.

The other marked trends are in mid-level altostratus, which is 7% of cover and increased sharply by 1.5–2% in 2002; high cirrostratus at 6%, which also increased sharply by 0.5% in 2002; mid-level nimbostratus at 3% cover, increasing earlier in 2000 by less than 1%; and deep convective high-level cloud at only 2.5% cover, increasing by 0.5%.

The data show that cloud changes occur not in a gradual decline but in cycles and there is a break-point around the year 2000–01, where continued declines in low-level cloud are offset by rises in mid- and high-level clouds. The overall effect would appear to be a lower input of radiation at the surface in a steep change at the beginning of the millennium but potentially balanced by long-wave radiation held by the cloud increases at higher levels.

In the above data, the overall percentage cover is given below each type together with any shift in percentage from 1983 to 2005. The main changes are in cumulus (decline 4%); stratocumulus, a recent decline of 2%; altostratus, recent (2002) increase of 1–2%; stratus dropped by 0.5% in relation to the late 1980s but has remained stable since 2000; nimbostratus has risen since 1999 by 1%; cirrostratus up 0.5% since 2002; and deep convective cloud up 0.5% since 2001; there is a sudden drop in cirrus by 2% in 2005.

Cloud over land and ocean: the Atlantic, Pacific and Arctic regions

The ISCCP data is separable with regard to land and water surfaces. In the land data (Ethos A4.1), low-level cloud began to decline markedly in 1998 and shows two clear correlations of increased cloud with the low points of the solar cycle in 85 and 95 and then a 4% drop at 2001, which is since maintained; mid-level cloud is constant until 2002 then up 2–4%; high-level cloud is fairly constant but since 2002 is about 1% above mean.

Thus in the land data the effect of increasing high- or mid-level cloud is not very marked until 2002 and thus is unlikely to have contributed to the late centennial warming signal, whereas there is a clearer decadal decline in low-level cloud which would warm the surface. However, the low-level changes at a 3% decline over the two decades also show a marked phase change in 2000, after first displaying a remarkably good correlation with the solar cycle peak of 1990 coincident with troughs in the cloud data, and peaks in the cloud data coincident with the solar cycle troughs in 1997 and 1987. This data emphasizes the importance of looking at pattern rather than trends.

We can look also at cloud cover over oceans generally (Ethos A4.3). Here we can see a similar pattern but with no obvious correlation to the solar cycles. There are less-marked increases for high-level cloud, but with a noticeable break-point in 2002; similarly mid-level cloud increased sharply at this time by nearly 2% and low-level cloud also dropped by 2% at the same time during a more gradual decline of about 5%. There is no obvious correlation with the solar cycle for the low-level cloud data, no marked long-term trend for high-level cloud, and changes in mid-level cloud do not begin to provide a significant warming addition to the low-level cloud changes until 2002.

We can see a sharp drop in 2002 in the low-level cloud; high-level cloud jumped nearly 4% at the same time; there was also a steep rise in mid-level cloud in 2002 and a large 4% anomaly in 2006. Thus, the oceans indicate a sharp phase change in cloud cover – at the same time as the NASA data (regarded as unreliable) documents a large increase in outgoing LW radiation.

I have briefly looked for any regional differences between the Atlantic and the Pacific, examining only the low-level cloud, which is the main interest in relation to varying inputs of heat to the oceans (Ethos A4.3). There was no marked trend until a transition in 1998–2000 when there was a 4–5% reduction of low cloud over both oceans. In the Pacific there is a sharp decline in 2002 and it would be of great interest to be able to locate this sharp fall in relation to latitude. This fall is mirrored in data for mid-level cloud (not shown), which shows a 2% rise in 2002 but with little inter-decadal variability before then. High-level cloud likewise shows no overall trend.

The Atlantic and Pacific data do not provide much support for the contention that increasing high- and mid-level cloud has contributed to warming trends — there is little increase until after 2002 when ocean temperatures had already peaked. Further, the decline of low-level cloud cover is not marked until late in the 1990s and most marked after 2002. If there had not been the major influences of Pinatubo in 1991 and ENSO in 1998, the low-level data would likely show a stronger correlation with the solar cycles. Thus, these data sets confirm the long-term decrease in low-level reflective cloud and most of this decline relates to the Pacific region. This would imply continued warming whereas they show no additional warming after 2002. A possible explanation would be a spatial shift with the decline centred in northern regions where low cloud has an insulating effect.

However, in the Arctic region (A4.4), which we shall examine in more detail in Chapter 6, as it is currently causing so much attention, there is a clear phase change around 1997 in low-level cloud which declined to 2005 by 5%, after which it rose slightly; mid-level cloud increases slowly after 1997 and maintains an excess of 5%; the high-level trend is flat, apart from a major 'blip' of 40% in 2001 (the year of other major shifts around the globe).

In the Arctic Ocean region therefore, warming in the early part of the decades from 1980 to 2000 is unlikely to be due to cloud changes and in particular not from a decline in low-level or an increase in high-level cloud as evident in other regions. However, there is a marked decline in low-level and rise in mid-level cloud after 2000. In other regions this would contribute to a warming but in the Arctic much depends upon the season and location in relation to water. There is a clear phase change in 1997, which may be linked to the Arctic Oscillation of cyclonic pressure systems, and a rise in mid-level cloud by several percent after 2000, accompanied by a 5% fall in low-level cloud from a break-point in 1999–2000 with what appears to be a rise at the end of this data set. In general, increased cloud in the Arctic will mean warmth because the clouds originate in sub-polar waters, radiate heat to the ice surface and insulate open waters from heat loss to space. There is data reviewed later when we look closely at the Arctic 'meltdown' that shows a 14% build-up of cloud in the central Arctic region above the sea ice. However, the longer-term Arctic warming must originate elsewhere and we shall see that it relates to the movement of warm water and air masses from the Pacific and North Atlantic.

I have looked further south in the 60–90°N band of data (not illustrated) to see if there is a different pattern, but low-level cloud does not decline significantly until 1999 with about 3% loss thereafter; mid-level cloud is constant until 2002, when a sharp rise is noticeable, up 5%; and high-level cloud is constant though there are spikes in 2001 and in 1993. This data would not support the hypothesis that cloud cover changes and increased insolation led to the formation of warmer water in these sub-Arctic regions, at least not before 2000. In the 30–60°N band (not illustrated) low-level cloud does not begin to decline until after 1999, then by 4%; high- and mid-level clouds do not increase until 2001–02, also by 4–5%. So, it is again an issue of break-points rather than overall trends that is important, and again, there is no clear indication of a solar source of energy input to these northern regions in the period 1980–2000, though increasing cloud over ocean areas might then enhance any warming which, as we shall see, is likely to originate in ocean current systems further south.

Cloud over the tropical region

The tropical region (Ethos A4.6) is the main source of energy input to the planet surface; at this latitude it is 85% ocean. In the data for 15°N to 15°S there is a decadal decline of about 4% for low-level cloud (briefly inter-rupted by Pinatubo), but only markedly since the mid-1990s. Mid-level cloud changed dramatically in 1992–93 by 2% at the time of Pinatubo and stayed lower but appears to have recovered in 2005. High-level cloud shows a small 2% increase in 2000. The effect of the 1991 volcanic eruption can clearly be seen – an increase in low-level and decrease in high-level and mid-level clouds, but it is curious that this data does not show any marked effects from the 1998 El Niño. The low-level decline of 4–5% is however long term and at almost 20% cover this will have had significant effects upon heat inputs. There is no obvious evidence for pulses linked to the solar cycle.

In the southern hemisphere 30–60°S (Ethos A4.7) low-level cloud shows a big shift in 2001–02 with a 5% drop, which is maintained. This is also reflected in mid-level data with a sharp rise in 2002, also maintained. There was no obvious trend in high-level until a similar sharp rise in 2002, which was then maintained at 2–4% above the mean. And the Pinatubo effect is much more marked in this hemisphere. I at least managed to locate the origin of the major shift in 2001–02 global data sets – it occurs in the southern hemisphere oceans at 30–60° latitude. There is also just a hint of a solar pulse in the low-level data before a sharp phase change in 2002, which is also reflected in the mid- and high-level cloud data. Whether the 2002 shift would constitute a warming or cooling effect over the oceans is open to discussion; the increased high-level would be expected to produce a warming and the decreased low level also, but this would depend upon the spatial distribution in relation to stored ocean heat. As we shall see later, southern ocean temperatures have not shown as strong a global warming signal compared to the northern hemisphere waters, but this data, though in line with that trend for 1980–2000, would suggest a warming should occur after 2001. I have looked further south at 60–90° to see if there is any change in pattern (Ethos A4.8). Here, the VIS-IR data shows a 5% decline in low-level cloud after 2000, and on examination the IR data detects a significant drop in 1999 that appears to mark a major transition. This is further picked up in the IR data for mid-level where a 5% increase is equally sharp – and this is not obvious in the VIS-IR data and there is no obvious change in the IR high-level data. This data also shows little change in high-level cloud, major shifts at 1999 for mid-level cloud increases (most marked in the IR data) and late post-1999 shifts in low-level cloud.

Thus, the southern hemisphere from lat 30–90° shows a significant phase change around 2000–02 for low-level cloud, a sharp decrease of 5%; and for middle level cloud, an equally sharp increase of the same order. These will have had a combined warming effect unless the middle-level cloud types were strongly reflective, in which case they may have compensated for the low-level effect. This phase is so marked and so significant – it is also illustrated in the other data sets – over land for Eurasia (Ethos A4.9), land generally, the North Atlantic and Pacific regions that it requires detailed investigation. Such a major phase change and percentage shift in low and middle cloud cover relative to each other may also have been accompanied by abrupt changes in spatial distribution. In the light of albedo data and GISS radiation flux data, which also indicates a shift at the same point in time, further work is required to correlate cloud changes and fluctuations in the spatial distribution of ocean heat storage.

Summary of cloud data

In the above data banks, we can readily see the global pattern of decline in reflective low-level cloud. This confirms the satellite data on radiation flux indicated by the work of Wielicki, Wild, Pinker and Pallé. There is a correlation of low-level cloud variability with the solar cycle 22 (1986–96), but not obviously with cycle 23 (1996–2007). However, there is evidence of an unusual and as yet unexplained step-change that seems to have begun in the Arctic in 1997 and around 2001 elsewhere. This may have obscured the relationship to the solar cycle.

The bulk of the 'global warming' signal actually occurs between 1997 and 2006, and if we take out the unusually high peak of the 1998 ENSO it forms a plateau rather than a rising curve; the other down-sloping edge of the plateau was not evident until all the data for 2007 were plotted. The following twelve months of 2008 will likely keep the temperature in a trough.

The explanation for these cloud changes and temperature patterns must lie in the source of the cloud thinning from 1980 to 2000, the nature of the pulsing effect, the warming patterns as the oceans absorb this periodic heat pulse, and the subsequent redistribution of that heat on a regional basis. Much of the discussion relating to the persistent warming, whilst cloud and other indicators such as the reducing solar magnetic field would indicate a fall, is made redundant by the nature of the time lags which are now evident.

The contradictions between Pallé's earthshine data and Wielicki's satellite studies have now been resolved, with the balance of evidence from the ISCCP data sets supporting Pallé. It would appear from the ISCCP data that the sudden shift in albedo must be due to mid- or high-level cloud contributing to reflectance. Thus, the cloud data have shown that the decline of low-level cloud, particularly after 1997, is part of a long-term trend since cloud data began to be collected in 1982 and that this decline, coupled with radiation flux data at the surface and some time lag in relation to oceanic responses, is adequate to explain the late twentieth-century warming and the plateau between 1997 and 2005. These cloud patterns are not predicted by the global circulation models and cannot be claimed as feedback due to greenhouse gas increases.

In relation to the unexplained step-change, almost all data sets show a shift between 2001 and 2002 in the proportion of low- and high- to mid-level cloud and this is most marked in the southern oceans. The consistency and scale of these changes of phase have significant consequences for the surface heat budget, particularly over oceans. In this respect, further study of the spatial variation of cloud cover would help to clarify the impacts. There are also recent papers showing that most of the land temperature increases in the late twentieth century can be accounted for by the influence of winds from warming oceans and, in particular, ice-mass and glacier changes in both Antarctica and Greenland. (**Compo & Sardeshmukh, 2008**). When we examine ocean heat storage patterns in the next chapter, spatial shifts in winds and cloud cover will be of great significance in affecting the heat budget of ocean waters and land regions downwind.

Note

1. Takmeng Wong at NASA's Langley Research Centre has led a team that has re-analysed the various instrument data on radiative fluxes (mostly at the top of the atmosphere in the tropics, **Wong, 2006**) and reports in 2008 that two inter-pretations are possible: that cloud thinning is a consequence of ocean warming (implying a greenhouse-gas feedback and confirmation of models), *or* that cloud thinning is a primary effect that warms the oceans, and that the state of the science cannot discriminate between the two (http://eospo.gsfc.nasa.gov/eos_observ/pdf/Jan_Feb08.pdf).

5

Ocean Cycles

What happens in the oceans is central to the global warming thesis. The IPCC estimate that 80% of global warming has been stored in the oceans. And we saw in the previous chapters that the only significant global warming signal occurred between 1980 and 2005 – a mere 25-year period that is regarded by the Panel as 'anomalous' and without explanation in relation to natural causes. In a 30-year period prior to this signal, global temperatures dipped and this necessarily included ocean surface temperatures, as they make up 70% of the surface. Prior to that time global (and ocean) temperatures had been rising steadily since the end of the Little Ice Age, and the trend from 1900 to 1950 simply continued this long-term trend. We also saw that the satellite data clearly indicate a significant thinning of cloud over the 1980–2000 time period and an increased flux of solar short-wave radiation to the surface of the planet.

Surface waters absorb solar heat

Only surface water can absorb and store significant amounts of the energy coming in from the sun. In contrast, land surfaces cool down rapidly at night unless insulated by cloud, and the atmosphere itself holds only one-thousandth of the global heat store, mostly in water vapour and cloud. Energy from incoming visible light is stored in the top one hundred or so metres of surface waters. This absorption of light in the short wavelengths effectively accounts for the natural warmth of the planet. Geothermal heat from beneath the crust is regarded as negligible. The amount of heat transferred by long-wave (infrared) radiation from clouds, water vapour and greenhouse gases back to the ocean is also very small – it has minimal capacity to heat water directly as only the first few millimetres of surface water can be penetrated by infrared wavelengths of light.

As we saw from the satellite data, the net energy flux to the oceans occurs in pulses of short-wave radiation in periods of about a decade, with some evidence that these pulses are timed by the solar 11-year cycle. The oceans then lose this heat to the atmosphere by evaporation and the action of winds, which transfer heat to the air. It is now known that the past warming of land masses is mainly due to this transfer of heat from warming seas. In a major examination of global modelling, Gilbert Compo, at the Climate Diagnostics Center in the University of Colorado, and Prashant Sardeshmukh, at NOAA's Earth System Research Laboratory at Boulder, concluded:

> ... that the recent worldwide land warming has occurred largely in response to a worldwide warming of the oceans rather than as a direct response to increasing greenhouse gases (GHGs) over land. Atmospheric model simulations of the last half-century with prescribed observed ocean temperature changes, but without prescribed GHG changes, account for most of the land warming. The oceanic influence has occurred through hydrodynamic-radiative teleconnections, primarily by moistening and warming the air over land and increasing the downward longwave radiation at the surface. The oceans may themselves have warmed from a combination of natural and anthropogenic influences.

These authors also make several criticisms of the IPCC's use of models:

> Although not a focus of this study, the degree to which the oceans themselves have recently warmed due to increased GHG, other anthropogenic, natural solar and volcanic forcings, or internal multi-decadal climate variations is a matter of active investigation {Stott et. al. 2006; Knutson et al. 2006; Pierce et al. 2006}. Reliable assessments of these contributing factors depend critically on reliable estimations of natural climate variability, either from the observational record or from coupled climate model simulations without anthropogenic forcings. Several recent studies suggest that the observed Sea Surface Temperature (SST) variability may be misrepresented in the coupled models used in preparing the IPCC's Fourth Assessment Report, with substantial errors on interannual and decadal scales {e.g., Shukla et al. 2006, DelSole, 2006; Newman 2007; Newman et al. 2008}. There is a hint of an underestimation of simulated decadal SST variability even in the published IPCC Report {Hegerl et al. 2007, FAQ9.2 Figure 1}. Given these and other misrepresentations of natural oceanic variability on decadal scales {e.g., Zhang and McPhaden, 2006}, a role for natural causes of at least some of the recent oceanic warming should not be ruled out.

These are powerful criticisms by a team of well-supported scientists in a key programme concerned with critical examination of computer models. Much of the references are very recent and I have not had time to explore them in detail. These conclusions very much support my own analysis.

Clouds play a major role in the transfer of this heat from oceans to land in the cycle of evaporation and precipitation. The mapping of relatively stable climate zones shows that these patterns of heat transfer are well established – although they are also known to shift, sometimes dramatically on longer timescales. For example, a major change in pattern occurred about 6000 years ago when the Sahara, formerly a lush wetland and savannah, ceased to receive rain from wind patterns in the Atlantic and dried out. Another major shift at this time affected northern Britain, which became wetter and cooler with large areas of forest dying and turning into acid-grassland and bog. We shall see in this chapter that during the Maunder Minimum, about 400 BP, when Europe became cold and wet and harvests failed – with famine also in China – there was a major shift in the storm tracks that cross the Atlantic. Normally these storm tracks transfer the subtropical heat of the Gulf Stream north and westwards, but during this solar minimum, something happened to shift the normal pattern of transfer of warmth.

In the current pattern, there is a high degree of expectation with regard to the regional weather in each season. Certain regions are regularly cloudy, others clear, and given what we have seen regarding the heating power of even small percentage shifts in cloud cover it follows that any major shifts in either the percentage cover or spatial distribution of clouds can dramatically affect the heat budget of surface waters. The temperature of surface waters also affects air temperature and pressure above the water mass and by a series of 'teleconnections' changes pressure, wind patterns and heat transfer around other ocean basins.

The oceans oscillate

When studied in recent decades, all the major ocean basins have proven to exhibit oscillatory patterns, usually mapped by changes in sea level air pressure at different ends of the basin. For example, the North Atlantic Oscillation shifts from high to low pressure over Iceland and low to high pressure over the Azores on a timescale of 10–20 years. The cycles alternate between periods of warmer and cooler surface waters for the North Atlantic, and there also are longer-term cycles – known as 'low frequency oscillations' – of about 60 years which lead to peaking temperatures. The NAO pattern is accompanied by changes in winds, storm tracks and landward precipitation.

A similar process operates in the Pacific Decadal Oscillation where the northern waters go through warm and cool phases of 30 years duration. Eastern and western equatorial Pacific regions undergo a high frequency oscillation known as the El Niño Southern Oscillation (ENSO), which has profound impacts on precipitation in South and Central America and Australia. The amplitude of the peaks in ocean temperatures, and hence the severity of the landward impacts, also varies over longer timescales in what appear to be regular cycles. In addition to these decadal, multi-decadal and centennial shifts, the ocean basins are also subject to very long-cycle, large-scale circulation patterns with connections from one ocean basin to another.

We shall see that much of the research on ocean cycles was just breaking when the IPCC models were first constructed and commitments made to the dominance of greenhouse gases such as carbon dioxide. As we saw in Chapters 1 and 2, there was considerable uncertainty as to the power of carbon dioxide to drive global temperature cycles and much depended upon the nature of water vapour and cloud feedbacks. In recent years, it is oceanographic data that have initiated a drive to revise the global warming models. Firstly, between 2006 and 2008 there was a major revision of *past* estimates of ocean heat storage, by 200%, but the previous estimate is still factored in to predictions of future global warming, as well as having been taken as an apparent validation of the global warming models themselves. Additionally, oceanographers are now predicting decades of future cooling as the PDO shifts phase (noting that the previous cold cycle from 1947 to 1977 coincided with the global dimming episode hitherto erroneously ascribed to sulphate pollution). Thus, the role of oceans and the new science of ocean cycles is proving crucial in understanding the natural predominant forces of climate change.

Oceans circulate and redistribute heat

The ocean depths are not uniform, but behave rather like a stack of plates down to the abyssal depth of about 4000 metres, with thickness varying over hundreds of metres and each layer moving independently of the layer above. Thus, for example, in the North Atlantic, the surface layer of a few hundred metres moves northwards towards the Arctic with minimal transfer of heat below, whereas the bottom layer in the western Atlantic moves south, and in the eastern moves north, forming a loop beneath the huge 4000 m rise that separates the shallow Arctic seas from the Atlantic deeps. South of Greenland, Arctic and Atlantic surface waters begin to sink to the bottom and join the abyssal circuit.

This 'meridional' overturning is part of a global conveyor system, with the southward moving bottom water travelling down to the Antarctic and eventually northward into the North Pacific before it up-wells to the surface. The North Atlantic part of the conveyor is thus of global significance in that it draws warm surface waters down into the cold abyssal depths and hence acts as a giant heat pump. This pattern can be seen in Fig. 11.

The dynamics of this gigantic conveyor is not well understood. The abyssal circulation may take 500–1500 years for one circuit and the flow is far from constant. The science of ocean currents is bringing continual surprises. I will detail later that it was some of the breaking science in the early 1980s that undermined the credibility of the rather simplistic computer models that attempted to show radioactive waste could be happily removed from the rest of the biosphere. However, much as the global conveyor may be involved in very long-term cycles, it is an unlikely source of variability in the shorter term in which global warming has been measured. If nothing else had changed, I would be inclined to look more deeply at the science of these 1500-year cycles because there is clear

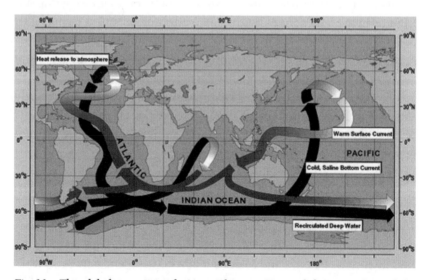

Fig. 11   The global ocean circulation pathways. (Derived from graphics of the Woods Hole Oceanographic Institute)

evidence of 1500-year periods of climate change, but it is also clear that the electromagnetic field of the sun and interplanetary space has also changed over the global warming time-period. And the two may be linked.

It is onto this dynamic that the apparent signal of global warming has been projected. That signal has been detected, as we saw in Chapter 1, in both land and ocean surface temperature data sets, as well as in the atmosphere itself by statistically examining thousands of data sets in a global grid. It follows, therefore, that to reliably identify that signal as unusual with respect to natural cyclic variability and hence conclude that there is a human imprint on the climate one would need a very good understanding of the cyclic patterns of the oceans.

In actuality, these ocean cycles are only recently becoming clear, and up until very recently there had been little attempt to build these cycles into the predictive models. The exact causes or drivers of the cycles remain unknown but there is mounting evidence through correlation with solar cycles that variations in the sun's energy play a key role. This mechanism may be direct in the form of sunlight – it is now known that the small amount of variability observed over the 11-year cycle is reflected in ocean surface temperatures (we will look further at this work). Or it may be indirect in the form of a solar-cycle influence on cloud patterns, and we shall look in more detail at the recent breaking science in this field.

These patterns of oscillations, cloud cover, precipitation and the links to solar pulses of radiation will create a pulsed pattern in the surface waters of the ocean. These waters are in continual motion around the globe. Thus we can expect landward time lags in the impact both of precipitation and heat transfer. This area is little studied.

Polar amplification and peaking cycles

When I first began to look for an alternative explanation for what I saw as the rather too early onset of global warming, I was intrigued by the patterns of heat transfer from equatorial regions to higher latitudes. The equatorial region receives the greater part of the sun's energy input and is a region of permanent heat gain – that is to say, the daytime input is not balanced by night-time losses. In higher latitudes there is a transition to areas of permanent heat loss, and by the time the polar regions are reached the tilt of the earth ensures long frozen months of 'night'. Thus, the planet operates a constant heat pump and the transfer is affected by surface currents, wind, evaporation, condensation and cloud movement.

It is curious that in most presentations relating to global warming – for example, the Hadley Centre information packs distributed to schools – the oceans and their cycles are not highlighted. The main focus is upon land temperature patterns and atmospheric heating coupled to futuristic maps of temperature distribution and the expectation of impacts such as rainfall, drought, and heatwaves. Long-term ocean cycles and shorter-term oscillation are not mentioned, nor is the phenomenon of harmonic interactions in teleconnected ocean basins.

When several cycles (or waves) are in phase with each other they amplify temperatures, and when out of phase they can form interference patterns that cancel their effects. If cycles of 4, 10, 20, 30 and 70 years are in motion and they are teleconnected through time-lagged effects on each other, then there will be complex harmonic interaction that may defy analysis. Occasionally there will be the equivalent of the surfers' 'seventh wave'.

The science of such ocean patterns is only just breaking. The PDO cycle was not elucidated until 1995, and studies are incomplete on its impact upon the well-known 70–80-year Arctic Oscillation, itself known to be connected to the North Atlantic Oscillation. The global reach of the ENSO can also affect these other cycles, and the PDO may modulate the peaks of the ENSO. For example, the ENSO of 1998 followed a period of rapid warming in the equatorial Pacific caused by changes in wind and surface water circulation, and it led to the peak of twentieth-century temperatures in the atmosphere. This carried over to land masses as well as other ocean basins during that year. ENSO peaks irregularly on a four-to-five-year cycle. It is often followed by a period of much cooler surface waters known as La Niña years. This cycle affects rainfall, in particular the occurrence of torrential rains and dangerous floods and landslides in South America, as well as drought or flood conditions in Australia. Prediction of this event, other than in the first few months of the year in which it happens, has proven elusive, despite decades of modelling. The ENSO can also go into longer-term modes of suppression or higher intensity and the controlling factors are not well understood.

I have been discussing my own sense of how ENSO affects the global temperature with a number of oceanographers and hydrologists. It is apparent that a large pulse of warm water at some depth is created very rapidly in the central equatorial region, and yet the very next year the region's waters apparently cool. It seems unlikely that the very same body of water is so suddenly cooled, and more likely that this body of warm surface water is taken by currents either northward or southward, to be replaced by cooler water either from deeper levels or from higher latitudes.

In the last major ENSO, central Pacific temperatures rose 3°C above normal and I think this created a warm pulse of tropical water at depth which then took several years to be redistributed by ocean currents. Charles Perry, a hydrologist with the US Geological Survey, is studying the linkage of ocean oscillatory patterns and river basin flow patterns in the mid-west of the USA. He has found a time-lagged linkage of rainfall patterns, ocean temperatures and solar cycles (paper presented at the International Geophysical Congress, Oslo, 2008, and see **Perry & Hsu, 2000**).

In an attempt to understand what might happen to bodies of water that were periodically heated — as the GISS satellite data from NASA show would be the case — I first looked at the difference between the southern and the northern oceans. In the north, warm water that moves from the equator is trapped between two continents and forms circulating gyres in both the Pacific and the Atlantic, whereas in the south, surface waters are free to move and get spun around the globe in a continuous belt of ocean several thousand kilometres wide. Most crucially, this belt enters the zone of permanent heat loss below 60°S and thus little heat can accumulate. The southern oceans show much less of a warming trend since the mid-1980s at 0.77°C per decade in the extra-tropical zone compared to 0.276°C in the northern hemisphere. The south polar seas began cooling at the end of the period, whereas the north polar seas and north extra-tropics display the bulk of 'global warming' with the polar region the most marked at 0.497°C per decade. There is a phase change in the north polar region in 1995. The large ENSO event in 1998 can clearly be seen in the tropical and north extra-tropical record but much less so in the south extra-tropical. There is actually no trend over this period in the tropical zone (see Ethos A.6 for detailed graphs of sea surface temperatures).

This data shows the importance of the northern region temperature anomaly as the main signature of 'global warming'. It is also clear from regional data that the northern extra-tropics warmed at exactly the same time as the 1998 ENSO peak and continued to show a 0.4°C anomaly thereafter. Curiously, the *north* polar region warmed *before* the ENSO event and showed two cooling spikes at 1998 and 2002 coincident with the ENSO of those years. Such sudden spikes in surface water temperatures might be a signal of unusually clear skies in the far north as a teleconnected atmospheric response to the equatorial high temperature regimes.

These ocean data would support the hypothesis that a large body of warm tropical surface water travelled northward following the ENSO event in 1998. As we shall see, this can be tested by examining the records of depth average temperature (DAT), a system of measuring variation of temperature with depth below the surface. This is a much better indication of the pattern of global warming since it represents accumulated heat and is not subject to the same variability as surface clouds and wind.

The accumulation and distribution of heat in the upper ocean waters

The two major ocean basins behave differently, as we can see from Fig. 11, where in the North Pacific between 30° and 60°N the surface waters are warmed by the sun but move south and westward, eventually to join surface currents that take a proportion far into the southern oceans and a proportion of that joins the tortuous journey of surface waters into the Atlantic conveyor; in the North Atlantic, the warm surface waters move north as far as Greenland. In both cases, prevailing westerly winds driven by the rotation of the planet and directed in their track by the waves of the jet stream extract heat from the surface, form clouds and dump that heat in rainfall landward of the major gyres. Thus, in the northern Pacific, any accumulated heat will be transferred to Alaska and the Pacific coastal regions of Canada. In the northern Atlantic, the accumulated heat will be transferred to northern Europe, in particular northern Scotland and Norway which are on much the same latitude as southern Greenland and Hudson's Bay in Canada.

I have tracked the expected build-up of heat in the DAT record. There have been two recent analyses — one in the UK at the Hadley Centre in Exeter, and one by John Lyman at the Scripps Institute of Oceanography in San Diego in the USA. They use differing methods but derive much the same patterns. The data are usually shown with regard to trends over particular periods — for example, at the Hadley Centre they have compared the period 1985–2004 with the period 1961–1980. Lyman compared 2005 with 2003, and there are now data for each year available from NOAA (see Ethos A.7). In Fig. 12, I have attempted to show the main zones of heat build-up and loss in the global oceans.

In the graphic there are specific areas where warm water has built up

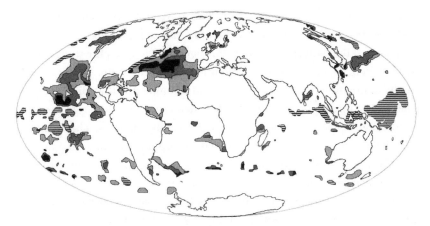

Fig. 12   *The main zones of heat gain (blocked shading) and loss (hatched shading) in the upper oceans. (Note: these zones are approximate and based largely upon the data sets 1985–2004 minus 1961–1980 generated by the Hadley Centre, with some additions in the southern oceans based upon the work of John Lyman at NOAA)*

over this period — in the North Atlantic and north-west Pacific particularly. We can clearly see a hemispheric difference with some cooling in the southern oceans. Of great interest would be to interrogate the Hadley monthly data sets of surface temperature with respect to the post-2000 period relative to the long-term mean because this is the period of the step-change noted in the cloud data. Lyman studied this period, as we shall see, and found at first a rapid fall in heat content, and then when the data were revised on account of instrument error he concluded there was at least a period of no extra heat accumulation between 2000 and 2008 and this is a major challenge for global warming theory.

From the preceding chapter we can conclude that variation in the ocean heat budget is primarily mediated by clouds, and as we have seen there are inter-decadal changes in cloud cover with significant phase shifts. Thus, we expect to find that the accumulated heat of the 1980–2000 period will find its way northward where it will be hemmed in by continental masses, and southward where it will be circulated freely in the southern oceans and more readily dissipating its heat to space.

The heat content patterns are reflected in ocean surface temperatures records. The annual variation in surface temperatures measured as anomaly used to be readily accessible on the Hadley Centre website as well as NOAA, but in recent times only the monthly series are easily accessible. In the Hadley maps the relationship of warm land masses downwind of large warm ocean masses is readily apparent (and now shown to be the main driver of land temperature increases).

I followed the patterns of change through 2000–05 and the transition to 2006, when the large region of anomalous warm water in the North Pacific suddenly changed to a large cold anomaly. In that time also the extent of southern ocean cooling also increased, but the large areas of North Atlantic warm anomaly remained unchanged. These changes may be followed in colour on the Ethos website supporting material.

We are observing bodies of warm water formed in the regions of permanent heat gain which are then transferred by ocean currents to areas of permanent heat loss. The rapidity of cooling is then determined by how much they are exposed to night skies or the polar winter as well as prevailing storm tracks (directed by the jet stream) which suck heat out of the reservoir. Changes in spatial patterns in cloud cover could enhance cooling or insulate the warmth, as well as transfer heat landward in the form of warm air masses and precipitation. There is also a complex interaction with long-term oscillations within the ocean basins and I am hoping to stimulate others better placed than myself to investigate this relationship by tracing the warm water anomalies through the records, as well as take a more detailed look at the spatial distribution of these cloud changes.

I have recently been watching animations of the global weather system drawn from satellite observations over three-day consecutive periods (at *www.intelliweather/imagesuite_speciality.htm*) and it is very instructive to follow the tracks of the cloud trails as they are driven by the jet stream and trade winds. The animation for the warm period in the northern Pacific clearly shows the persistent winds blowing warm moist air into Alaska and depleting the Pacific heat store. Similarly, storm tracks cross the warm water pool in the North Atlantic which has yet to show depletion. What is striking is that any significant southerly movement of the jet stream and any change in the standing waves will move the zones in which heat is extracted as well as where the moisture is eventually precipitated. In the Pacific cold phase the winds circulate in a cyclone in the northern Pacific, and Alaska becomes less cloudy, drier and colder. Likewise, recent changes in 2007–08 show the North Atlantic part of the wave now extracts heat closer to the American seaboard; the wave loops further north, sending drier and colder winds into Britain and Europe on the downward loop and warm air from Arabia, Iran and south central Asia into Siberia on the upward loop.

The decadal oscillations of the pressure systems in these two ocean basins will be associated with very different patterns both of warming and cooling, wetter or drier, and differing zones of cloud cover and heat extraction from the oceanic surface waters. Moreover, were such shifts to become entrained for a longer time period they could significantly alter the global heat balance. In my view, these two regions will become central to understanding the shifting global patterns of the interglacial period.

Ocean oscillations: the North Atlantic, or NAO

It can be seen from the data on upper ocean heat storage that the North Atlantic is a key region of warming over the past two decades. This region affects the whole of western Europe and Greenland, and, as we shall see in the next chapter, is responsible for the incursion of warm water deep into the Arctic Ocean. In Fig. 13 we can see that the North Atlantic region was affected by a cycle that involved a positive phase of the NAO between 1900 and 1940 and a negative phase between 1940 and 1990.

It can also be seen from the above that the North Atlantic is subject to a low frequency cycle with peaks separated by about 80 years, as well as smaller cycles of 10–20 years duration that seem to alternate. The global warming decades of 1980–2000 can clearly be seen as part of a repeating pattern. The NAO high beginning in 1995 also coincides with a linked Arctic Oscillation which we shall look at in some detail.

The negative phase of the NAO corresponds to a weak subtropical pressure system and weak higher pressure in the Arctic, and this directs storm tracks on a flat west-east trajectory, bringing moist air to the Med-

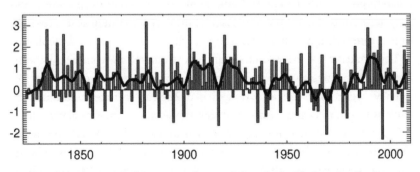

Fig. 13   The positive and negative phases of the North Atlantic Oscillation over two centuries. The index is based on pressure differentials between Iceland and the Azores. (Source: Lamont Docherty Earth Observatory, University of Columbia, New York)

iterranean and cold weather to northern Europe. The eastern seaboard of the USA has more snow and wintry cold, but Greenland has milder winters. In the positive phase, higher subtropical pressures and deeper Icelandic lows result in stronger storms tracking further north into Britain and western Europe. This results in warm and wet winters in Europe, but cold and dry conditions in Canada and Greenland, with the US eastern seaboard becoming milder and wetter and the Mediterranean becoming drier.

Thus we can see that the NAO entered a positive phase at the same time as the apparent global warming signal became evident and at the end of this period it has entered a negative cool phase. We can now look at the major oscillations in the Pacific Basin – the Pacific Decadal Oscillation in the northern part and the El Niño Southern Oscillation in the equatorial region.

The Pacific Decadal Oscillation, or PDO

This major oscillation was identified rather late in the twentieth century with decade-long signals (**Trenberth & Hurrell, 1994**) and longer cycles of 50–70 years (**Latif & Barnett, 1994**). Further, in a study using tree-rings in Baja, California at about 30°N, Franco Biondi at the University of Nevada together with Alexander Gershunov and David Cayan at Scripps Institute of Oceanography in San Diego, California identified variations in amplitude of the cycle in a period from about 1700 to the late twentieth century. There were periods of shorter frequency and high amplitude and teleconnections to the El Niño oscillations such that ENSO was magnified when in phase with the PDO (**Biondi et al., 2001**). Tree-ring growth in the American south-west reflects the oscillation very accurately as it correlates to precipitation in the mountains.

This pattern of constructive and destructive interference is very important in relation to teleconnections, as it can give rise to lower frequency cycles. We can also note that a 23-year recurrent pattern was evident in the tree-ring data – and that this is also a major solar cycle.

The interaction of PDO and ENSO led to a clearly defined cooling in 1947 and a shift back to warming in 1977, and from the graphs in Chapter 1 we can see that this had global implications. The graph in Fig. 14 shows where the amplitude of decadal level swings is increased when in phase.

The longer-term oscillatory pattern appears as a 30-year phase change,

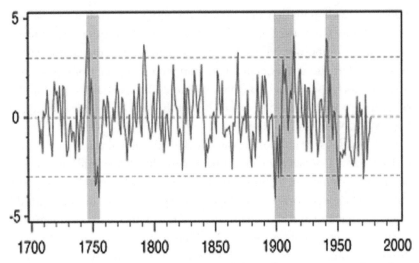

Fig. 14   The interaction of the Pacific Decadal Oscillation (PDO) and Southern
Oscillation Index (ENSO), 1700–1977. (Source: Biondi 2001)

where in the warm phase there are warm waters in the eastern Pacific off
Alaska and cool waters in the west off Russia and Japan, with large areas of
warm water in the central equatorial region; in the cool phase, the western
Pacific off Japan warms and the waters off Alaska cool, with the central
area also cooling but the southern region warming. The graph in Fig. 15
shows the last century – the index is calculated to reflect the differing
atmospheric pressures in the regions where the dynamic is strongest.

In the graph below, the negative troughs are the cool phases. The correlation between the PDO and the global temperature pattern is very clear
– with the trough from 1945 to 1975 coinciding with the period formerly

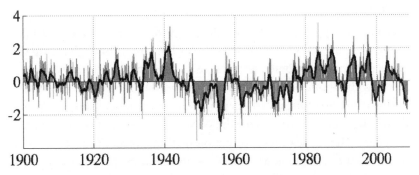

Fig. 15   The PDO Index, 1900–January 2008. (Source: Joint Institute for the
Study of Atmosphere and Oceans, University of Washington)

associated with pollution and 'global dimming'. There are two strong warm phases from 1925 to 1945 and from 1977 to 2006. The strong variation in amplitude in 1998–2000 coincides with the major ENSO event of 1998 and subsequent La Niña. It would appear from this series that the 30-year warm period from 1977 to 2007 may have shifted to the cool phase – it remains to be seen whether this will be a similar trough to that of the 1950s and 1960s.

If we recall discussions of the satellite data, the global warming decades of 1980s and 1990s were marked by a naturally clearer and more transparent atmosphere in relation to cloud cover and aerosols in oceanic regions. The close correlation of global temperatures to the PDO would suggest that cloud and aerosol patterns related to the Pacific Ocean are the main driver of global temperatures. What happens in the Pacific feeds through to other ocean basins either through atmospheric pressure patterns, cloud movements or ocean currents, and the latter are likely to have time lags associated with their influence.

As we noted in Chapter 2, those scientists researching the radiation fluxes to the earth's surface make guarded criticism of the global circulation models that couple atmosphere and oceans, commenting as late as 2005 that the GCM models do not replicate the periodic patterns. Recently, a research team lead by N.S. Keenlyside at the Leibnitz Institute of Marine Sciences in Kiel, Germany, made headlines with a study that sought to incorporate ocean cycles into the coupled GCMs (**Keenlyside et al., 2008**). The study integrated the oscillations of sea surface temperature for the Pacific and Atlantic in a simple model that also incorporated the projections of future global warming due to carbon dioxide and concluded that the meridional overturning – the conveyor in the North Atlantic – would slow to its long-term average and that both Pacific and North Atlantic waters would go through a cooling period strong enough to override the apparent global warming signal. They expected warming would recommence in 2015 when the North Atlantic Oscillation switched phase again and combined with the rising greenhouse effect (which they accept without question and without reference to other explanations).

This is the first time that a scientific team using model simulations has shown that natural cycles have the power to overwhelm the greenhouse gas effect. They repeat the assumption that the greenhouse gas model is correct (oceanographers are not likely to question the reliability of the atmospheric physics), but they do not then point out that just as the oceans have the power to dampen the signal they must also have had the power to amplify it in the first place. They make no calculations as to the quantitative relations of the supposed greenhouse effect with the suppressive or amplifying power of the natural cycle.

Clearly, this team is still working with an unrevised physical model for the carbon dioxide effect and if this is, as I suspect, erroneous by a factor of 2 or 3 then the future cooling might be more prolonged. There is a clear correlation between the longer-term cycle of the PDO and the lower amplitudinal cycles of the NAO, as also with the ENSO. We will return to the issue of cycles when we look at the relation of these oscillations to the Arctic and Antarctic temperature records. We turn now to the issue of upper ocean heat content as a signal of the human-induced global warming.

The anthropogenic signal in the oceans

Up until 2005 there had been some debate as to whether the heat content of the oceans showed a significant change as would be expected by orthodox global warming theory. In that year, Tim Barnett of the Scripps Institution of Oceanography Climate Research Division summarized the issue in *Science* with 'Penetration of human-induced warming into the world's oceans' (**Barnett et al., 2005**), where he concluded that recent warming of the oceans cannot be explained by natural factors. This paper was widely reported in the media in arguments underpinning the greenhouse-gas hypothesis.

The authors used two independent climate models and recently updated ocean data sets of the penetration in depth and time of the global warming signal to attribute causes. They found 'the warming signal is far stronger than would be expected from natural internal variability, as estimated by the models', and state 'the control run (of the model) variations are a reasonable representation of the natural internal variability, at least on the decadal timescale of interest here'. They go on to model solar inputs and volcanic aerosols according to recent analyses of irradiance, as well as greenhouse gas (GHG) forcing, concluding that only GHG forcing can explain the recent warming signal. They state that they could find no difference between runs for all natural variability and the runs using only solar variation.

Thus once again this conclusion is based upon the results of computer models that aim to simulate the natural world. The solar inputs relate only to acknowledged small variations in visible light output at source, and there is no reliable model for the cloud variations recorded on these timescales. Furthermore, the science of oceanic cycles is still in its infancy and characterized as uncertain by the IPCC.

As we shall explore in the next chapter, there is other evidence correlating solar cycles with sea surface temperatures but they do not rely upon a virtual reality in a computer model. If there are other factors involved in the solar cycle, such as UV variability or electrical effects upon cloud seeding and aerosols, then these will only be picked up in the real world data.

The modellers also talk of a 'reasonable' match of the computer model with the past natural variability, and this is used as a 'validation' of the model. But 'reasonable' is a subjective term – the models are known not to replicate the decadal variability, whereas they more easily replicate the overall trend. Barnett's analysis was published in the same year as the satellite data reviews and thus there is no excuse for such important findings on ocean heat storage to be reported out of context with other research on the flux of energy that creates such storage. At the very least scientists should refrain from comment on causal relationships that they have not specifically assessed. This is an example of the institutional inertia and authority structure of science – none of these scientists are examining the atmospheric model with a critical eye, despite the existence of high-level debate on the nature of the ocean-atmospheric feedback at the very outset of the modelling exercise (**Lindzen et al., 1982, 1991, 2001; Mitchell et al., 1989**).

There is another dynamic at work in relation to prior commitment. Oceanographers have worked for the last 50 years on the assumption that a natural *internal* variability exists within the oceans – that is, the oscillations are a mechanical phenomenon, sometimes in phase where they amplify each other and sometimes out of phase where they interfere in what is called a 'destructive' manner. Yet in the past 20 years many studies using more sophisticated techniques have shown correlations between ocean cycles and solar cycles – that is, the cycles may be *externally* driven. We will review some of these key papers below.

This narrow focus is a common enough occurrence but it reflects badly upon the limited referee process of even major journals such as *Science*. It may also illustrate another process; in the competition for funding, mention of global warming and greenhouse gases is a sure way to convince funders that data collection is worthwhile, whereas reference to natural drivers – which government can do nothing about – may not attract such funds.

Errors in the estimation of ocean heat content

The Barnett review relied upon ocean heat content data and analysis from other studies that were available to the Scripps team at the time and which showed that 84% of the excess heat of global warming was contained in the upper oceans (**Levitus et al., 2005**). This underpins the IPCC's statements on global warming signals. Barnett's conclusions about the anthropogenic signal made headlines in science reports around the world and were widely reported in the media. Less well reported was the critical review of the methods of data collection used by Levitus by Gouretski and Koltermann published as 'How much is the ocean really warming?' in *Geophysical Research Letters* (2007), which found that due to instrument bias the global heat content had been overestimated by a factor of two (**Gouretski & Koltermann, 2007**). A recent paper by Matthew Palmer and colleagues at the Hadley Centre, 'Isolating the signal of ocean global warming', using a different method comes to the same conclusion (**Palmer et al., 2007**).

Further to this complication in what is an essential area of global warming science, great interest was generated by an announcement in November 2006 by John Lyman and colleagues at NOAA's Pacific Marine Environmental Laboratory in Seattle, and published as 'Recent cooling of the upper ocean' in *Geophysical Research Letters*. The team studied a 13-year period from 1993 to 2005 and discovered a downturn between 2003 and 2005 which 'could be the result of a net heat loss from the earth to space' (**Lyman et al., 2006**):

> ... this variability is not adequately simulated in the current generation of coupled climate models used to study the impact of anthropogenic influences on climate. Although these models do simulate the long-term rates of ocean warming, ***this lack of inter-annual variability represents a shortcoming that may complicate detection and attribution of human-induced climate influences.***

These guarded comments are a major criticism of current climate models upon which the detection of the anthropogenic signal depends. Moreover, the key changes, which are here described as 'inter-annual' and are not amenable to modelling, refer to very recent changes that correlate with other data from other disciplines which the NOAA oceanographers – if they are aware of them – do not mention, such as the satellite data on solar flux to the ocean surface and to earthshine data on albedo, as well as the apparently faulty NASA data on a sudden out-flux of long-wave radiation at this same point in time.

In a later paper, the team reported an instrument bias in the new system of floats and revised their original estimate of the cooling. This work was widely reported in science journals, perhaps less widely in the press, but even so, several researchers in other disciplines related to radiation fluxes and cloud cover failed to notice the connection – this ocean cooling followed reports of increased albedo and increased global cloud cover from which cooling would be expected. One conclusion is clear: no consensus now exists on the oceanic signal of global warming and this issue is not flagged by IPCC's 2007 Report.

The importance of El Niño

In the southern hemisphere, ENSO is the strongest cyclic phenomenon and another measure of global periodicity that is not well replicated by current climate models. The intensity of the effect varies on millennial timescales and can be traced in its impact on river runoff, lake sediments, tree-rings and various other indices related to precipitation throughout the southern hemisphere.

There are dozens of studies showing this cyclic pattern. Christopher Moy and colleagues at Stanford University published in *Nature* an analysis of ENSO activity derived from proxies in sediment over the last 10,000 years (the Holocene period, since the end of the last Ice Age), 'Variability of El Niño/Southern Oscillation activity at millennial time scales during the Holocene epoch' (**Moy et al., 2002**). This paper studied the intensity of the ENSO events as recorded in debris deposited in sediments in an Ecuadorian lake high in the Andes. Under normal conditions offshore winds drive surface currents westward across the Pacific and this allows colder nutrient-rich waters to up-well from depth. During El Niño conditions these winds fail and warmer surface water spreads eastward, shutting off the up-welling (and collapsing local fish populations which rely upon the nutrients) but also increasing convection and precipitation over South America and in Australia. This phenomenon often peaks at Christmas, hence the title of El Niño as 'The Child', i.e. Christ.

The authors find that ENSO activity in the early Holocene was weak but began about 7000 BP and intensified to a peak at about 1200 BP, after which it has been declining. The variance shifted from a 1500-year period to a 2000-year period later in the Holocene. They report that an ENSO computer model (referenced to Zebiak and Cane in their paper) has been able to replicate this longer-term pattern using changes in insolation resulting from variations in the earth's orbit of the sun. These changes induce more persistent easterly winds which drive the system towards La Niña as temperatures and pressure gradients increase across the Pacific. These Pacific oscillations parallel 'Bond' events (periodic rafts of ice) in the North Atlantic at 1800-year periodicity – with the Bond events (see below) tending to occur during periods of low ENSO activity followed by high activity. This variability shares similar periodicities to the Atlantic subtropical 1500-year cycle identified by deMenocal, which we shall look at below – noting also that the Bond cycle average was 1470 years but varied from 1300 to 1800.

My own view is that such variability, long regarded as 'internal' to the planetary ecosystem, if it is not linked eventually to solar activity could only be ascribed to long-term oscillations in inter-ocean current systems, such as the global circulation – but intuitively, I think this an unlikely driver. There is also a potential lunar cycle factor, proposed recently by Charles Keeling and Timothy Whorf at the Scripps Institution, where the moon's declination cycles of 1800 years and 5000 years may be involved in tidal forces amplifying oceanic processes (**Keeling & Whorf, 2007**). As we shall see when we look further at solar cycles, there is much yet to be discovered in relation to what drives these long cycles.

There is, however, a good deal of evidence that the shorter cycles, at least, *are* correlated to solar activity. As we shall see below, Warren White at Scripps Institution argues that the inter-decadal ENSO variability can be excited by shorter-term and quite small insolation changes. However, we should point out that even oceanographers are not yet able to explore the potential science of a solar magnetic cycle amplifier – through cloud seeding, which would also correlate with the visible radiation flux.

The intensity of ENSO is expressed as the number of events per 100 years and shows late Holocene peaks of 30 events per hundred at 1600, 1400 and 800 years BP, and troughs at 2000 BP (major), 1500 BP (major), and minor troughs as ENSO drops in intensity at 500 and 300 BP. The latter dates correspond to the two troughs of the Little Ice Age (from temperature proxy data) and to two troughs in the beryllium-10 record (the proxy for solar magnetic activity). We should note that low ENSO intensity is therefore associated with falls in solar magnetic activity and periods of cooling. The longer-term periodicity of 1500–2000 years may be associated with orbital changes (according to the model) rather than the intrinsic energy cycle of the sun. There remains a possibility that the internal oscillations of the deep ocean conveyor system are also involved.

As we can see, a complex of factors may be involved and although the mechanisms are not yet clear there is strong evidence that the ENSO strength is linked either to peaks in solar output or to orbital changes affecting insolation to the surface – either of these sources of heat to the ocean surface waters would intensify the ENSO peaks and perhaps add to their frequency. There are currently no major orbital shifts, which are in any case slowly changing. But as we shall see in Chapter 7, the earth has experienced a recent Solar Grand Maximum as well as the strongest ENSO event in a two-hundred-year record.

This 1998 major ENSO appears to be the result of a steady increase in amplitude over several decades coincident with the pattern of global warming between 1980 and 2000. However, the 2002 and 2006–07 ENSO events were small and La Niña or neutral conditions have domi-nated the last ten years. This coincides with a waning of the PDO, a lower peak in the solar cycle (2000–02) and the decline to a recent two-year period of very low solar magnetic activity. From Moy's data, it is clear that this ENSO pattern is a cycle and that previous intensity correlated with solar activity.

Thus, we have one more factor indicating a natural and cyclic peaking capable of enhancing global temperatures worldwide – over the same time period that attribution studies purport to be able to identify the relative strengths of the two processes – and as we saw at the outset, the IPCC models using natural factors *alone* predicted a flat global tempera-ture pattern from mid-century. Once again we have ample evidence not so much of the inadequate knowledge of cycles, but of the inability of computer modellers to translate ecological data into predictive pro-grammes. There is no doubt that the cycles are there but their complexity defies simple mathematics and effective incorporation into the models.

The ocean's response to global warming

We need to look more carefully at the ocean's response to global warming. According to GHG theory and models, the oceans will retain more heat and the climate system will recirculate this heat poleward. As we will see when we look at the work of Warren White, the upper one hundred metres of the ocean retains this heat with relatively little transferred to greater depth. The oceans can only lose heat by radiative transfer (long-wave radiation back to space) or by latent heat of evaporation and the convective forces of wind. Cloud will affect this process by insulating warm waters and blocking the losses to space.

Thus as the oceans warm due to SW radiation input they radiate LW up from their surface – and this is measured by satellite (they are not warmed significantly by LW back-radiation from cloud or by warmer winds because infrared radiation does not penetrate beyond a few millimetres of the surface). The LW return flux from cloud is much greater than the radiative forcing calculated for GHG and hence no signal can be detected above the background noise (Lindzen calculates that of 340 watts/sq metre downward flux of infrared 290 is due to water vapour, 40 due to cloud, leaving 10/340 or 2% from the natural reservoir of carbon dioxide (which has now been increased by 35% to 2.7%).

We saw that the heat inputs of solar SW flux account for the 1980–2000 warming period and the largest region of heat gain is in the tropics. It is therefore instructive to look at tropical waters, and in particular the western Pacific zone that generates the ENSO. Recent global temperatures have followed a pattern dominated by the two ENSO events of 1998 and 2002, and in the central Pacific region, apart from these peaks (see Fig. 16), there is no evidence of continued temperature rises – only the periodic cycles.

As is clear from this data set, these oceanic surface waters show no overall pattern of global warming other than in the amplitude of the ENSO peaks, which rose through the 1980s and 1990s but peaked in 1998. The last two El Niño years, 2002 and 2007, have been small by comparison. To my knowledge, no research teams have proffered explanations for this recent dampening of the El Niño. One of the largest and best informed multi-disciplinary centres (Hadley in Exeter, England) predicted exactly the opposite for 2007 with a major ENSO event and world record temperatures, thus indicating how far out of touch such a world-renowned climate modelling centre can be with what is rather common knowledge among oceanographers.[1]

The black vertical line in 2006 represents the point at which the Hadley Centre made their prediction, which the Centre announced was based upon an in-depth analysis of ENSO dynamics and increased confidence in the models of global warming.

The data set clearly shows the periodicity in this equatorial region of the Pacific where the major influx of solar energy occurs. The immediate

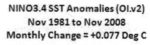

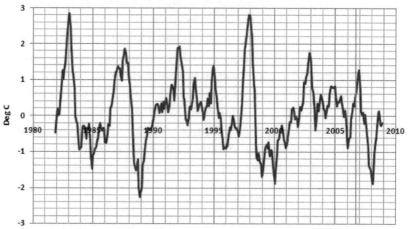

Fig. 16   *The variability of El Niño. The vertical line indicates the prediction point discussed in the text. (Source: Bob Tisdale blog)*

decline in surface temperature after the peaks either indicates that the surface has cooled due to clouds, or that surface waters have been replaced by cooler waters. If, as I suspect, large bodies of sub-surface tropical water move into the extra-tropical zone in cycles, then the ocean current patterns outlined above will determine their further evolution.

Longer-term cycles of change

Having established that the Holocene has seen significant cyclic variation, we are left with the still open question of whether these changes can be explained by *internal* variability or whether they correlate to solar cycles. The existence of a 1500-year cycle, as noted by deMenocal below, implies some form of periodic change, but there remains a possibility that inherently internal oceanic cycles may operate over this time period. Recently Y. Wang at Nanjing University reported work using a proxy record of temperature from stalagmites in China: 'The Holocene Asian monsoon: links to solar changes and North Atlantic climate', published in *Science* (**Wang, 2005**).

This paper demonstrates the link between events in the North Atlantic and the Asian monsoon and shows periodic cycles linked to solar magnetic activity proxies; the data are gathered from oxygen isotopes in cave stalagmites. This is one of many recent studies that shows correlations between oxygen isotope proxies of temperature fluctuations and solar magnetic cycles. It also demonstrates the connectivity of the northern hemisphere environment, with the Chinese record also correlating with major fluctuations recorded in the Greenland ice cap. A major dry event is recorded by both at 4400 BP.

This work and similar work in Oman (**Fleitmann et al., 2003**) support the idea that solar changes are partly responsible for changes in the Holocene monsoon intensity, and Wang's paper references about a dozen similar data sets that correlate changes in temperature and precipitation with solar cycles.

The fine-tuned stalagmite record in Oman

> ... confirms the correspondence of statistically significant solar cycles at 205 (de Vries cycle), 132, 105, 90 (Gleissberg cycle), 60, and 55 years. Whether variations in solar output affect the Indian Ocean Monsoon indirectly, by internal forcing mechanisms, or more directly, by external forcing mechanisms, is not yet fully resolved.

Interestingly, in the discussion over internal and external drivers, one of the external mechanisms referred to was the work of Drew Shindell at NASA on variability of UV radiation from the sun. There is clearly no consensus on the relative roles of the two. For our present purposes we note that firstly the solar magnetic cycle signature is present in the temperature proxy record, and hence the (apparently) small changes in solar activity over these cycles are adequate to cause climatic and ecosystem changes; secondly, this cycle is imposed upon larger orbital and inclination cycles, and thirdly, these effects are felt right across the northern hemisphere.

There is a growing body of scientific literature on past climate fluctuations and on the presence of cycles of change. These cycles are not limited to the large-scale changes attendant upon the onset and retreat of ice ages, which have a periodicity of 100,000 years – many studies look in detail at recent millennia when the earth's climate has been regarded as 'stable'. We will not take time with the issue of ice ages except to say that, though correlated with orbital changes in the earth's relation to the sun, for some climatologists there is still a 'missing link' with regard to causation, particularly in relation to the sudden endings; most would also accept that the changes are not initiated by changes in carbon dioxide levels, though they may be amplified by them, because in the ice-core record temperature rises precede carbon flux changes by several hundred years. Here we are more concerned with identifying significant change during the Holocene period (the last 10,000 years) and correlations to solar cycles.

In this respect, it is noticeable that in the work of the IPCC and other scientific bodies that present information to policy makers there has been a tendency to focus attention on the last 150–200 years and only within a context of the last 1000–2000 years, a period of *relative* stability (i.e. between 0.5° and 1°C above or below the mean). The marked warming of 0.6°C since 1980 is not unusual, even in the rate of change.

The Little Ice Age and the Medieval Warm Period

There is a detailed recent review of the late Holocene pattern as a whole by Willie Soon and Sally Baliunas at Harvard University's Department of Astrophysics. In 'Climatic and environmental changes of the past 1000 years' published in *Climate Research* (**Soon & Baliunas, 2003**), they summarized the palaeo-environmental evidence for temperature and precipitation change over the last 1000 years. It is doubtless stimulated by the controversy regarding the 'hockey stick' graph used by the IPCC and the Panel's 2001 conclusion that the recent temperature rise was unprecedented.

The paper shows that recent Holocene temperature fluctuations were significant and that there is no reliable evidence that twentieth-century changes are beyond the level of variability in the record. They conclude:

> . . . the Little Ice Age exists as a distinguishable climate anomaly from all regions of the world that have been assessed with only two regions, Tasmania and Antarctica failing to show this depression.

Similarly the Medieval Warm Period is demonstrated except for some conflicting results in South America, and they conclude:

> most of the proxy records do not suggest the twentieth century to be the warmest or the most extreme in their local representations

The paper gives a region by region account of the MWP and the LIA – and this is important reading considering the predictions of a coming Maunder-type minimum by some solar scientists. Such scholarly treatment in a reputable journal ought therefore to counter arguments in the media and from environmental groups that have consistently downplayed the two temperature anomalies – on the grounds that they were localized and not global phenomena.

I will add to this point by selecting two recent papers that go into greater depth on the issue. There may well be good papers that I have not seen which do not find evidence for the warm or cool periods but, given the number that do, the absence of evidence in a few studies is not compelling. In any case, my task is not advocacy but to show that on these crucial issues a significant body of science exists and at the very least there is no consensus.

Firstly, on the 'coherence' of data showing that the two periods show up in the temperature proxy records of both hemispheres: Peter deMenocal and colleagues at the Lamont-Doherty Earth Observatory at Columbia University published 'Coherent high- and low-latitude climate variability during the Holocene warm period' in *Science* (**deMenocal, 2000**). This paper deals with the central issue of whether the Little Ice Age (LIA) and the Medieval Warm Period (MWP) were extensive phenomena. It demonstrates that subtropical Atlantic sea-surface temperatures varied in phase with the North Atlantic region records of the LIA and the MWP.

The LIA is recorded off the coast of West Africa with a temperature drop of 4°C and 2°C in two marked periods of 300 and 600 BP respectively. The MWP is evident at 1°C above the recent average. These data are also in phase with Bermuda Rise data which show fluctuations of 2°C from the MWP to the LIA.

Both sets of data (derived from isotope studies of Foraminifera shells) are in phase with a 1°C MWP rise marked in boreholes from the Greenland ice sheet, which also showed a 0.5°C drop over the LIA. The paper also shows very clearly the correlation with the drying out of the Sahara at 5000 BP after a wet period caused by an enhanced African monsoon. This wet period lasted from 14,000 to 5000 BP and coincided with a peak in summer insolation in the subtropics caused by the orientation of the earth's orbit at 10,000 BP (an 8% increase of about 20 watts per square metre caused by the orbital precession aligning boreal summer solstice with the perihelion). There is reference to a computer model that replicates these changes (Claussen, *Geophysical Research Letters* 26, 2037, 1999).

The faunal changes noted in the sediment studies point to periodic changes in ocean currents; during cool phases there is a suppression of summer season warming and northward migration of warmer tropical waters, coupled with increased up-welling and movement of sub-polar waters into the region.

The solar-driven changes at 10,000 BP are thought to be amplified by albedo changes in the Sahara which was heavily vegetated until the climate shift at 5000 BP.

The causes of the LIA and MWP are not so clear:

> Whatever their ultimate cause, these millennial scale Holocene SST (sea surface temperature) variations appear to involve the entire North Atlantic, recurred with a 1500 period throughout the glacial and interglacial intervals, were accompanied by terrestrial climate changes and involved large scale ocean and atmospheric re-organizations that were completed within decades or centuries, perhaps less.

The authors note that the recent warm period *is part of a fluctuation of increasing amplitude* (my emphasis) and supports the view that Holocene variability has increased in recent millennia with the LIA the largest amplitude in 20,000 years.

The crucial importance of this paper (and there are perhaps a dozen parallel references to other work) is that oceanographic sediment studies show that the current warm period is not unusual – and not the warmest – and can be seen as potentially still in recovery from the largest amplitude cool period in the record. In other words, natural factors are quite capable of causing the scale and rate of change that we see today, although they note that the recent warming is unprecedented for the last 1000 years (after the MWP). The authors word their conclusions carefully:

> ...these results underscore the need to understand anthropogenic warming within the context of rates and amplitudes of natural late Holocene climate variability.

This is another restrained 'scientific' way of saying that the past picture of relative stability during the Holocene (as for example given in IPCC's 'hockey stick' graph) must now be discounted and that any claims for an anthropogenic signature must be made against new evidence that climate has warmed and cooled on a large scale over the past 1000 years, and has been doing so at 1500-year intervals for 20,000 years. The authors cite about twelve papers dealing with similar data sets from *Science*, *Nature*, *Geology*, *Quaternary Science Review* and *Palaeoceanography*.

Timing the cycles – the possible role of the sun

Thus far, we have seen how new science undermines the models based upon a previous understanding and we have seen evidence of resistance to revising models, particularly when it undermines the global warming thesis. One of the key areas of resistance relates to the long tradition of regarding ocean cycles, and hence climate variability, as essentially mechanical properties of an oscillating system. They are thus classed as *internal* variability. Climate models do not have to replicate these cycles exactly – that is, predicting the peaks and troughs, to be regarded as reliable, as long as the variability looks reasonably similar. There has been very little work on correlating ocean cycles with solar cycles – and we turn now to some key papers.

Warren White and a team at Scripps Institution of Oceanography in San Diego have been working on how decadal mode changes such as ENSO can be timed by the solar cycle. A paper delivered at the 1st Solar and Space Weather Euro-Conference on the Solar Cycle and Terrestrial Climate, 'Global average upper ocean temperature response to changing solar irradiance: exciting the internal decadal mode', provides critical evidence for understanding several elements in climate data and global patterns. Firstly, the authors are looking at depth average temperature (DAT) as a more relevant measure than ocean surface temperatures, and secondly they are looking at solar irradiance and solar cycles as a mechanism for timing decadal variation hitherto thought to be the 'internal' oscillations of the oceanic system.

The authors refer to earlier work which found:

> ...the sun's irradiance and global average sea surface temperature records significantly correlated with each other over the 95 years from 1900 to 1994.

Subsequent work showed that the decadal and inter-decadal peaks displayed temperature signals with time lags of 30–45 degrees of phase 'in apparent response to the sun's surface irradiative forcing of about 0.1 watts/sq m'. These correlations were maintained to depths from 40 to 120 m but not deeper, demonstrating that the inter-decadal variability was held in the upper ocean:

> ...the excess heat storage associated with changes in the sun's radiative forcing on decadal and inter-decadal period scales was found trapped in the upper layer of the global upper ocean ... presumably balanced there by radiative-plus-turbulent heat loss to the overlying atmosphere.

The researchers also point out that :

... changes in global average sea surface temperatures are geophysically irrelevant when considering the earth's global average oceanic response to changes in the sun's surface radiative forcing.

This is an important conclusion because virtually all discussion outside of the oceanographic science of 'global warming' relates to surface temperature records. Whilst these are doubtless influenced by what happens in the ocean, the focus upon surface records means that correlations with solar cycles may be obscured and the broader pattern of heat transfer is under-researched.

The Scripps team found that the phase-locked patterns could be explained by a 'delayed action oscillator model' and this model also displayed the amplified peaks of the ENSO periodicity. They conclude that the relatively weak solar signal is amplified by white noise and phase-locked to the solar cycle.

This work demonstrates the potential role of white noise and stochastic resonance in producing periodic patterns that hitherto were explained as 'internal' oscillations. However, the authors appear not to have considered the potentially larger wattage changes that would accompany a 3% change in cloud cover as indicated by Marsh and Svensmark's work (see Chapters 7 and 8). This would produce a much stronger signal and less necessity to invoke stochastic resonance phenomena. The peaks and troughs of the NASA GISS and ISCCP data show clear patterns of the order of several watts/sq metre and we shall see from recent work by Camp and Tung, detailed below, that correlations of solar cycles are associated with larger surface temperature changes than previously thought and that they also have a spatial component.

These surface flux changes over the solar cycles — we can see that satellites record cyclic variations of 7–11 watts per square metre of net radiation anomaly (averaged over the globe and hence greater in the tropics) and this would represent a powerful periodic pulse to the oceans (many times greater than any carbon dioxide input over the same period). The apparent trends in the albedo and cloud cover figures from ISCCP, when broken down to regions, as we have seen, display decadal phase changes rather than smooth trends. The changes documented are, as we have noted, quite enough to drive the global pattern of warming that we have seen and provide no obvious signal for a carbon dioxide component.

Recent work by Charles Camp and Ka Kit Tung at the department of Applied Mathematics at the University of Washington in Seattle (**Camp &**

Tung, 2007a,b) suggests that on average the 11-year solar cycle leads to a 0.2°C signal in near-surface air temperatures and that surprisingly (they state), this signal is present in the global unfiltered data (see Ethos A6).

What is surprising is that this simple relationship has not been high-lighted before. The authors note that they are the first to publish this correlation, which survives simple statistical tests of significance. They then go further and look at the *spatial* pattern of temperature changes – perhaps motivated by the fact that previous correlations of surface tem-perature with the 11-year solar cycle had used primarily regional data.

Here they found that the original data yielded a time-dependent index when temperatures at the solar maximum were compared with those at the solar minimum (the years of perturbation by volcanoes were left out) and they found a significant correlation to latitude (see Ethos A6.2). The pattern is of great interest in that it shows a higher effect in the Arctic regions, and as we are looking at surface temperature responses, we can assume that this spatial pattern would also be influenced by any cloud and water vapour feedback.

When this work was first published, *New Scientist* commented that as the researchers had discovered a larger than expected response to the solar cycle it underscored the greenhouse gas global warming model! (*New Scientist*, 11 August, 2007, p. 14.) At that time I had not seen the original paper – it took several months to view the work. In fact the authors do not claim to support the greenhouse gas global warming model. On the contrary, they comment:

> Our work establishes that the surface-temperature response is correlated with the solar-cycle forcing at over 95% confidence level. For compar-ison, a similar relationship between response and forcing has not been statistically established for the greenhouse global-warming problem.

This is one of a number of examples of *New Scientist* adopting a 'global warming' spin in its journalism. My initial response to the reports of their paper has proven correct: that their work further established the power of solar cycles to drive global temperatures. The fact that they have found double the expected temperature response, and a poleward amplification, should direct research towards further spatial analysis of cloud patterns.

My expectation is that the heat input of several watts/sq metre for a few years at the top of the solar cycle is circulated by the oceans over decadal timescales in the first 120 m of the ocean. Oscillations and currents that circulated these waters would have the potential to release heat in regional locations far removed from the main input areas and with time lags that would then distort the relationship to the solar cycle.

As I was drawing these conclusions from the oceanographic data, I came across the work of Charles Perry. He has elucidated time lags of up to 34 years in the relation of Mississippi Basin hydrology to the jet stream and North Pacific Ocean temperatures and correlates these patterns to cycles in Total Solar Irradiance and the geomagnetic index – all of which demands a mechanism for the amplification of the solar cycle as proposed by Svensmark. His work is published in leading science journals yet draws little attention (**Perry & Hsu, 2000; Perry, 2005, 2007**).

There has been recent work on spatial patterns at the Centre for Marine Science at Bremen University. Gerrit Lohmann and colleagues published 'Climate signature of solar irradiance variations: analysis of long-term instrumental, historical and proxy data' in the *International Journal of Climatology* (**Lohmann et al., 2004**). The team looked at the patterns of ocean surface temperature in gridded data sets to detect influences of solar variability. The index used for this variability was solar irradiance but I would add that this would act at least as a proxy for whatever other effects might be associated with the same periodicity of activity, such as solar-cloud fluctuations or UV induced atmospheric effects. They conclude:

> Taking the hemispheric temperature average, one can find a good match between the long-term average surface temperature and the solar cycle ... this could partly explain the cold periods around 1700, 1810 and 1900, the anomalous warm 1940s and the slight cooling between 1950 and 1975.

However, the observed temperature variations are rather high and

> ... cannot be explained by a direct radiation effect. Indeed our analysis suggests that the influence of the solar forcing onto climate is via atmospheric modes of variability. We detect the Schwabe (11yr), Hale (22yr) and Gleissberg cycles (80yr) in the solar irradiance forcing.

and that this

> ... suggests the warming trends since the 1970s and prior to 1940 are in accord with the Gleissberg cycles.

> ... the multi-decadal variability associated with solar irradiance is not confined to the northern hemisphere ... raising possible links with the Pacific Ocean.

Although the authors go on to acknowledge that the recent 30-year trend of temperature is larger than the trend of solar forcing and hence likely to indicate a greenhouse gas component, they are still not introducing (nor do they reference) a solar-cloud factor that might amplify the solar irradiance changes, nor do they reference data showing the trend in thinning clouds. Once again we see researchers paying deference to greenhouse gas components that they themselves cannot critique but not to solar-cloud research which might better explain their trends.

These works highlight regional anomalies, such as the warming Arctic, and we shall deal with this in some detail in the next chapter. We may conclude here that such anomalies are not likely to be solely a direct response to solar irradiance. The decadal integration referred to by Lohmann and the lack of equilibrium in surface waters referred to by White suggest ocean currents and atmospheric heat transfer will be key elements. As we have seen from Camp and Tung's work, the solar pattern of warming is amplified in the Arctic region, but the mechanism is not elucidated. It is likely a combination of the movement of warm water into the Arctic Ocean, warm air, cloud and precipitation from atmospheric circulation; and as we shall see when we consider the role of UV radiation, the polar vortex responds to higher atmospheric heating phenomena related to the absorption of UV light by ozone.

Thus we may conclude from this review of recent oceanography that the oceans play a crucial role in the absorption and dissipation of heat over decadal and millennial timescales and with distinct cyclic patterns. These patterns are poorly understood and not replicated in global warming models, and any conclusions drawn with respect to those models being able to isolate an anthropogenic global warming signal must be regarded as unproven and unlikely.

Note

1. The Met Office at Hadley issued *Global Temperature Anomaly for 2007 from dynamical and statistical method* by Chris Folland & Andrew Colman, Hadley Centre, Met Office on 3 January 2007 with the key message: ... we forecast that our best estimate of global surface temperature in 2007 will be a record value of 0.54°C above the 1961–90 average. This just exceeds the 1998 record of 0.52°C, taken from a new Hadley Centre and Climatic Research Unit data set being used by the IPCC Fourth Assessment Report as the key data to assess observed global warming. This is appreciably warmer than the expectation for 2006.

6

Poles Apart

As we saw in the last chapter, from a climate perspective, Planet Earth ought perhaps to be named Planet Ocean. Its two poles are then quickly seen as very different places, despite the superficial similarities of hosting ice caps several kilometres thick with calving glaciers and large areas of sea ice. The Arctic is essentially a relatively shallow ocean with only one major ice cap, on the island of Greenland, whereas the Antarctic is a frozen mountainous continent with two extensive ice caps and large thick ice shelves reaching out into the shallow shelf seas. The geographic North Pole consists of a few metres of sea ice, whereas the South Pole lies atop three kilometres of ice grounded on the mountainous continent. The key difference climatologically is that Antarctica is isolated from other land masses by two thousand kilometres of ocean, in contrast to the Arctic which is surrounded by and connected to the major land masses of North America and Eurasia via shallow straits. The two poles have shown marked divergence in their response to global warming with the Arctic registering the extremes of warming and the Antarctic experiencing little change.

Despite the media concern about cracking ice shelves in Antarctica and the potential for catastrophic sea-level rise, the continent as a whole has barely responded to the global warming pattern. The small percentage of land (less than 15%) in the western section and the Antarctic Peninsula has seen a rising trend of temperature and break-up of ice, but the rest of the continent has experienced no significant rise. This is in apparent stark contrast to the Arctic, where the most obvious impact has been the acute loss of sea ice, with much media and even scientific talk of an ice-free North Pole within the next decade and of accelerating ice loss from the Greenland ice cap.

The Arctic Polar Environment

There is no doubt that the Arctic is experiencing a greater degree of warming than other parts of the globe. In the years from 2003 to 2007 the summer sea ice around the North Pole declined by 50%, with a 20% fall in 2007 alone. If that trend were to continue at the same acceleration the Arctic would be ice-free (in summer time) by 2012. Other signs of unusual warming range from melting permafrost in Alaska and Siberia to massive lakes and cracks appearing atop the Greenland ice cap. Wildlife campaigning organizations have used the iconic polar bear and its need for sea-ice hunting ground to spearhead their campaigns for public awareness of climate change.

In all of this, media interest has been intense, with much focus upon threatened habitats and extinction for the bears. News teams show pictures of glaciers accelerating and calving off huge chunks into the sea and there is alarm over potential 'tipping points' where the Greenland ice cap undergoes rapid disintegration, perhaps then affecting the North Atlantic circulation and paradoxically plunging western Europe into freeze-lock. Just in the course of writing this chapter, over a dozen environmental organizations have taken out a full-page ad in British national newspapers arguing for accelerated carbon dioxide reductions to avoid a 'tipping point' catastrophe which they argue is only 120 months away. All of their references relate to papers that project the trends of the past and take no account of the power of cycles to turn.

Tipping points?

In the last months of 2008, the New Economics Foundation ('Green New Deal: Joined up policies to solve the triple crunch of the credit crisis, climate change and high oil prices', NEF, 2008) and the World Wildlife Fund ('Climate Change: faster, stronger, sooner: an overview of the climate science published since the UN IPCC 4th Report') both issued reports using the latest scientific references to underpin a claim that a tipping point was only 120 months away. What they mean, of course, is that at some point several decades from now there may be a tipping point unless we *begin* to make serious reductions in greenhouse gas emissions. If the current Arctic ice-melt is related to greenhouse gas emissions (which very few Arctic specialists would argue), then a tipping point 120 months away cannot be avoided because greenhouse gas concentrations will not fall by that time. As the latter report makes great play of recent science showing *accelerating* climate change, I deal with this under 'tipping points' later in this chapter.

In many ways, the Arctic is a test case for the 'global warming' thesis. All the maps of excess temperature show that the warming is concentrated here, and this is held also as an observational validation of the computer models that also predict a north-polar amplification of the warming. The apparently obvious nature of the meltdown grabs media headlines, but it is also one of the major focuses for campaign groups raising awareness for their carbon reduction objectives. The Arctic thus carries many lessons related to the way the climate science is interpreted, and the way that politics and campaign agendas use information.

Repeating cycles

The first question in the mind of an ecologist is whether such an Arctic meltdown has happened before, and if so under what circumstances. It doesn't take long to find an answer. The Arctic is subject to cycles of warming and cooling. In fact, the pattern is much clearer for the Arctic than perhaps any other region *because* the global climate patterns are amplified here. However, it is also clear that each peak or trough in the cycle can vary with regard to the pattern in the Arctic, with see-saw effects in the eastern and western regions and sharply within a region such as Greenland.

In the following analysis, we shall see that many scientists and virtually all of the media and campaigners share the same blindspot – they focus upon trends and ignore cycles. As I will detail, this is not entirely a matter of ignorance, because several Arctic experts have complained that their response of 'nothing that unusual going on here' gets no press. The Arctic has experienced heatwaves before, and appears to do so on a regular cycle of about 70 years. Arctic specialists are well aware of this, but their warnings go unheeded. In the following, we will look at the data and their predictions. We will see that the scientific case for an Arctic 'amplification' of global warming (i.e. the human signal) is by no means agreed among experts. We shall also see that campaign groups have wilfully ignored what these scientists have been saying – the information is readily available, published by the leading institutes and in the peer-reviewed science journals.

From the preceding chapters we have seen that the late twentieth century certainly experienced a warm period and, given the nature of the global climate system, the increased warmth will have been transferred to the Arctic. That is the nature of the ocean current systems in the Pacific and the Atlantic. However such global warm periods have happened before during the stable periods of this interglacial, and as we shall see there is evidence that the Arctic Ocean has been free of summer ice. Some of those warm periods are more readily explained by long-term orbital and inclination changes in the earth's relation to the sun; the recent twentieth-century warming is clearly not an orbital phenomenon, but it is linked to extraordinary solar activity. There is now evidence that the Arctic warm period may be ending, but this receives virtually no media attention, despite several key science reviews stating this clearly.

The heatwave of 1920–40

When I first became aware of the periodic nature of the Arctic warming (early in 2006), I consulted the data sets held by NOAA (and readily available on John Daly's website). I was shocked by what I found. Of 32 Arctic data sets running in many cases for over one hundred years and ending variously between 1996 and 2004, I could find only one that showed higher temperatures compared to the last warm phase around 1940 – and that was for one site on the west Greenland coast, and even then not until 2004. In later years, particularly during 2005, some of these records may have been exceeded as 2007 is regarded by some as the warmest year in the Arctic region – but the increase over the previous warm period is marginal.

In Fig. 17 it can be seen that for Greenland on the eastern seaboard at Godthaab annual mean temperatures in 2002 were still below the 1940s records, whereas in Angmagssalik on the west coast the temperature peaked and surpassed 1940 only in 2002, having remained below the 1940s records throughout the century.

This pattern is typical of all the Arctic data sets. We can see here some evidence of a roughly 70-year cycle. It is also worth noting for future reference that if an annual mean is expressed as an anomaly to a reference

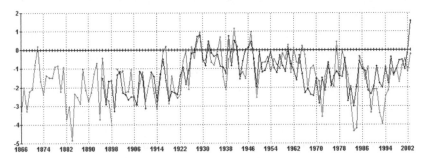

Fig. 17 *Annual mean temperatures at two stations in Greenland, 1866–2002. Grey line with longer-term data: Godthaab on the east coast. Darker line: Angmagssalik on the west coast.*

period of lower temperatures, as is the case with many data sets using 1961–1990 (Hadley, for example), then twenty-first-century temperatures will show a three-degree 'warming'. Another way of expressing the data would be to compare the twenty-first-century peak (temperatures peaked in 2005–07) with the 1940s peak and note that it is about 15–20% higher. If there is nothing else unusual about the late twentieth century, then we could assume this is an upper bound for the greenhouse effect and the human signature. But, as we have recorded, there *are* other factors that are unusual, in particular the previous decades of increased solar activity and the peaking of several oceanic warm cycles.

The other Arctic data sets I looked at failed to show recent higher peaks. The data from Norway, for example, had the highest records in 1931–40. A similar pattern exists for the Jan Mayen Islands further north between Norway and Greenland, where the 2002 temperature matched the 1939 peak, but throughout the 1980s and 1990s temperatures were not higher than the 1940s and 1950s. In the western regions of the Arctic at Barrow, Alaska, the 1940s heatwave is marked by record winter temperatures not equalled until 2001, though summer and autumn temperatures were a little higher in the 1990s. And in the Yukon of Canada the highest recent temperature (the data I have seen ends in 2002) was 1987, which was slightly less than the 1944 record.

The turning point: 2007?

There is no doubt that overall Arctic warming continued through 2007 (Arctic Report Card, 14 October 2008) and particularly over most of the Greenland ice cap, but few real experts on the Arctic system would base a prediction of trends upon the last decade of warming when it is clear that the main drivers of that warming – ocean temperatures and cloud patterns – showed significant changes. It will take a few years for these changes to work through. For example, the record 2007 summer ice-melt was not repeated in 2008.

The media seldom report the previous Arctic heatwave from 1920 to 1940, nor do they consider that in Viking times (about 1000 years ago) cattle could be grazed around their Greenland settlements. In the latter case, no one monitored polar sea ice thickness, the melt on the ice caps or polar bear numbers. But there is evidence, though disputed, that the Vikings made sorties to Canada, and a few centuries later a Chinese naval expedition navigated the Arctic Ocean. In the 2008 Arctic Report Card, the authors show a map of the retreating Jakobshavn glacier and note that it has retreated to a point that was also clear of ice during the Medieval Warm Period (**Box et al., 2008**).

The current propensity to illustrate the Arctic 'meltdown' by spatial maps of temperature anomaly referenced to the 1961–90 mean – as in Hadley Data Centre maps – obscures the cyclic nature of the warming. The period for calculating this mean is dominated by the low temperatures from 1961 to 1980, and this can be clearly seen in Greenland data sets among others. Thus the Arctic looks very warm on the spatial coloured maps, usually red, of the globe (see Ethos A5).

If, however, we take a closer look at the cycle pattern revealed over the twentieth century, as evident in the surface air temperature (SAT) record, we can see that the Arctic is subject to a roughly 70-year periodic oscillation. This is illustrated in Fig. 18 taken from the *State of the Arctic Report*, a major international monitoring exercise (**Richter-Menge, 2006**).

The propensity of the Arctic climate to oscillate is well studied, and one of the foremost experts, Igor Polyakov, leads an international team at the University of Fairbanks in Alaska. In a paper published in 2000 in the

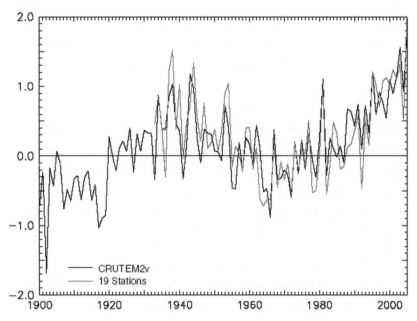

Fig. 18 *Arctic-wide annual averaged surface air temperatures (60–90°N) over land for the twentieth century. Based on the CRU TEM2V monthly anomaly data set. (Source:* State of the Arctic Report, *Richter-Menge, 2006)*

Journal of the American Meteorological Society entitled 'Variability and trends of air temperature and pressure in the maritime Arctic, 1875–2000', he and his colleagues at the Institute of Marine Sciences, collaborating with the Arctic and Antarctic Research Institute in St Petersburg, Russia, analysed the trends over the century. These are their conclusions (my emphasis in bold):

> Air temperature and pressure display strong multi-decadal variability on timescales of 50–80 yr [termed low-frequency oscillation (LFO)]. Associated with this variability, the Arctic Surface Air Temperature (SAT) record shows two maxima: in the 1930s–40s and in recent decades, with two colder periods in between. In contrast to the global and hemispheric temperature, the maritime Arctic temperature was higher in the late 1930s through the early 1940s than in the 1990s.

> ...the large-amplitude of multi-decadal climate variability impacting the maritime Arctic may confound the detection of the true underlying climate trend over the past century. LFO-modulated trends for short records are not indicative of the long-term behavior of the Arctic climate system. The accelerated warming and a shift of the atmospheric pressure pattern from anticyclonic to cyclonic in recent decades can be attributed to a positive LFO phase. It is speculated that this LFO-driven shift was crucial to the recent reduction in Arctic ice cover.

> The multi-decadal variability (LFO) is evident in various instrumental and proxy records for the Northern Hemisphere. **This variability appears to originate in the North Atlantic** *and* is likely induced by slow changes in oceanic thermohaline circulation.

> We speculate that warming alone cannot explain the retreat of Arctic ice observed in the 1980s–90s. Also crucial to this rapid ice reduction was the low-frequency shift in the atmospheric pressure pattern from anticyclonic to cyclonic. Positive and negative LFO phases of the SAT are shifted by 5–15 yr relative to those in the SLP (sea-level pressure) record. **The complicated nature of Arctic temperature and pressure variations makes understanding of possible causes of the variability, and evaluation of the anthropogenic warming effect most difficult.**

Fig. 18 is a more recent source than Polyakov's paper which only extended to 2000, and shows the cyclic nature of the surface air temperatures (SAT). It is clear that the Arctic is recovering from the nineteenth-century low period of temperature after which there was a major peak in the 1920–60 period, followed by a trough in 1971 within the 1960–90 period. The recent rise can be seen as not unusual and part of the long-term low-frequency oscillation (LFO). This graph references the anomaly to 1961–90 and for all regions poleward of 62°N. As can be appreciated, if the sampling is limited to the second cycle, as many pre-sentations do (especially those of the Hadley Centre website), recent temperatures will appear very high because the 1960–90 base period samples the trough of the cycle!

In a further collaboration of the Institute of Marine Science and the International Arctic Research Center at Fairbanks and the St Petersburg Arctic and Antarctic Research Institute, titled 'Observationally based assessment of polar amplification of global warming' (available from the Center), the researchers concluded:

> If long-term trends are accepted as a valid measure of climate change, then the SAT and ice data do not support the proposed polar amplifi-cation of global warming.

Thus, contrary to a great deal of misinformed comment, the Arctic warming is not unusual and is clearly affected by a long-term oscillation pattern. The apparently unusual warming and 'amplification' seen, for example in the Hadley spatial data sets for temperature *anomalies*, are the result of 'incomplete sampling' – in other words, the selective use of a 1961–90 base period from which to compare recent 1990–2005 tem-peratures. As Polyakov notes, these temperatures are not anomalous when compared to earlier cycles. He further argues that this oscillation is driven by the Arctic Ocean's relationship to the North Atlantic. However, since Polyakov's papers were published in 2002, Arctic sea ice has retreated spectacularly and we look at this in more detail.

Arctic sea ice decline

Since 2002 Arctic sea ice has undergone an accelerated decline – of about 50% in extent, and 20% of that in the last year to September 2007. This may not be unusual in terms of the expected pattern (the recent peak associated with the LFO would be expected to go higher than the previous peak because the globe has become warmer both over a long-term cycle from a global low in 1800 and from the pulse of warmth from 1980–2000, which we have argued was due to increased flux of sunlight to the oceans in that period), as well as from whatever contribution carbon dioxide is making. Nevertheless, the loss of Arctic sea ice has been startling to observe – see Fig. 19.

We can see that there is a 22% loss of sea ice in the eastern Arctic in just two years. The narrow black line denotes the median ice-edge for September (the usual low for summer ice). Further, in the summer of 2007 a large body of open water could be seen in the middle of the sea ice – a quite unprecedented sight in recent times, indicating how thin and unstable much of the remaining ice is. However, in the summer of 2008 the rate of change reversed such that the September low was 9% above the record in 2007, but the Laptev Sea ice-tongue disappeared in a storm and the Arctic made news again as for the first time in human history, according to much of the press, the Arctic was circumnavigable.

Despite the media concern at the ice-melt, Arctic specialists were already warning that conditions were very unusual – a combination of pressure systems, winds and ocean currents was thinning the ice and driving it out of the Arctic seas into warmer waters. However, many commentators simply extrapolated the trends which might mean an ice-free North Pole by 2013 and, of course, put it all down to carbon dioxide emissions.

The very small percentage increase in carbon dioxide levels over the past five years (about 3%) could not cause such accelerated change, which

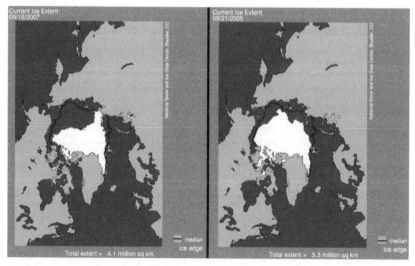

Fig. 19   Rapid loss of Arctic sea ice from 2005 to 2007. (Source: National Snow and Ice Data Center, Boulder, Colorado)

must be driven by the remixing of the accumulated warmth of the previous century coupled with whatever caused the unusual atmospheric conditions over the polar region. We should note that the long-term build-up of carbon dioxide is not a likely contributor to the acceleration because the models based upon greenhouse gas theory do not predict a summer ice-free Arctic until much later in the twenty-first century. Either carbon dioxide is more powerful that thought or something else is driving the ice loss.

This amount of ice-free water is not likely without longer-term precedent – as we noted before, Greenland was much warmer in about AD 900, and there are reports (much disputed) of the Chinese fleet circumnavigating the Arctic in the late sixteenth century. However, very recent geological research on northern Greenland beach structures shows that the region from there to the North Pole must have been ice-free during the Holocene Optimum of approximately 6000 BP. (Astrid Lyså, Geological Survey of Norway). This period was however marked by very different conditions for solar visible light falling in the 60°N latitudinal zone.

The complexity of Arctic cycles

In the previous warm phase the eastern Arctic was warmer than the western, and in the 1990–2000 warm phase there is a strong warming of eastern Greenland that was not seen in the previous cycle. As we shall see from the discussion below, Arctic temperatures see-saw between eastern and western regions. We shall look at Greenland in more detail considering its importance in the global dynamic of a warming world, particularly with regard to sea-level rise (the 'meltdown scenario') and also the freshening of the North Atlantic and potential effects on the ocean conveyor system.

However, this recent Arctic high is certainly affected by the warmer oceanic waters further south, as in Polyakov's supposition that the warming has its origins in the North Atlantic. There is also an influx of Pacific warm water through the Bering Strait and it is apparent that conditions in this region are sensitive to the Pacific oscillations. As we noted with the PDO, it is now at the end of a 30-year warm phase in this part of the Pacific and it turned negative in late 2006.

The very rapid thinning of sea ice in the last few years has a more immediate explanation – also connected to Polyakov's thesis. On 1 October 2007, NASA issued a press release to announce that a research group had determined that unusual winds caused the rapid decline (23% loss) in winter perennial ice over the past two years in the northern hemisphere, stating that 'this drastic reduction is the primary cause of this summer's fastest-ever sea ice retreat in recorded history which has lead to the smallest extent of total Arctic coverage on record'. According to the NASA study, the perennial ice shrunk by an area the size of Texas and California combined between the winter of 2005 and the winter of 2007. What they found was the Arctic Ocean north of Siberia and Alaska was dominated by thinner seasonal ice that melts faster compared to the thicker ice confined to the Arctic Ocean north of Canada. The thinner ice is more easily compressed and responds more quickly to being pushed out of the Arctic by winds.

'Unusual atmospheric conditions set up wind patterns that compressed the sea ice, loaded it into the Transpolar Drift Stream and then sped its flow out of the Arctic,' said Son Nghiem of NASA's Jet Propulsion Laboratory and leader of the study. When that sea ice reached lower latitudes, it rapidly melted in the warmer waters.

Discussion among meteorologists centred upon the Arctic Oscillation (AO) and the North Atlantic Oscillation (NAO) with the AO and the NAO having been predominately in the positive phase between 1989 and 1995 and again from 1999 to the current time. The positive phase of the (AO) typically leads to milder than normal winters over Scandinavia and Siberia, while colder than normal conditions prevail across Greenland. The positive phase of the NAO again leads to colder conditions over Greenland, while much of the eastern US is warmer than normal. There were also reports from NASA and NOAA of unusually clear Arctic skies during the summer of 2007.

In relation to this argument, we should note that Greenland is not then following the normal pattern; it is currently warming, particularly on the eastern seaboard and in the south-west, and this is undoubtedly due to warmer North Atlantic water penetrating further north than usual.

This period also corresponded with a southward shift of the jet stream that caused widespread rainstorms and severe flooding in western Europe throughout the summer of 2007, an event that also coincides with the lowest recorded measures of solar electromagnetic activity. And this has been repeated during the summer of 2008.

These apparently unprecedented atmospheric conditions have coincided with a deepening of the solar cycle. The summer of 2008 was marked by an absence of sunspots and the unexpected lengthening of the current cycle number 23 and delayed onset of 24. My own sense is that atmospheric research has yet to elucidate the key mechanisms relating these long-term oscillations and solar electromagnetic energy cycles, which I suspect will eventually prove to have a primary role – whether by UV modulation, the flux of ions and particles into polar regions, ionization and cloud condensation or some as yet undisclosed mechanism that affects cloud properties. There is already strong evidence that UV fluctuations can affect global wind patterns. We shall turn to this in greater detail in the next chapter.

We can consider the following potential causes for this meltdown:

- warmer water is penetrating from further south in the Pacific and Atlantic;
- warmth is being carried by wind, cloud and precipitation from warmer ocean regions;
- carbon dioxide levels are trapping heat that would normally escape from the surface;
- cloud cover changes have insulated ocean and land areas and prevented heat loss;
- UV fluctuations (or other forms of solar energy) have caused stratospheric warming and interfered with high-level wind patterns.[1]

All of these factors are likely to play a part. What is required is a detailed investigation and appraisal. Polyakov's analyses were written before the exceptional ice loss of 2005–07, but the *State of the Arctic Report* – an international effort from all the Arctic research institutes, summarized the situation in 2006 (**Richter-Menge et al., 2006**), my emphasis in italics:

> During 2000–2005 the Arctic system showed signs of continued warming. *However, there are a few indications that certain elements may be recovering and returning to recent climatological norms* (for example, the central Arctic Ocean and some wind patterns). These mixed tendencies further illustrate the sensitivity and complexity of the Arctic physical system. They underline the importance of maintaining and expanding efforts to observe and better understand this important component of the climate system to provide accurate predictions of its future state.

As we shall see when we consider the Greenland ice sheet, Arctic trends have to be seen within a longer-term context and with regard to recent techniques of measurement and data series. During the 1920–40s warming period, there were fewer data sets and many measurement techniques were not then available – particularly those derived from satellite observations. Furthermore, this latest Arctic cycle, though it shows differences from the previous cycle, may be part of an even longer-term oscillation related to the North Atlantic. The leading research on longer-term cycles and connections between ocean temperatures and solar factors is carried out in Russia and is not readily accessible.

Such connections are indicated in the Report's conclusion regarding atmospheric northward energy transport (ANET) and the persistence of cyclones in the Arctic Sea region:

> About 25% of the Arctic surface air temperature trend from 1979 to 2001 in winter and spring is related to an increase in the ANET. The strongest linkage is in the Atlantic sector. Both the general trends of an increase in cyclones and an increase in the ANET suggest increased linkages of the Arctic to the mid-latitude atmospheric circulation in recent decades.

The persistence of cyclones during 1989–1996 indicated a phase change in Arctic weather systems but this changed again between 2000 and 2006, when anticyclonic conditions returned. Recent work at NOAA's Earth Systems Research Laboratory in Barrow has identified a close correlation between surface air temperatures in the western Arctic and the down-welling of long-wave radiation from clouds:

> ... variations in western Arctic temperatures during the cold season are associated with cloud-radiative effects that in turn relate to advective processes. While not a direct consequence of greenhouse warming, the changes may relate indirectly to global warming through physical teleconnections that link the Arctic to the tropical and north Pacific. Warmer conditions at Barrow prevail when cyclonic activity increases in the north Pacific resulting in the advection of clouds northward. If instead, outflow from anticyclones centered in the Beaufort Sea is dominant, colder, drier air influences the continental regions to the south. Because GCMs do not always simulate the positions and/or intensities of the Aleutian Low, Siberian Anticyclone or Beaufort Sea Anticyclone correctly, and these are primary centres of synoptic activity in the region, predictions of Arctic climate based on such models must be viewed with caution. Moreover, the teleconnections between tropical and polar regions are not yet understood fully (**Stone, 1997**).

Thus cloud patterns in the Arctic, which are also linked to cloud patterns in the Pacific, are central to western Arctic temperatures. The Beaufort Gyre is known to oceanographers as the main driver of surface sea currents across the Arctic Ocean and controls the influx of warmer waters. As such it drives the whole of the Arctic pattern of wind and cloud systems (see below); as the Alaskan temperatures rise the Gyre has less power and warmer waters enter the Arctic from the Atlantic. We have also seen that there is a clear teleconnection between ENSO events and North Pacific sea surface temperatures, and this may also carry over into Arctic patterns of temperature, wind, and cyclonic and anticyclonic flows.

Additionally, the latest reports from NOAA's Arctic Climate monitoring website (*www.arctic.noaa.gov*) show a 14% increase in cloud cover over the central Arctic region (**Schweiger, 2004**), Fig. 20.

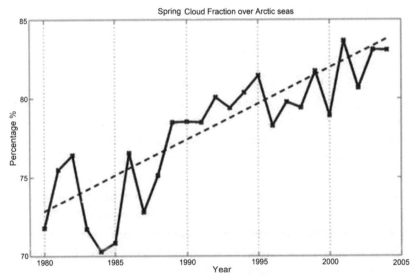

Fig. 20 Increased cloud cover over the central Arctic in spring 1982–2004. (Source: NOAA Arctic Change website)

Cloud and ocean circulation patterns explain the warming

We can conclude therefore that the recent 'heatwave' in the Arctic is caused by a combination of changes originating in the Pacific and North Atlantic Oceans (a late centennial warming due to changes in insolation at the sea surface) that have led to changes in the major cyclonic and anticyclonic oscillatory patterns in the western Arctic with the result that the last two decades have seen a 14% increase in cloudiness over the central Arctic region, as well as more recent strong winds and pressure systems breaking up and moving the thinner ice, hence reducing ice cover.

There are signs, however, that colder Arctic weather systems are re-establishing themselves and we shall see that this may be linked to a recent cooling of the oceans and reduction in heat storage in the key areas of the North Pacific that may ultimately drive the Arctic system. Alaska has experienced a much cooler summer in 2008. My expectation is that the cool phase of the Pacific oscillation – and we noted the sudden exhaustion of the warm water body in the north-west off Alaska in 2006 – will regenerate the very cold winds off the Alaskan shelf, recharge the clockwise circulation of the Beaufort Gyre and ultimately refreeze the Arctic. We turn now to the issue of the Greenland ice cap melting down and potential 'tipping points' in the North Atlantic.

Greenland ice mass changes

The response of the Greenland ice sheet (GIS) to global warming is an issue of great concern to some scientists – mostly those working with virtual reality simulations of the GIS response to warming. A complete melting of this ice mass would raise sea levels by about 7 metres with clearly disastrous consequences for low-lying land and coastal cities. We can also see from the NOAA temperature data from Angmassilik and Godthaab that temperatures have risen to match and in some cases exceed the previous records in the 1940s warm period. However, the cyclic nature of these changes would lead to an expectation of cooling at some point in the near future – not that this appears in media coverage, and neither is it incorporated into predictive models.

Much attention has been drawn to accelerated runoff and increased speed of glaciers moving to the sea. However, there is also data that show increased snowfall. A recent review of the Greenland ice mass data has been carried out by a team led by E. Hanna at the University of Sheffield as part of an international multi-agency surveillance of the Arctic (Arctic Report Card, 2007). The team found that although there was an undoubted acceleration of glaciers and melt-water loss at the margins of the ice sheet there were also compensatory increases in snowfall in the centre and at higher altitudes (**Hanna & Cappelen 2003; Hanna et al., 2005, 2006**). Overall, recent decades had seen an increase in ice loss from about 60 km^3 to about 200 km^3. However, when the long-term temperature trends of four research stations on the ice sheet were examined, recent years were regarded as not exceptional:

> In this longer perspective, only 2003 at Tasiilaq is outstanding in recent decades. Over the past century, years in Greenland that register as abnormally warm, 1929, 1932, 1941, 1947, and 1960 are outstanding, having temperatures warmer than observed recently. Increases in GIS melt and runoff during this past century warm period must have been significant and were probably even larger than that of the most recent last decade (1995–2006).

Furthermore, in the 2007 Arctic Report Card, Hanna notes some glacier speeds returned to normal in 2006–07. And in the same report, Grebmeier and co-workers draw attention to the sudden cooling of the Bering Sea over a two-year period, potentially reversing major ecosystem changes that had been noted for the previous decade. From my reading of the dynamics, I have no doubt that the warming in Greenland is primarily driven by the proximity of the warm body of water that has accumulated in the North Atlantic. The warmer coastal waters have reduced ice cover and accelerated the flow of glaciers. Additionally, warmer spring temperatures have led to enhanced melt-water penetration through fissures, which further accelerates flow. In part compensation, increased snowfall in the interior demonstrates an enhanced dynamic; but as several Arctic experts point out, there is no evidence that this dynamic is exceptional because no one was measuring these parameters during the 1920–50 warming period.

However, James Overland of NOAAs Pacific Marine Environmental Laboratory, reporting in the 2008 Arctic Report Card, leads a team that interprets the most recent twenty-first-century changes in the Arctic as indicative of 'global warming' with a continued rise in surface air temperatures to 2008 (**Overland et al., 2008a**) and repeated high melt-rates in Greenland which rose again after the lull reported by Hanna in 2006. This team argues that the continued rise and shifts in pattern mark out an unusual 'Recent Arctic Warm Period' (**Overland et al., 2008b**). The data could appear to support this conclusion as the most recent surface air temperature charts of CRUTEM show swings that go higher than the graph in Fig. 18 (moving to an anomaly of 2°C). However, this data may be unduly influenced by the high autumn temperatures of 2007 and 2008 over the ice-free areas of Arctic Ocean.

Overland sees this as a result of the absorption of solar energy. Another explanation could be the release of heat from the exposed and normally ice-insulated surface waters that still carry the warm signal from the North Atlantic incursions of the previous decades' build-up. The late summer and autumn sun is at an angle that would not favour a great deal of absorption. Thus I would expect the atmospheric warming to have peaked in 2008 and to follow the downward shift in ocean temperatures in the Norwegian and Bering Seas.

The role of the North Atlantic Oscillation and recent global warming

As we noted in the developing argument of previous chapters, the warming of the Arctic can only be caused by mass heat transfer from further south and the main routes are via the atmospheric circulation of warm air masses, development of cloud, and via warm water incursion from the Atlantic or Pacific. I have looked at only two data sets for these waters. The Hadley data in Fig. 21 are expressed as anomalies referenced to the 1961–90 mean.

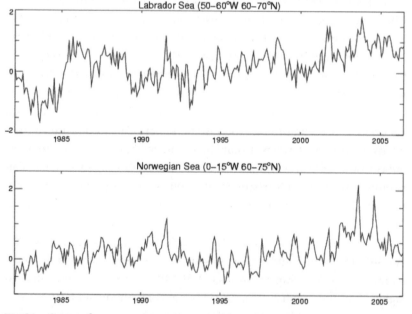

Fig. 21 Sea surface temperature in two Arctic Sea regions east and west of Greenland. (Data kindly provided by John Kennedy at the Hadley Centre)

These data plots show interesting patterns: in the Labrador Sea, where water is flowing south out of the Arctic, there is a rise of about 2°C from 1985 to 1995, another 2°C from 1995 to 2002, and a slightly higher peak in 2003–04 after which temperatures formed a plateau. There is an indication of roughly 10-year phases in the troughs. The peaks and troughs correlate with the solar cycle – with the troughs roughly corresponding with the *high* points, and peaks with the *low* points in the 11-year Schwabe cycle. This anti-correlation to solar radiation peaks could mean that during the peak of the solar cycle (1980, 1990, 2000) there is less cloud and greater loss of heat to space; during the solar minima (1986, 1996, 2006) there is more cloud and more heat is trapped in the surface waters. There is a clear double peak between 2001 and 2004 and then a recent fall. In the Norwegian Sea, where warmer surface waters are travelling northward, there is no clear trend between 1985 and 1997 and the pattern of troughs is reversed except for the last cycle which has two major peaks slightly later in 2003–04 and then a recent fall similar to the Labrador data. How far these surface temperatures represent a surface response to wind, cloud or solar irradiation, and how much a response to movements of warm water bodies is difficult to tell.

If we consider the previous chapters on the heat flux to the oceans and subsequent heat storage, we can begin to construct a hypothesis to explain the current pattern of global warming and its impact upon the Arctic. Firstly, we have seen that the equatorial regions will have absorbed the greater part of the enhanced solar flux noted in the period 1980–2000, but as these waters appear not to *store* that extra heat it follows that surface currents have carried bodies of warm water both north and south. In the south these waters are circulated freely in the circumpolar current in a region of overall radiation deficit, hence losing their heat (we saw earlier a compilation of sea surface temperature data which shows only very small rise in the southern hemisphere). This effectively isolates the Antarctic continent climatologically and as we shall record below there is no significant warming trend over Antarctica apart from the Peninsula region which juts out into the warmer waters of the circumpolar current.

In contrast, northward travelling waters carry the heat stored within them into gyres bounded by land – such as between Alaska and Russia south of the Aleutians, and south of Greenland between the USA and northern Europe. These gyres will have accumulated heat over decadal timescales by increasing the depth of the isotherm, and they will release it slowly from the surface to the northern skies. This release will be pronounced in areas of permanent radiation deficit as the warmer waters penetrate into the shallow Arctic seas. The degree of heat loss will be strongly influenced by cloud cover. Thinning cloud or a spatial redistribution southward would lead to higher heat losses, thicker cloud would conserve the heat. Transfer of heat landward would occur through wind and moisture transfer. If this heat store is not augmented, it will gradually diminish as the depth of warm water becomes shallower, and then suddenly it will turn cold. In the 2007 Arctic Report Card it was reported that the accumulated warmth of the Bering Sea had taken only two years to dissipate (as part of the 2006–07 shift in North Pacific temperatures).

As we shall see in more detail, the western Arctic has warmed most, particularly in Alaska, Canada and Greenland. In the western region this could be explainable by heat transfer from North Pacific waters to Alaska. A warming Alaska would reduce the power of the surface winds that drive the Beaufort Sea Gyre which is wind-driven by the deep cold region that normally develops across the Alaskan shelf. A reduction in the power of this gyre then leads to greater incursion of warmer North Atlantic waters into the Arctic basin.

When we examined trends in cloud cover in Chapter 4, the papers we looked at dealt with broad percentage changes over decadal timescales. It would be equally important to look at spatial changes in relation to ocean currents and stored heat. Cloud cover in oceanic regions of the northern hemisphere is confined to certain regions and as ocean waters move north they will be susceptible to the effects of cloud cover, either insulating or exposing to space. A spatial shift of cloud southward *without* any overall percentage reduction could still have enormous consequences for oceanic heat budgets and there has been a southward shift in the jet stream in 2007/2008 which affects storm tracks and the cloud cover associated with them. Furthermore, it is clear that low-level reflective cloud is more important than higher-level cloud in regions of heat gain because the cooling effects are so large, whereas both high-level and low-level cloud would be important in regions of heat loss as they provide insulation.

I have been able, in a relatively unsophisticated way, to monitor the development of the heat store in the Pacific waters off Alaska and to observe that late in 2006 that store became exhausted. It is my supposition that over the previous two decades onshore winds had taken warmth deep into Alaska and gradually reduced the heat store. The resultant cold patch is the hallmark of the PDO reversal.

As noted, much of the changes in the Arctic will be driven by the behaviour of the Beaufort Gyre. This gyre dominates ocean current structure in the Arctic region where the main source of heat comes in via the North Atlantic surface currents, which being saltier undercut the westward flowing, fresher and less dense surface currents. According to recent reports, the Beaufort Gyre has slowed significantly in the last decade, thus allowing a greater penetration of warm North Atlantic water. The ultimate cause however will be the strength of winds off the Alaskan shelf, which can be expected to have lessened with the past 30-year warm period.

This significant shift in the gyre is confirmed in the State of the Arctic Report. The cyclonic pattern dominated during 1989–96; the anticyclonic pattern has prevailed since 1997. It is therefore clear that the Arctic region is affected by several factors combining to produce a strong temperature anomaly relative to the second half of the century mean, but not necessarily in relation to the previous period of warming. There are major differences between the two warming periods, but these may be accounted for by changes in the Pacific and Atlantic Oceans which are subject to oscillations, as well as the overall centennial rise in heat content of the oceans.

Variations in the North Atlantic climate: the ocean conveyor

We dealt earlier with the recent variability of the North Atlantic Conveyor and it is instructive to look at Holocene records of longer-term cycles. Giancarlo Bianchi and Nicholas McCave at Cambridge University reporting in *Nature*, 'Holocene periodicity in North Atlantic climate and deep-ocean flow south of Iceland', analysed long-term variations in the deep water current (flowing south in the North Atlantic and part of the global 'conveyor' system) by means of sediment grain size and origin in an area south of Iceland (**Bianchi & McCave, 1999**).

The paper's significance lies in detecting a 1500-year periodicity as well as correlating the speed of the conveyor with temperature conditions. They also show fluctuations which confirm both the Medieval Warm Period (MWP) and the Little Ice Age (LIA). Their observations show that even during the generally stable Holocene there is an underlying fluctuation in the strength of the what is termed the 'Iceland–Scotland Overflow Water' (ISOW):

> . . . we infer that periods comparable to the Little Ice Age and Medieval Warm Period were a recurrent feature of earlier parts of the Holocene

climatic history, with the warm intervals coinciding with faster near-bottom water flow.

The authors pick up the signal for the Roman Warm Period at 2000 BP followed by cooling in the Dark Ages. They make the point that the 1500-year periodicity that they observe is not present in either beryllium-10 or carbon-14 ice-core records and hence solar variation is unlikely to be the forcing mechanism, which they attribute to an 'oceanic internal oscillation'.

But as we have seen, beryllium-10 and carbon-14 data *do* show correlations with the LIA and MWP according to other work and this area of science clearly needs further work. Also, oceanographers may be wedded to the idea of 'internal' oscillations. It appears to me a relatively common occurrence that researchers from a particular discipline are not up to date with key research in other disciplines – and also, at times, with their own, especially where some kind of 'paradigm' shift is occurring. In the latter case, as the work of Warren White in the USA and Gerrit Lohmann in Germany has shown, shorter-term oscillations formerly regarded as internal appear more likely to be timed by solar factors. And it may well turn out that the longer periods are also timed by harmonics of the solar and orbital cycles.

Very recent work by Lohmann in collaboration with Mihai Dima in Bucharest, which I have not had time to review, investigates the combination of ocean and solar cycles (**Dima & Lohmann, 2007; 2008**). In the abstract of their 2007 paper they outline a model of the Atlantic Multidecadal Oscillation (approximately 70 years), linking it to variation in the thermo-haline circulation and, through atmospheric pressure waves ultimately connected to the North Pacific, subsequent wind patterns, ice export through the Fram Strait and effects upon the freshwater balance in the northern Atlantic. In the later paper, they provide a conceptual model for longer global climate cycles that link to solar and thermo-haline variability.

There has been recent work correlating solar cycles and ice-rafted debris which would be indicative of ocean temperature changes. The pattern of Bond cycles, Dansgaard-Oeschger and Heinrich events identified by study of ocean sediment debris that falls out from melting ice-rafts argues for solar influences (**Bond et al., 1997; 2001**). But there is no clear consensus on the Bond cycles, with recent work by Muscheler, a researcher much favoured by the IPCC, failing to confirm a solar correlation. For our current purposes it is important only to note that there are long-term cycles in sedimentary data that are as yet not well explained and that there is no clear consensus on the cause.

In seeking to understand the fluctuations in the North Atlantic deep water currents of the conveyor, Bianchi and McCave state 'there is no doubt the earth's climate is highly unstable on millennial timescales', and on forcing mechanisms for the deep water variability 'more than one process could be responsible – the salt or freshwater flux on convection in the Nordic Sea or a shift in the density of shallow and intermediate water masses entrained by ISOW after overflowing the Scotland Iceland ridge'. However, although they cannot pinpoint the mechanisms, they state that the sensitivity of the system is greater than thought and that if 'anthropogenic' melting from the Greenland ice sheet reached a critical level the conveyor could 'flip' and plunge northern Europe into extremes 'exceeding those of the Little Ice Age'.

The authors do not detail the scale of changes required to cause a flip, nor do they identify the mechanism of melt-water as responsible for the fluctuations that they note. There are clearly other mechanisms for significant variability of the flow that do not entail Greenland ice-melts but may simply be related to oceanic surface water temperatures and densities or to freshwater inflow to the Arctic Ocean from increased river flow and precipitation. It is perhaps more likely that we could be faced by another Little Ice Age and a slowing conveyor well before we could be faced by an ice cap meltdown.

Tipping Points: the *Day After Tomorrow* doomsday scenario

The issue of a shut-down of the North Atlantic conveyor has been uppermost in the public mind since the Hollywood blockbuster *The Day After Tomorrow*, which depicted a sudden freeze in consequence. Whilst the ice record (and fossil beetle remains in British peat) show that very severe temperature fluctuations of 16°C occurred within a matter of a decade in the northern hemisphere, it is not at all clear how the thermohaline circulation (THC) 'shut down' in such a short period of time. In his classic work *The History of British Mammals*, Derek Yalden uses the peat record of climate-sensitive beetle fauna in Britain to illustrate the rapid warming, referring to Peter Osborne's painstaking work showing a transition that spanned the 16°C over a period dated from 10025–9970 BP, or 55 years (**Osborne, 1980**). This is salutary material for those concerned with rapid climate shifts and their impact upon biodiversity, but it shows that such shifts can occur naturally and on a large regional scale.

There is little doubt that a major shift in the THC would have enormous consequences across the northern hemisphere – at least equivalent to the Little Ice Age, and in the case of a complete shut-down far more severe (**Broecker, 1997**). The matter is not simple, however, nor is it well reported in the popular media which usually refer to a potential failure of the Gulf Stream. Firstly, the THC is not quite the same thing as the Gulf Stream, though the two are linked. The Gulf Stream is a west-to-east surface current that is driven by winds and the rotation of the planet; it does not 'stop' or 'shut down'. On the occasions when the THC has apparently shut down, the Gulf Stream still flows but on a more southerly trajectory. In normal circumstances its waters are drawn further north by the massive down-welling between Iceland and Scotland. This down-

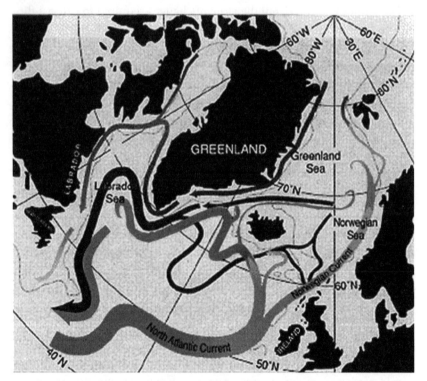

Fig. 22   *The North Atlantic part of the global conveyor (from a graphic of ocean-world at Texas A&M University, courtesy of Woods Hole Oceanographic Institute). Note: the Norwegian current carries warm water in, the Labrador current cooler water out of the Arctic.*

welling is a function both of topography – there is a very steep drop of 4000 metres from a depth of only 200 metres in the space of 10 km – and of the density of very cold Arctic water. The North Atlantic part of the conveyor is shown in Fig. 22 (and the global system in Fig. 11 in the previous chapter).

It takes around one thousand years for bottom water travelling from the Arctic to complete a circuit after resurfacing in the zones of up-welling in the North Pacific, thus providing a potential for long-term cycles or low-frequency pulses of warmer or colder water to emerge and potentially influence global temperatures. It may be possible for a signal to develop in the form of a multi-decade warming of surface waters in the North Atlantic which is then entrained in the conveyor. It would need to maintain the integrity of the signal for the long journey to up-welling sites (for example in the North Pacific), and although bottom waters are insulated from surface waters and are held not to mix greatly with the layers above, my own sense is that the maintenance of such a signal is unlikely. I would expect long-term variations in solar output to have a much greater effect on surface waters.

The shut-down of the conveyor

Most scientific commentators assume that a shut-down of the 'conveyor' can only be caused by a large influx of fresh water into the North Atlantic. This is because the last known shut-down, evidenced from bottom sediment records, was at the outset of the Younger Dryas cold period of 11,000 BP when warming up after the ice age caused glacial melt-water lakes in Canada to burst their banks and flood into the regional ocean. There are, however, other potential causes of disruption of the circulation. If Arctic waters warm significantly they will also be less dense and this would affect the down-welling, possibly even curtailing it. The conveyor also captures saline Gulf Stream surface water in the gyre south of Iceland as it cools and this also contributes to the down-welling. Any warming of this gyre would counteract that force. As we have seen, there is a body of warm water that has built up in the North Atlantic in recent years, as well as signs of freshening in the region.

Observations over the past decades show a gradual warming of the North Atlantic, particularly in this region *and* a freshening of the Arctic overflow (**Dickson, 2002; Petersen, 2002; Karcher, 2005**). Further, both sets of available data on the strength of the THC conveyor – one covering the period 1945–95 and the other 1992–2004 – show a slowing of the current speed. The first data set from 1945 to 1995 indicated a 20% slowing of the down-welling off Iceland (**Bacon, 1998; Hansen et al., 2001**). The second data set was centred on another part of the conveyor in mid-Atlantic at 25°N and showed apparently little change until 1992 (though data were sporadic) but a marked slowing down between 1992 and 2004 of about 30% (**Bryden, 2005**).

This potential slow-down had been predicted by earlier Norwegian work summarized by Bacon at Southampton's Oceanographic Institute and added to by his own analysis:

> It appears that changes in the Arctic and North Atlantic atmospheric and oceanic circulation are linked, and are at present weakening the Atlantic (and so the global) thermohaline circulation. This, ultimately, could alter the way that heat is redistributed around the globe by both ocean and atmosphere. For the Greenland Sea, the accumulated heat could at present be removed by just two years of intense winter cooling. But how far does Arctic warming need to progress before the oceanic circulation changes catastrophically in the manner suggested by recent coupled ocean–atmosphere modelling efforts, which represent climate transitions between 'normal' (present) and ice-age states? We cannot yet tell whether, or how much of, the variability reported here is normal or unusual.

I contacted Harry Bryden to discuss the two data sets, because if they were combined they would show evidence of an exponential curve in the slow-down. He was of the opinion that the two data sets, though spanning consecutive time periods, should *not* be combined. If combined this would have indicated a 50% slow-down over 60 years *and* an exponential rate of decline. Furthermore, other commentators have remarked that any such major slow-down would have had a noticeable cooling effect on the whole North Atlantic region where, instead, a major warm cycle has occurred.

I am not sure that this conclusion is correct. A slow-down of the down-welling section ought to cause a *build-up* of warm water in the seas south of Iceland, which has been observed in the depth average temperature measurements we examine below, and this would allow a *greater* output of heat into the North Atlantic weather systems that warm northern Europe and Greenland. Eventually, of course, this heat store would diminish and a cooling take place, but this has obviously not happened yet.

The issue has attracted more detailed scientific attention in the USA, where there have been congressional hearings with evidence from Woods Hole Institute of Oceanography, perhaps the foremost laboratory in the region (Joyce & Keigwin, 2002). As I noted before, US oceanographers are less sanguine about the THC potential for surprises compared to those in the UK, where Jeff Knight and colleagues at the Hadley Centre, working with computer simulations and the pattern of sea surface temperatures, have argued that the THC may have actually speeded up (Knight, 2005).

The situation is far from satisfactory with some commentators arguing that the work on the 'speed' of the conveyor can only make use of brief 'snap-shots' and hence Bryden's work may pick up a brief slow-down within a general speeding up. The freshwater incursion reported by Karcher for the mid-1990s (and perhaps a consequence of the more intense North Atlantic Oscillation from 1960s to early 1990s) may have been responsible, and with Arctic conditions now returning to normal this freshening will cease. There is ongoing work in the USA with Carl Wunsch and Patrick Heimbach at MIT analysing data from a variety of sources and reporting that the dramatic change of 1993–2004 is exaggerated and that changes in the transport of heat are negligible. There is clearly no consensus among the experts on these issues.

Very recent analysis of the waters of the Norwegian Sea shows, however, that the freshening trend has reversed in this region (Hátún et al., 2005). This analysis shows that the record highs of temperature and salinity are due to a decade-long reduction in the extension of the cold and fresh waters south of Iceland and Greenland, which allow for an enhanced Gulf Stream transport of warm and saline waters into the Nordic Seas west of the Faeroes and along the Norwegian coast. These findings are important since they indicate a reversing of the gradual freshening of the North Atlantic and the Nordic Sea that characterized the period since 1960 and which it had been assumed could weaken the THC system.

The researchers comment on the record high ocean temperatures and salinity which they ascribe to changes in the circulation of the North Atlantic and they emphasize these changes have important implications for the climate in northern Europe: 'On the one hand, the high salinity of the water masses will secure that the strength of the Gulf Stream system is maintained in the upcoming decades. On the other hand, the high temperatures will enhance the impacts of global warming on the climate of the northern hemisphere. The team regard it as 'unclear whether these changes are due to human influence on the climate system or if they are simply a consequence of natural variability of the North Atlantic climate system'.

If the recent reversal of Atlantic polar sea freshening is maintained then this should allay any fears of the conveyor slowing significantly. However, if the North Atlantic Oscillation moves into a deep negative phase, this could also slow the conveyor and deepen the cooling. Whatever the outcome, and I rather suspect that we will see the North Atlantic cool significantly within the next few years, there seems a very low risk of the conveyor actually shutting down.

Climate change and WWF: faster, stronger, sooner?

It follows from the preceding analysis that the main error of all commentators and many scientists who should know better is the projection of Arctic *trends* that have been established over the past 30 or 50 years and failing to sample the longer-term cycles of 70–80 years studied by Polyakov and others. The World Wildlife Fund report quotes a series of papers on Arctic sea ice decline, in particular those of Julienne Stroeve of the National Sea Ice Data Centre supported by NASA, which compare the observed decline with that predicted by computer models. As will be obvious, the decline is much faster than the models predict, but the equally valid scientific conclusion that this shows natural forces operating rather than greenhouse gases is not made by Stroeve or those who use her as a reference. The points we made about recent phase changes and potential reversal of trends are not made by Stroeve (**Stroeve et. al., 2007**).

The WWF report also refers to accelerated change in Antarctica, in particular the disintegration of ice shelves (**Pritchard & Vaughan, 2007**). But no mention is made of the opposite trend in sea ice extent compared to the Arctic, with 2007 recording a maximum extent for the satellite record since 1979 (*nsidc.org/sote/sea_ice.html*). Antarctic specialists do not attribute the break-up of ice shelves to global warming, because there has been very little in the Antarctic apart from the Peninsula region (see below). Instead, reference is made to unusual stratospheric heating and shrinkage of the polar vortex due to solar activity, with possible links to the ozone hole (now also thought to have a significant natural process as a co-factor with CFCs).

The WWF report also makes reference to accelerating sea-level rise projections (**Rahmstorf, 2007**). However, this paper deals with projections based on past trends and assumed future greenhouse gas levels (and a higher projection of 3.4 mm per year for the century). There is no mention of the cooling oceans and recent reduction of sea-level rise (see Fig. 29, Chapter 10).

Further reference is made to recent work appearing to show that the southern ocean carbon sink is declining, with implications for an accelerated build-up in the atmosphere (**LeQuere, 2007**). However, the paper refers to a period from 1981 to 2004 before the current levelling of ocean heat content. The study also compares observed measures with computerized expectations, and makes a number of entirely unsupported assumptions that recent increases in southern ocean winds (held responsible for the reduced carbon uptake) are 'resulting from human activities'. In this case, the inference relates as much to UV light and ozone depletion as to greenhouse warming.

Antarctic warming or cooling?

Just as the Arctic 'meltdown' has been hyped by media and campaign groups, the layperson could be forgiven for thinking that Antarctica was suffering the same fate, with stories of massive ice shelves cracking and falling into the sea. In fact the continent as a whole has seen virtually no change either in temperature or overall ice loss, with what changes there have been limited to the area of the Antarctic Peninsula.

It has long been an observation of palaeoclimatologists that the polar regions warm and cool like a see-saw – when the Arctic is warmer than normal, Antarctica is cooler and vice-versa (**Svensmark, 2007**). In the Antarctic and southern oceans global warming is not as evident as in the Arctic and northern oceans. On the continent itself, 60% of the surface shows either no trend or a slight cooling. The only pronounced warming is in the Peninsula regions. This temperature distribution was reported by Peter Doran, an Antarctic specialist at the University of Chicago, in 2002 and so much was made of it by 'sceptics' that he issued a statement to clarify that the area of cooling he studied – the dry valleys of the McMurdo Sound – was a relatively minor region and he stressed that the Peninsula had warmed by several degrees, although he confirmed that 58% of other stations on the continent showed a cooling trend.

A recent report (January 2008) from Bristol University's glaciology unit, headed by Professor Jonathan Bamber, gained over 10,000 references on Google and a whole-page spread in several national newspapers when it showed that the West Antarctic Ice Sheet (WAIS) had increased its melt by 75% in the decade 1996–2006. This translated into headlines that implied the Antarctic as a whole was melting down. But the WAIS is only 20% of the ice mass — the main East Antarctic Ice Sheet surveyed by the same team showed no trend. Furthermore, the team based its conclusions on the analysis of glacial movement rather than change in the mass balance of ice. The issue of whether Antarctic ice mass balance is changing is not yet resolved, but it would appear there is little overall change.

The warming of the Antarctic Peninsular region is due to warmer southern ocean waters flowing from the Pacific. These waters have begun to cool and it would be instructive to look at the pattern of cooling and whether there is any downturn in glacial ice movement towards the end of the data period that Bamber deals with. As we have noted for the Arctic, any data sets based on anomalies referenced to 1961–90 will still show excesses of several degrees, and even a recent downturn of one degree might not negate the decadal trend and go unreported. Likewise, trends quoted for only two decades will obscure any indication of cycles.

In this latter respect, breaking news (21 January 2009) that NASA satellites have confirmed Antarctic warming over the satellite monitoring period needs to be greeted with some scepticism. Seasoned Antarctic meteorologists point out that surface stations show a cooling; and although they do not cover the remote ice mass of the high continental interior, they indicate important aspects of the region's climatology. In actuality, despite the media hype, the NASA team emphasize the uncertainty of their methods, which appear to detect a 0.6°C rise over 25 years for the whole continent, despite the fact that most surface stations report little change. There is also no indication of how the last 25 years relates to the previous one hundred years of the records which are based upon surface station data alone. Further, as we noted earlier, satellite data show record ice extent for 2007, the reverse of what is happening in the Arctic.

Ozone changes and polar climate

In addition to the potential effects of short-wave radiation flux to the oceans and the changes we have discussed relating to warmth entering the Arctic from southern latitudes, we shall see in the next chapter there is also a potential for the flux of ultraviolet light (and recently reported cosmic radiation) to affect the polar vortex. The work of Drew Shindell at NASA has shown that as ozone absorbs the UV flux and heats up the upper atmosphere, energy is transferred to the lower levels and can affect high-level wind patterns such as the jet stream (**Shindell et al., 1999;**

2001). The upper atmospheric warming that took place between September 2001 and April 2002 (at the height of the solar cycle) caused the northern winter polar vortex to shrink, resulting in an extremely warm winter in low and middle latitudes; and the southern summer vortex contracted and for a time even broke up into two centres for the first time ever observed.

In the Arctic, these vortex changes may have contributed to the unusual weather patterns that affected sea ice cover, and in the Antarctic, to the brief summer break-up of the Larsen ice sheet observed at that time. This area of science appears to me to be badly neglected. During the current prolonged solar minimum the jet stream has shifted south during the summers of 2007 and 2008, bringing torrential rain to western Europe. This directional shift is opposite to that predicted by IPCC models for a warming world. It could be important to understand the dynamics of this relationship because several solar scientists are predicting a prolonged solar minimum with low UV activity, and this would be expected to impact upon jet stream dynamics and regional if not global climate. Cosmic ray flux will be significantly higher under these conditions. We have learned that what happens in the Arctic and in particular the North Atlantic has a global reach. Shindell believes that the prolonged solar minimum, the Maunder Minimum, which corresponded with the Little Ice Age, experienced a prolonged southward shift in the jet stream.

We have seen that the Arctic region has cycles of about 80-year peaks in temperature. Other ocean basins have cycles ranging from 4, 10, 23 and 30 years; and there are longer irregular cycles of 400 years and 1500 years, with much evidence relating many of these periods to cycles of solar activity. In the next chapter we shall take a look at the potential for solar activity to be driving the cycles of ocean temperature on these various timescales.

Note

1. There is some new science reported in January 2009 that cosmic ray fluxes are correlated with sudden stratospheric heating in the north polar regions (see Chapter 8).

7

Solar Source

In our review thus far we have seen that the global warming pattern of the late twentieth century can be explained by a combination of variations in cloud cover and the consequent amount of sunlight reaching the earth's surface. This pulse of radiation over two decades will have created a 'warm pool' of surface water in the oceans. This warm pool is then subject to currents and eventual redistribution of the heat to land, and especially in the northern hemisphere and Arctic regions. Such pulses will have happened before in regular cycles of irregular length as each of the ocean basins interacts with the others, with time lags and varying interference and amplifying patterns that defy modelling and prediction. In this chapter we will see that such pulses of warmth are not likely simply to be the result of random fluctuations or internal oscillations, as often characterized, but also subject to longer-term cycles of solar activity than the 11-year sunspot cycle. The science of these cycles is not well understood but, as we saw in the previous chapter, there is a great deal of evidence for correlations of solar activity to ocean temperatures, ENSO activity, North Atlantic sediments, continental watershed, tree-rings and stalagmite indicators of monsoon intensity.

We saw also that several ocean basins exhibited warm periods that have peaked 'in phase' with each other. The global warming signal – that part attributable to human actions – is dependent upon computer simulations that claim to have isolated a signal from the period 1950–2000 that can only be explained by the rise in emissions of greenhouse gases. We have seen how that claim cannot be substantiated when natural cycles of varying frequencies are so poorly understood and not replicated by the global circulation models. We now examine a large body of evidence to suggest that much of this natural variability is not random nor is it entirely internal, but correlates to the solar magnetic cycle.

The 11-year and 22-year cycles

Solar magnetic cycles have periodicities of about 11 years (the Schwabe cycle), 22 years (the Hale cycle) and the Gleissberg cycle (about 80 years).

The first of these is the basic sunspot cycle known since the invention of telescopes from the number of small dark spots that traverse the solar disc. The cycle rises to a peak over a period of 4–5 years and declines over a period of 5–6 years, and in the troughs in between there may be an absence or very low numbers of spots for a few months. The number of spots at the peak count has been known to vary, a low count being about 45 and a high as many as 200. During the space age and the deployment of satellites, two key observations were made. At the peak of the 11-year cycle the sun reverses its magnetic polarity, thus showing a 22-year cycle from one polar phase to another. It was also discovered that there was a small variation in the amount of sunlight (short-wave visible spectrum) from trough to peak of about 0.1%. Hitherto, science had thought the visible light output of the sun to be constant.

I use the term magnetic cycle to refer to the changes in the pattern of sunspots. At what is called solar minimum these spots disappear, or there are periods of overlap between spots of the former cycle and the new cycle which will have an opposite polarity. The spots appear dark against the solar disc because temperatures are cooler beneath the thin plasma sheath surrounding the solar mass known as the photosphere, which is the origin of the solar 'shine' in the visible light spectrum – too cool, in fact, to shine at all.

Modern physicists work from a model known as the solar dynamo, where the outer mass of the sun rotates through its own magnetic field. The coronal holes or sunspots are assumed to be created by vortices developing in the mobile zones beneath the surface. They regard the mechanisms whereby the sun winds itself up to a peak of sunspots and then down again as internal to the sun. Likewise, there is a generally accepted model of the solar interior as essentially a fusion-reactor generating energy from the reactions of hydrogen and helium.

However, standard theory does not entirely explain the mechanism. Firstly, the sun generates the 'solar wind', a flux of high-speed electrons and protons that accelerates away from the surface. This dilute plasma is channelled by magnetic field lines into a narrow twirling 'skirt' that spirals out from the solar equator in the plane of the ecliptic across the whole planetary system. It continues to accelerate past the earth and eventually reaches a point where the pressure of the galactic wind of dilute plasma creates the 'heliopause'. The solar system thus exists as an invisible bubble travelling through space. This bubble is criss-crossed by varying magnetic and gravitational effects caused by the interaction of the larger planets. Standard theory cannot account for several key features of this system, chief of which is the sudden rise in temperature of the coronal sheath or photosphere, and also the acceleration of the solar wind. The Nobel laureate, Hannes Alfven, the founder of magneto-hydrodynamics, developed an alternative theory in the 1920s, since out of fashion, that the sun's shine was created by an electric arc, and that the solar wind was accelerated like an electric current in a magnetic field with a back-current of dilute electrons from the solar system to complete the circuit.

Failed models and predictions

I make these points about fashionable but inadequate theories because, as computer models of the mechanisms became ever more sophisticated, NASA in particular sought to predict the amplitude of the sunspot cycle and the speed of the solar wind. The main reason for this was that damaging flares of solar radiation are variously correlated with parts of the cycle (indicated by the number of spots). These predictions have begun to fail spectacularly as we end cycle number 23 and await the beginning of cycle 24.

For most of the time that we have known of their existence the spots were an enigma and no one suspected they could have significance for the climate on earth. This view changed radically in the decade of the 1990s when correlations were found between the variable *length* of the sunspot cycle and the surface temperatures of the northern hemisphere (1991), and when correlations were found between low cloud cover and the peaks and troughs of the 11-year cycle (1997). The cycle varies from 9 to 13 years and is usually measured from trough to trough. The shorter cycles entail more spots, a stronger solar wind and stronger flares.

The length of the 11-year cycle varies in a periodic fashion known as a Gleissberg cycle of about 80 years. There are also longer-term variations but no general agreement on there being further regular cycles, though some scientists see a periodicity of 400, 800 and 1500 years. These longer-term cycles can only be inferred from proxy data relating to the effect of the solar wind on the earth's atmosphere, as we saw illustrated in Fig. 6. Another method lay in extrapolating data from similar stars that suggest sunlight can vary by small but significant percentages over millennia.

Solar radiation: visible light

At first, the variable light of the 11-year cycle at 0.1% from trough to peak was not regarded as significant for climate. The cyclic pulses were initially regarded as stable and thus any effect over the sun's repeating cycle would be evened out. This assumption began to change in the late 1990s when various studies, ranging from comparisons with distant sunlike stars and readings from proxies of the solar wind indicated that the sun had varied in its electromagnetic power over longer-term cycles. By correlating visible light output with magnetic data during the modern instrument era, it is possible to infer changes in solar light before that era. The key proxies of solar wind activity are the isotopes of beryllium and carbon created when cosmic rays bombard the upper atmosphere. When the solar wind is strong at the sunspot peaks the geomagnetic field gains power and deflects cosmic rays, leading to a fall in the rate of production of these cosmogenic isotopes.

The timing of this new knowledge is important for our argument, because most of this new evidence has come *after* the first computer models were constructed and the equations for carbon dioxide's effect were locked in. It rapidly became apparent that these models would have to be revised and an increasing solar component added. This work was carried out largely between 1995 and 2000, and it became obvious that the rise in temperature from 1800 to 1950 was more likely due to an increase in solar energy, though there was some dispute about the exact proportion. This would then be the period of recovery from the Little Ice Age now referred to by several leading researchers (**Akasofu, 2008**).

However, as solar science progressed, the effect of the sun became more evident. New techniques of assessing the sun's past activity were being devised, based upon the observed relationship of solar visible emissions to its electromagnetic field strength, and proxy evidence for a strengthening of that field, particularly since the trough of the Maunder Minimum between AD 1400 and 1800 and the recovery period leading to what is now known to be a Solar Grand Maximum. This latter phenomenon was not fully appreciated until 1999 and then mainly in terms of its implications for the additional output of visible light. However, a paper published in 1991 by researchers at the Danish Meteorological Institute first alerted the world to another possibility: that the earth's climate could be influenced by some component of the electromagnetic field itself. This work showed a correlation between the *length* of the magnetic cycle and the pattern of northern hemisphere temperature fluctuations, work that later led to study of correlations of the solar cycle with clouds.

Solar magnetics and cosmic radiation

We can recall from Chapter 1 that global temperatures did not rise in a straight line through the last century – there was a large dip between 1950 and 1980. The Danes correlated this dip with longer-cycle lengths. The longer-cycle lengths coincide with a weaker solar wind. The discoverers of this link, Eigel Friis–Christensen and Knut Lassen, proposed that the lessening of the solar wind weakened the earth's magnetic field, and that as this field deflected ionizing radiation (cosmic rays) it might lead to an increase in cosmic ray ionization and more cloud seeding – hence the cool period. Conversely, as the sun increases its power there are fewer clouds and more sunlight available for warming the earth (**Friis-Christensen & Lassen, 1991**).

This line of reasoning was followed up by physicists at the Danish Meteorological Institute, and in 1997 Friis-Christensen and Henrik Svensmark published the now seminal paper 'Variation of cosmic ray flux and global cloud coverage – a missing link in solar climate relationships' in the *Journal of Solar-Terrestrial Physics*. They demonstrated a link between the sun's field, cosmic ray fluctuation and percentage changes in cloud cover (**Friis-Christensen & Svensmark, 1997**). Their reference to the 'missing link' relates to the inability of climatologists to adequately account for past climate swings that correlated to solar cycles but where the amount of solar energy could not have varied enough to explain the changes. If the magnetic field variance caused cloud changes, the climate effect would be amplified. Thus, a potential link between the solar wind and climate had been discovered, and Svensmark went on to publish more detailed correlations, in cooperation with Nigel Marsh, at the Danish Space Centre in 2000 (**Marsh & Svensmark, 2000**).

The implications of this work were not lost on the international climate science community. Firstly, it coincided with other ground-breaking work on the past variations of the solar wind. Mike Lockwood's team at the Rutherford Laboratory in England published work in 1999 in *Nature* that showed a 230% increase in the strength of the solar wind from 1900 to 2000 (**Lockwood, Stamper & Wild, 1999**). And further, continued research using the proxies for the strength of the solar wind were demonstrating long-term cyclic patterns over thousands of years, with the most recent corresponding well with environmental data on past temperatures such as the Little Ice Age and Medieval Warm Period.

It had become clear that as the solar magnetic field and the geomagnetic field had strengthened throughout the century the flux of cosmic rays would have fallen, potentially leading to less cloud and more warming sunlight reaching the heat stores of the oceans. If it could be shown that cloud cover had varied over these timescales then the mystery of past solar correlations would be solved. But of course there are no past data on global cloud cover! Apart from relatively sparse and localized weather records, there were no reliable data on clouds until the satellite era, although meteorologists specializing in the study of atmospheric oscillations, such as ENSO, the NAO, the PDO and the Arctic Oscillation, were well aware that changing cloud patterns played a vital role in the temperature dynamic. Such data did not become available until relatively late in the satellite era, with reliable records going back only to 1983. And as we have noted, trends in the data were subject to some uncertainty due to the short life of each instrument set and the difficulties of calibrating from one satellite to the next.

However, the Danish work appeared to show a clear correlation over the solar cycle from 1983 to 1996 – a 3% increase in cloud cover correlated with the drop from the peak of sunspot activity when the solar wind is at its strongest to the trough when the wind weakens. This relationship is shown in Fig. 23. The correlation shown after 1994 has been disputed by others but Svensmark is supported by his colleagues in this assessment – we revisit the issue in Chapter 8.

Further, data on sunspot activity going back to 1610 (following the invention of telescopes) shows a clear pattern of increasing activity throughout the twentieth century. By implication, the steadily increasing strength of the solar wind would have led to a steady percentage decrease in cloud cover. It would then be a simple matter to extrapolate the 1983–96 data to the century-long rise and calculate the amount of extra sunlight available to warm the earth.

As one might imagine, this new research presented a problem for the emerging consensus that carbon dioxide had been the main cause of the late twentieth-century warming. Svensmark had made a rough estimate of the amount of extra sunlight that would have warmed the planet during the two decades of warming and in a carefully worded statement he noted that it was of a similar amount, in watts per square metre, to that ascribed

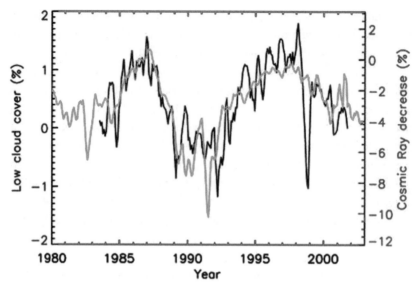

Fig. 23   Low cloud cover and correlation with cosmic ray flux. (This diagram is based on the updated work of Svensmark)

to carbon dioxide in the models. To the initiated this meant that both theories could not be correct.

The challenge to the consensus on carbon dioxide

Svensmark's work was published in the run up to the IPCC's Third Assessment Report and at a time when environmentalists were exerting considerable political pressure upon governments to implement the protocols of the Kyoto Convention on Climate Change – agreed in 1997, but not yet ratified by the USA and Russia. The IPCC, of course, also had its own internal pressures of prior commitment to the carbon model. When it came to considering this new research, it merely noted its existence, the fact that it was controversial and that the jury was out on any implications for modelling.

At first, Svensmark's team found it virtually impossible to gain funding for experimental work to confirm their thesis. In a recent account of Svensmark's efforts and difficulties in *The Chilling Stars: an alternative theory of climate change*, co-authored by the science journalist Nigel Calder, it is clear that the scientific establishment made life very uncomfortable for the Danish team. It was not until 2006 that they managed to get private funding for experimental work, the results of which they published in the *Proceedings of the Royal Society*, showing the clear potential for cosmic rays to create condensation nuclei and hence the seeding for clouds (**Svensmark et al., 2006**). This work is currently being repeated at the CERN laboratories under Svensmark's guidance and with European Space Agency funding.

These initial results were thus available to the IPCC in time for their 4th Assessment, as well as an expanded amount of research on cloud cover trends, solar visible radiation reaching the earth's surface, other workers confirming the cosmic ray correlations (see next chapter), and more detailed work on the solar wind. However, whilst the IPCC could no longer ignore this theory, it still concluded that the correlations were weak and, as we saw in Chapter 4 on clouds, that there were contradictory data on trends.

The science establishment also responded with what, in my view, were papers clearly aimed at debunking the solar-cloud theory. The main objection to Svensmark's hypothesis was that over the crucial period of rapidly increasing temperatures the critical solar factor of visible light and the cosmic ray flux had not shown any increasing trend. In 2007, Lockwood, who had repeatedly drawn attention to the lack of trend in the cosmic ray flux, and Claus Fröhlich, based at the World Radiation Centre in Davos, Switzerland, published a critique in *Proceedings of the Royal Society* (**Lockwood & Fröhlich, 2007**) which attempted to show the contradictory trends of the key solar factors in relation to rising temperatures – such as sunspot number, solar flux, cosmic rays into the earth's atmosphere and the total solar irradiance (visible and UV light). All show either no overall trend or a slight reduction over the period 1975–2005 – see Fig. 24.

This approach reflects the narrow focus of solar specialists. Firstly, the analysis is limited to the 30-year period and the authors do not consider ocean heat storage and inertia in relation to acknowledged time lags. Cycle 22 peaked in 1990 and ended in 1996 and the next cycle, 23, peaked in 2000, with cycle 24 not yet beginning (August 2008), thus the lengthened cycle is very recent. And although according to past patterns the lengthened cycle should correlate with lowering temperatures, there is evidence in the record which goes back to 1850 of time lags which could be caused by the ocean storing and later redistributing the incoming solar heat. If this paper had been published at the end of 2008, the 11-year running average for the temperature graph used in a later figure, and very often in media references, would register a downturn with a time lag to the

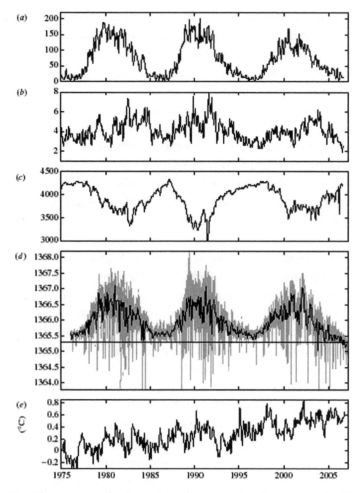

Fig. 24   *Observations of solar activity for recent decades compared with global mean temperature data.*

(a) The international sunspot index compiled by the World Data Center (WDC).

(b) The open solar flux derived from the radial component of the interplanetary magnetic field, compiled by NASA's Goddard Space Flight Center (GSFC).

(c) The neutron count rate due to cosmic rays (WDC, Boulder, USA).

(d) The TSI (total solar irradiation) composite compiled by the World Radiation Centre, Davos, Switzerland.

(e) The GISS analysis of the global mean surface air temperature anomaly DT (with respect to the mean for 1951–80), compiled by GSFC, primarily from meteorological station data.

(From Lockwood & Fröhlich, 2007)

solar indicators. By using the running average the authors also obscure the peak of 1998 and the flattening of the signal thereafter.

Furthermore, Lockwood and Fröhlich choose not to focus on the *aa* index – the geomagnetic field strength – which does show more of a rising trend than the solar flux data and until well beyond 1990 (see next chapter). The earth's field tends to store the magnetic flux such that the peak follows the peak of the sun's flux with a time lag of a few years. Additionally, although there is no obvious trend in cosmic ray flux, there are pulses and troughs of deepening amplitude, and the limited time period chosen does not show the longer-term rise of the solar flux and hence the expected fall in cosmic ray intensity. They also choose to show the GISS surface temperature record which is biased towards the northern hemisphere and has the global peak in 2005. The Hadley surface data set has a 1998 peak followed by a plateau (before the later drop), and tropospheric temperatures would also show the earlier peak and obvious recent falls. But the main criticism is that 30 years is simply too short a timescale to draw conclusions about trends and correlations, especially in a system subject to cycles. We deal with this issue in more detail in the next chapter.

The media response: the myth of solar influence

These criticisms of the solar theory have been carried over into national newspapers and the *New Scientist* ('Climate Myths', 19 May 2007) by commentators who do not examine the detailed literature and simply repeat comments made by others. For example, there has been no recent trend in cosmic rays and hence they cannot account for the temperature or cloud patterns. Whatever doubts there may be about the cosmic ray relation to cloud cover and temperature, my own feeling is that the focus should shift towards the relevant satellite data on radiation fluxes in the atmosphere – this confirms the effects of clouds and indicates periodic pulsing in relation to solar cycles. More detailed analysis should be performed on ocean surface temperature patterns and upper ocean heat content which, as reported in Chapter 5, are well correlated with solar cycles.

This criticism of Svensmark's hypothesis was already seriously out of date when published and reflects badly on the journal's review process. There was already growing evidence of oceanic responses to decadal scale variation. In their opening paragraph, the authors themselves make the point that: 'A number of studies have indicated solar variations had an effect on pre-industrial climate throughout the Holocene.' The authors note that the correlations of sediment debris by Bond and stalagmite isotopic indicators of temperature by Wang (which we looked at in Chapter 5) correlate with the cosmogenic isotopes beryllium-10 and carbon-14 'on all time scales between decades and several thousand years', but the authors appear uninformed on the obvious issue of time lags.

If clouds respond to the amplitude of the drop in neutron flux (or some other parameter linked to the depth of the cycle) then a pulse of short-wave solar energy to the surface will set up a pulse of warmed ocean surface waters and, as we shall see from reviews of the latest oceanographic work, these pulses may constitute the timing mechanism for oscillatory phenomena and account for a large part of the north polar amplification of sea surface temperatures seen by Camp and Tung in Chapter 5.

The continuing controversy over solar activity

This limited focus upon the second half of the twentieth century (and apparent lack of trend in solar activity) detracts from one of the key issues – the longer-term variation in the activity of the sun. There are no proxies for sunlight other than in the general dynamic of ocean currents and the rainfall and sediment patterns that record them, and thus the evidence for variable solar irradiance over long timescales is equivocal. Some inferences have been drawn from the study of cycles in similar stars to the sun, and some inferences are drawn from studying the recent magnetic cycles (for which proxies of the magnetic field show past cycles) and variations on light output.

There is extensive research on the issue with several leaders in the field arguing that there has been a significant rise in solar irradiance since the time of the Maunder Minimum. The graph in Fig. 25 is taken from work by Judith Lean and published on the official US Government website of NOAA. It clearly shows the Maunder Minimum and a 2.5 watts/sq metre rise since that time, and in particular the second rise since 1900 and the recent high plateau of solar activity. In another paper, Lockwood and Stamper in association with Sami Solanki at the Max Planck Institute in Germany regarded recent solar activity as 'unprecedented' over the past 8000 or more years (**Solanki et al., 2004**). This work has been revised by a NASA team (**Wang et al., 2005**) that has included Lean, although not all

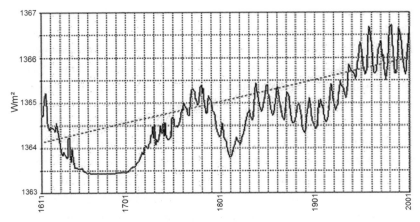

Fig. 25 Solar irradiance changes since 1611. (From J. Lean, 2004; NOAA/NGDC Paleoclimatology Program, Boulder, CO)

specialists agree with the revisions. I use Lean's graph as an illustration, but Wang's team would argue for much less of a dip through the Maunder Minimum.

In Lean's graph, the Maunder Minimum can be seen from 1651 to 1721 and the Dalton Minimum at 1811, but there is no dip to match the pattern detected by Friis-Christensen and Lassen from 1940 to 1978. It can clearly be seen in this work that the total additional wattage is of the order of 2.5 watts/sq metre. This is an increase in the total amount of solar energy available to the planet; clouds will then modulate how much of that reaches (and leaves) the surface, on average about 25% or 0.6 watts/sq m. This amount is 24% of the sum estimated by IPCC (with their computer assumptions) at 2.5 watts/sq m of all greenhouse gas effects over this period.

We should note that although many commentators regard the sun's output as stable since 1950, this is not the whole story. A glance at Lean's graph shows that the amplitude of the cycle has increased substantially and thus the 'pulsing' is as important as the inter-decadal lack of trend. Furthermore, there are unanswered questions regarding the ultraviolet component of the sun's radiation, which varies much more over the 11-year cycle − by about 8% compared to 0.1% − and which recent studies implicate in mechanism for climate change.

If we look at Lean's graph above, from 1910 when temperatures were 0.4°C below the 1961−90 average and from which they have risen to 0.4°C above, this parallels a 0.375 watts/sq m rise at the surface. Thus the long-term irradiance changes could have accounted for about 35% of the changes overall, and much more of the pre-1950 rise. These figures are hardly 'insignificant' yet this term has often been used in research papers that dismiss the variation of solar visible light as a significant climate factor.

Curiously, one such is the work of Sami Solanki, a specialist at the Max Planck Institute for Solar Research in Lindau, Germany, also working on the rise of the solar flux but more with respect to the evolution of solar visible light and reconstructing the past luminosity. Solanki has regularly stated that variations in solar energy cannot account for 'global warming' and that therefore greenhouse gases must be the cause. He has developed a model using the open magnetic field that extrapolates back from the availability of modern instrumental data using similar techniques to Lockwood's reconstructions and beryllium-10 data. The period 1900–2000 shows a rising amplitude peaking around 1960, falling in 1970 and peaking in 1980 and a little higher still in 1990, with the 2000 peak down on the previous high point (Ethos A6).

The sun is not constant after 1950

This analysis is important as it shows a *continuing* rise of the solar flux *after* 1950 and the peaks in the late 1990s are the highest since 1700. The data are not extended beyond 2000, but we can see from Lockwood's graph (Fig. 24) that the flux rose slightly in 2001, then to a smaller peak in 2004, and then declined sharply as did the total solar irradiance index. The cosmic ray count rose sharply in 2004, having been about 20% above the previous cycle's low point from 2000 to 2004. According to Svensmark's theory we would therefore expect a rise in low-level cloud cover compared to previous cycles and a cooling effect. Clearly, temperatures have begun to fall from their historic high points in 1998 and 2005, and as we noted in Chapter 4, there is evidence that cloud cover began to increase in this time period.

Solanki makes no exploration of the solar-cloud connection at all in this 2002 paper, merely referencing the earlier connection to solar cycle length with temperature and ignoring the work of Marsh and Svensmark published in 2000. He shows a graph depicting the late twentieth-century temperature rise coincident with the post-1950 plateau of the irradiance and concludes: 'hence the sun cannot be the dominant source of this latest temperature increase, with man-made greenhouse gases the likely dominant alternative'.

Solanki is an influential solar scientist and he repeats this opinion in other joint papers (**Solanki & Krivova, 2003**) but makes no allowance at all for Svensmark's theory, not even referring to it in the references of his 2002 paper. He makes no reference to the work of other scientists showing changes in cloud patterns and increased penetration of sunlight to the earth. He also uses 11-year running means for all curves and this smooths out the last two decades of temperature data, thus obscuring the post-1998 levelling off. Further, by not referring to the solar-cloud theory, he places no emphasis on the magnetic flux measurements; neither does he raise the issue of ocean cycles such as the 1998 ENSO peak, the PDO and time lags. This is yet another classic example of senior scientists publishing in the international peer reviewed literature and commenting on issues entirely outside of their field – such as carbon dioxide and atmospheric physics – without reference to other entire fields of relevant climatology. This work and opinion is seriously compromised either by the compartmented approach or political correctness in the face of 'controversial' science.

This and much other published work by solar specialists, such as Lockwood and Fröhlich, is highly unsatisfactory and shows how key scientists can, even at this late date, ignore or simply be unaware of important developments in the published literature of related fields. It also reflects upon the competence of reviewers in these key journals. Solanki refers to the solar cycle/northern hemisphere correlation and represents the graph, but simply states that some unknown mechanism or chance must be operating, making no reference at all to the published work of Svensmark and colleagues on correlating these cycles to cloud cover.

Thus although the solar irradiance (visible light or short-wave radiation) at the top of the atmosphere changed little since 1950, it had increased from 1850 to 1950 and then levelled off. After that time we see a series of pulses in the flux of light to the surface caused by thinning clouds, with the last decade in particular (1990–2000) indicating an amount that can account for the whole of the warming that has been observed in the late twentieth century.

The IPCC's 2007 Report and the treatment of this recent solar science

The IPCC review the work of Solanki and others with regard to long-term solar irradiance changes. In the February 2007 Report (Summary for

Policymakers) it was announced that recent research had shown that past solar influences had been overestimated. A 20–30% influence (which some would put higher at 40% overall) had been reduced to 12%. As I had seen no new science to justify such a move, I awaited the IPCC's 2007 detailed assessment with some interest.

It is clear from the 4th Assessment Report's Technical Working Group that IPCC have simply cherry-picked more recent literature. In one table (2.10 of Chapter 2, Changes in Atmospheric Constituents and Radiative Forcing) they provide a range of estimates of the TSI from 1700 to date but take a single new estimate (**Wang et al., 2005**) that might justify a downgrading and without referencing any discussion among the various scientists concerned. They state that a new model of solar irradiance uses recent flux patterns and a new model of solar activity cycles rather than cosmogenic isotope records, and that this lowers the solar influence by a factor of 0.27 over the previous assessments of Lean in 2000 which were largely based on comparisons of the sun with other similar type stars (see Ethos A.8). It is not clear how Solanki views this issue.

The Panel also refer to data which shows that there was a Medieval Maximum, and to the work of Solanki which reported 'exceptionally high levels of solar activity in the past 70 years, relative to the preceding 8,000 years', and they state:

> In contrast, when differences among isotope records are taken into account and the C-14 record corrected for fossil fuel burning, current levels of solar activity are found to be historically high, but not exceptionally so. (**Muscheler et al., 2007**)

Again, a new piece of research is used without reference to how other scientists in the field have responded. As we shall see in Chapter 8, Ilya Usoskin, working closely with Solanki's model, strongly disagrees with Muscheler's alternative model. Yet, these crucial disagreements disappear in the process of producing the Summary for Policymakers and the IPCC manages to prop up its prior commitment by selective use of the specialist literature.

Building in solar factors to the global circulation models

To some degree, solar factors are built into the GCMs that produced Fig. 1, but the methods are not transparent. They have assumed some small increase over the period 1900–50, based upon the work of Lean, and constancy thereafter. No solar-cloud relationship is accommodated. Recently, modellers tested the two solar projections, that of Lean, as in Fig. 25, and that of Wang (Ethos A8), and found that if the lower gradient from 1600 (Wang) is used, then the GCMs produce *higher* figures of up to 50% for the solar component in the period 1900–50 (**Scafetta & West, 2007**). These modellers have sought to reconcile the data showing higher variability of past temperatures (**Moberg et al., 2005**) compared to the projections of Mann and Jones (2003), a revised form of the 'hockey stick' which still shows much less variability. This study shows that various juggling operations can be performed with several valid data sets and the authors conclude that GCM modelling must begin to incorporate a range of assumptions regarding the influence of solar cycles in order to explain natural variability. However these authors restrict their inputs to variations in solar visible light and take no account, though mentioning its importance for future research, of the solar-cloud mechanism.

Other factors: the variable output of UV radiation and electrical effects

The variable output of UV light over the solar cycle is attracting more attention as potentially significant for climate. It has only recently become clear that the UV flux varies by as much as 10% between solar maximum and minimum compared to the 0.1% variability of visible light. And although the energy associated with this wavelength is absorbed mainly in the stratosphere, the recent work led by Drew Shindell at NASA is showing a potential for the UV flux to heat the upper atmosphere in photochemical reactions with ozone and for this heating to affect lower atmospheric wind patterns, especially in the polar vortex and the jet stream. There exists therefore another potential link between the solar cycles and oscillations in pressure and temperature such as the Arctic and North Atlantic Oscillations (**Shindell et al., 1999**).

The recent shift in the jet stream during the summer of 2007 brought widespread torrential rain and flooding to western and central Europe and occurred when the solar minimum dipped to an all-time recorded low point. Shindell and colleagues explored this connection between such solar lows (as in the Maunder Minimum) and shifts in the polar vortex and jet stream (**Shindell et al., 2001**). Such major movements of storm tracks and cloud banks will affect the absorption and distribution of heat in precisely those areas where the global warming signal is localized (as we saw in Chapter 5 where we looked at the spatial distribution of ocean heat storage). If the UV effect were to be correlated with solar lows in the Maunder Minimum, as seems likely, then this effect has the potential to raise or lower global temperatures via its impact upon such ocean heat stores and the major ocean oscillations.

I have a strong feeling that this will prove to be the case and that this factor is as important, if not more so, than the cosmic ray cloud correlation. The global warming signal is likely, in my view, to be multi-factorial, with TSI, the cloud element and the UV element all playing a role and peaking together under the unprecedented solar conditions of the late twentieth century. It is curious that Shindell's inquiry does not seem to have been continued considering its concluding paragraph emphasized the potential role of relatively small changes in solar output in large regional and even global change such as in the Little Ice Age.

There may even be a fourth factor. In addition to work on the UV flux and atmospheric change, there is long-running research on the global electric current, its relation to the solar wind, and effects upon cloud opacity (albedo) and cloud seeding. This work is lead by Brian Tinsley at the University of Texas at Dallas, and a review was published in *Space Science Reviews* in 2000 indicating that variations in the solar wind could have climatic consequences via the global electrical circuit, particularly in the north polar regions (**Tinsley, 2000**).

These are areas of scientific review that I wish to explore in future work, as it is clear to me that the north polar region plays a major role in the North Atlantic system and the global ocean conveyor. Changes in this region have the potential to influence temperature patterns on a global level. Considering the historical correlation between the fluctuations of the solar wind (correlated with sunspot numbers) and global climate, these areas of research may well hold the key to the 'missing link' mechanisms that Svensmark assumes relate to cloud condensation nuclei and cosmic ray flux. As yet, data on electric current variability and cloud cover is not advanced enough for correlative work.

Finally, the sun itself is not an isolated system – as modern astrophysics has discovered. Its cycles and periodic activity are unlikely to be as *internal* to the sun or even the solar system as often supposed by many solar specialists working within an older paradigm of dynamo mechanics. Given the recent failed predictions drawn from this body of science with regard to the start of solar cycle 24, perhaps the old paradigm will be revised and electrical factors revisited. The solar wind constitutes a powerful electrical current as well as magnetic flux. We now turn to the cosmic connections of solar activity.

8

Cosmic Rays

In the previous chapter we saw that contrary to many reports in the media and popular science journals there *has* been an increase in climate-relevant solar activity during the *later* decades of the twentieth century. In particular, there was an increase in the electromagnetic range of energy arising from a stronger solar wind and the transfer of magnetic field strength from the solar and interplanetary field to the earth's magnetic field. In this first decade of the twenty-first century we are witnessing a prolonged absence of sunspots and the lowest levels of geomagnetic field strength in the record. At the close of Chapter 7 we saw that recent research at NASA has established potential causal relations between low points in the solar magnetic field and climate changes through alterations in the upper atmospheric wind, sea-level pressure and the tracking of ocean storms. In this chapter we will take a closer look at the research on mechanisms linking the sun's magnetic status with climate on earth.

Clear evidence of long-term cycles

Evidence from proxies of the magnetic field's strength shows that the field varies over centennial timescales, with its low points correlating with the LIA cool period and its higher points with the Medieval Warm Period. Longer-term studies of these isotopes show cycles of several thousand years, indicating either that the solar flux varied, the geomagnetic field varied due to internal factors or that the flux of cosmic rays from inter-stellar space also varied. A combination of all of these factors could operate.

We have also seen that the solar magnetic cycle of 11 years modulates the influx of cosmic rays and has been correlated to variations in cloud cover over the same period. However, this interplay is not best repre-sented by the *length of the solar cycle* over short periods of a decade or two, or by crude sunspot numbers, or by variations of the interplanetary solar flux and strength of the solar wind, because the geomagnetic field cap-tures and stores this energy in a cumulative process. It is the strength of this field as well as the overall protection afforded by the sun's helio-spheric sheath which influences the penetration of cosmic rays and any potential they may have for affecting cloud cover.

Additionally, there are other qualities of the solar-terrestrial connection which may be important but under-researched, in particular those elements relating to the global electrical current, the ionosphere, the relation of UV light absorption, the polar vortex and the jet stream. We have seen that changes in the north polar environment can have profound consequences for the global climate through a series of teleconnections, with the North Atlantic Ocean being particularly sensitive to variations in the polar vortex and capable of transmitting this signal to the global environment.

We now turn to a more detailed study of the long-term cycles of several hundred years, the longer cycles of millennial scale and to the cycle of ice ages, which have now run for nearly two millions years, and explore the issue of whether these cycles are internal to the solar system and what driving and timing mechanisms may be at work.

The long-term rise of the solar magnetic flux, 1900–90

We can explore in a little more detail the long-term variation in the earth's magnetic shield and its relation to cosmic rays. This area of research is very recent. It has been pioneered by Mike Lockwood and colleagues at the Rutherford Laboratory in Oxford with a seminal paper in *Nature* and a presentation in 2000 at the 1st Solar and Space Weather Euroconference in Spain, 'Long term variations in the magnetic fields of the sun and possible implications for terrestrial climate'. This work confirmed and amplified data on the *long-term* decline in the cosmic ray flux, showing a decline of 15% for the lower energy rays that are more easily deflected by the magnetic shield compared to 4% for the higher range which suffer less deflection.

The centennial rise of the solar flux of 230% from 1900 to 2000 is derived by extrapolation from modern era direct measurement and correlation with the *aa* index, a measure of the earth's magnetic field that has been logged since the turn of the century. This rise and in particular the peaks of 1980 and 1990 can be seen in the Fig. 26 as well as the more recent direct measurements from satellites.

We can see that the magnetic flux continues to rise after the solar cycle peak of sunspot numbers in 1960, thus contradicting the impression from Lockwood and Fröhlich's 2007 paper. Lockwood, with Wild and Stam-

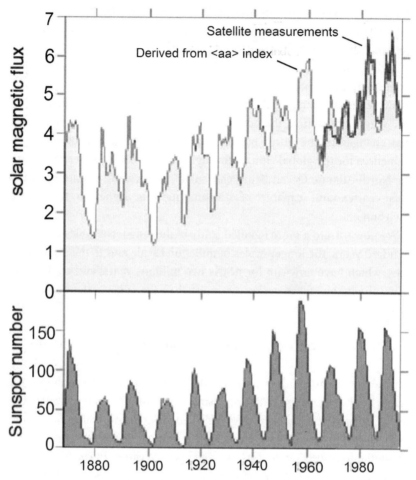

Fig. 26 *The long-term rise in the solar flux from 1860 to 2000. (Based on Lockwood, Stamper & Wild, 1999)*

per, first published the *Nature* paper 'A doubling of the sun's coronal magnetic field during the past 100 years' in 1999 and commented that 'changes in the heliospheric magnetic field have been linked with changes in total cloud cover over the earth'. This seminal paper showed how energy from the solar wind is transferred to the earth's geomagnetic field, and that the total magnetic flux leaving the sun has risen by a factor of 1.4 *since* 1964 and 2.3 since 1901. The figures demonstrate that although sunspot numbers decreased since the peaks of 1960 (highest), 1990 (second highest) and 2000 (two-thirds the previous), the magnetic fluxes continued to increase.

This paper should have alerted the world's climatologists to the importance of the magnetic flux which modulates the cosmic rays suspected of providing a mechanism for solar-cloud correlations. It could be more important than the measures of solar activity derived purely from sunspot numbers and it shows how unusual current solar activity is, which other work on beryllium-10 in ice-cores and carbon-14 in tree-rings confirms. There may have been recent spikes, for example, as argued by Muscheler for 1800, but these were of much shorter duration. Very recently, there has been some rectification of the earlier analysis based on the *aa* measurements (**Rouillard et al., 2007**), with specialists waiting 'for the dust to settle' on just how unusual is the sun's late twentieth-century peak. Leif Svalgaard presented a paper, 'Towards a consensus view', on this issue at the 2008 Fall meeting of the American Geophysical Union, and I recommend his assessment at *www.leif.org/research*.

The long-term rise in the coronal source flux from 1900 to 1995 is actually part of a longer-term fluctuation that can be tracked by the Be-10 record which, as we noted earlier, acts as a proxy for the flux of cosmic rays that are primarily modulated by the solar magnetic shield and the strength of the solar wind (see Fig. 27), where the inverse relationship between Be-10 and the solar flux can clearly be seen. Lockwood uses data from the deposition of the beryllium isotope to extrapolate the coronal source flux back to 1400 and it indicates two troughs of galactic cosmic ray activity at 1400 and 1700 (the latter being deeper). This double trough correlates with temperature profiles derived from oceanic sediment data. Thus we can see a clear correlation between increasing solar flux, the decline in cosmic ray flux and lower beryllium-10 with rising temperatures, and the inference is that the declining cosmic ray flux leads to less cloud cover and much sunlight reaching the oceans to be stored as heat.

However, despite the publication of Svensmark's work in 1997, when Lockwood's team consider the implications for the earth's climate, Lockwood and co-workers restrict their 1999 analysis to solar insolation in the visible spectrum and provide an estimate of the temperature change at 0.24°C since 1900 *due to brightness alone* (about 35% of the rise). They do, however, note that Svensmark's postulated cloud effects may be real and should be further investigated.

They point out that the rise in solar irradiance may have 'disguised the anthropogenic effect' which they think may have a 'later, but steeper, onset'. They provide no estimate for the cloud effect, however, merely noting the data of Svensmark and stating that it would be very important

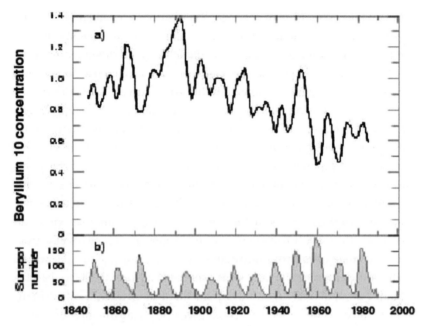

Fig. 27   Data on beryllium-10 in ice, supporting the assumption that galactic
cosmic ray flux has been declining for a much longer period. (Source:
Geography Department, University of Leeds)

indeed if a mechanism were shown and therefore that investigations
should be pursued.

We may note that as with many other comments in work that we
reviewed earlier (e.g. Wild, Solanki, Lohmann), there is a customary
deference to greenhouse gas theory. In this case Lockwood and Stamper
argue that solar effects may have 'disguised' the anthropogenic carbon
dioxide effect, an argument that had previously been applied to 'dimming'
from sulphur emissions. But this neglects the scale of *causal* linkage.
Rather than 'masking' an effect which will then 'steepen', the percentage of
forcing that can be ascribed to insolation and any added cloud effects
means that the hitherto assumed forcing of carbon dioxide is likely to
have been overestimated, leading to a *lower* than expected response to
rising carbon dioxide levels.

Furthermore, the much more recent decline in the solar field (the
sunspot peak of 2000 was lower than the 1990 peak) implies, according to
Svensmark's work, an increase in cloud cover and subsequent cooling,
which itself may be masked by the greenhouse effect (if it were strong
enough). As I will make clear as the argument develops, I have come to the

view that the carbon dioxide forcing effect has been overstated by at least a factor of three, and as much as four or five, in relation to the IPCC's computer attribution graph (Fig. 1), and that the main driver is solar with cloud and oceanic effects likely to prove the main mechanism. As we shall see when we look at predictions of future solar activity, I think there is a very real possibility of solar-induced global cooling and in this regard carbon dioxide and other greenhouse gas levels would act to ameliorate that cooling – but only slightly.

However, the pattern and amount of cloud is not going to be determined solely by GCR flux and is likely to be affected by ocean temperatures and influenced by ENSO and other cycles as well as intermittent volcanic eruptions. Such an eruption lowering global temperature occurred at the same time as the GCR trough and solar maximum in 1991, thus potentially obscuring a temperature rise. And El Niño peaked in 1998, at the same time as the GCR was high; further, the 11-year solar magnetic peak of 2000 was lower than the peak of 1990. This demonstrates that we should not be expecting direct decade-for-decade correlations and that the longer term picture will better reveal the mechanisms at work.

The current solar cycle: unprecedented over the last 8000 years

As we saw in the previous chapter, several experts regard the recent high output of the sun as unprecedented. More detailed modelling from Sami Solanki's team at the Max Planck Institute in Germany, published in *Nature* in 2004, 'Unusual activity of the sun during recent decades compared to the previous 11,000 years' (**Solanki, 2004**), shows that the recent high magnetic activity of the sun is the highest in 8000 years and the level in the last 70 years is 'exceptional' in both amplitude and duration. They further estimated on past patterns that there is only an 8% chance that the current high will continue for another 50 years.

But as we have seen so many times before, the authors then venture outside of their field of expertise and repeat the views of others that:

Although the rarity of the current episode of high sunspot numbers may indicate that the sun has contributed to the unusual climate change during the twentieth century, we point out that solar variability is unlikely to have been the dominant cause of the strong warming during the past three decades.

They provide no analysis to support this statement. Their comment relates entirely to the variations in solar irradiance and they do not refer to the solar-cloud theory, changes in cloud patterns or short-wave flux patterns, relying instead upon the assertions of predictive models that fail to incorporate these components. Neither do they refer to any oceanographic work on heat storage, ocean cycles and time lags, or the lack of a global warming signal from 1945 to 1978.

We cannot assume that in 2004 Solanki was merely ignorant of the debate, and must conclude again that an institutional bias is at work. The solar-cloud correlations had, by then, adequate representation in the peer-reviewed literature, at least to justify some more detailed treatment. However, the 2001 IPCC report had already labelled Svensmark's work 'controversial' and Solanki and colleagues perpetuate the orthodoxy and hence contribute to the apparent lack of scientific disagreement on the nature of global warming. This pattern repeats through many of the independent fields of enquiry with very few scientists attempting a cross-disciplinary analysis.

Cosmic ray correlation and cloud cover

A team led by Ilya Usoskin at the Sodankyla Geophysical Observatory at the University of Oulu in Finland looked at Svensmark's correlation of low cloud with cosmic ray flux and concluded (my emphasis in italics):

The recently reported correlation between clouds and galactic cosmic rays (GCR) implies the existence of a previously unknown process linking solar variability and climate. An analysis of the inter-annual variability of International Satellite Cloud Climatology Project D2 (ISCCP-D2) low-cloud properties over the period July 1983 to August 1994 suggests that low clouds are statistically related to two processes, (1) GCR and (2) El Niño–Southern Oscillation (ENSO), with GCR explaining a greater percentage of the total variance. Areas where satellites have an unobstructed view of low cloud possess a strong correlation with GCR, which suggests that low-cloud properties observed in these regions are less likely to be contaminated from overlying cloud. The GCR-low cloud correlation cannot easily be explained by internal climate processes, changes in direct solar forcing, or UV-ozone interactions. Instead, it is argued that a mechanism involving solar variability via GCR ionization of the atmosphere is consistent with these results. However, the results are marginal when including the recently extended ISCCP-D2 data covering the period until September 2001. This, we believe, is related to problems experienced with the ISCCP inter-calibration between September 1994 and January 1995 (**Usoskin et al., 2004**).

And additionally:

> A significant correlation between the annual cosmic ray flux and the amount of low clouds has recently been found for the past 20 years. However, of the physical explanations suggested, none has been quantitatively verified in the atmosphere by a combination of modelling and experiment. Here we study the relation between the global distributions of the observed low cloud amount and the calculated tropospheric ionization induced by cosmic rays. *We find that the time evolution of the low cloud amount can be decomposed into a long-term trend and inter-annual variations, the latter depicting a clear 11-year cycle.* We also find that the relative inter-annual variability in low cloud amount increases polewards and exhibits a highly significant one-to-one relation with inter-annual variations in the ionization over the latitude range 20–55°S and 10–70°N. *This latitudinal dependence gives strong support for the hypothesis that the cosmic ray induced ionization modulates cloud properties.*

Usoskin and his team are senior scientists supported by the Academy of Finland, yet this support for the solar-cloud theory received no media attention. It is particularly important because it draws attention to the latitudinal effect to be expected from the spatial distribution of cloud that we referred to in the previous chapter on cloud.

There was also collaboration between the Big Bear Observatory and the Armagh Observatory in Ireland. In 2000, E. Pallé Bago and C.J. Butler delivered a review paper at the 1st Solar and Space Weather Euroconference in Tenerife, Spain entitled 'Sunshine, clouds and cosmic rays' (**Pallé Bago & Butler, 2000**). This team corroborated the work at the Danish Space Centre, showing a modulation of low-level cloud cover by the GCR flux to 1994:

> ... we estimate the possible impact that such a solar-terrestrial connection may have on climate and find that much of the warming of the past century could be quantitatively accounted for by the direct and indirect effects of solar activity.

But the team noted that *overall* cloud cover appeared to have *increased* over the century and there were no direct measures of low-level cloud cover trends (an apparent contradiction later used, as we shall see, by the IPCC to dismiss the importance of the low-cloud correlation). In later analyses published between 2000 and 2005 the overall increase in cloud cover has also been noted for the latter part of the century covered by ISCCP data (see Wielicki's work on the 1960–80 period cited above). But in the period from 1980 that increase appears to have been a small percentage in high-level cloud (which warms the planet) with low-level cloud diminishing (which also warms the planet). And as we shall look at in more detail below, there then appears to have been a shift after 2000, when reflective cloud and albedo appears to increase.

Thus we can conclude that thinning *low-level* cloud is a likely consequence of the long-term build-up of the geomagnetic shield and some combination of effects upon cloud seeding, percentage cover and spatial distribution, particularly in north polar regions where the temperature anomaly is greatest. We can certainly conclude that there is support for Svensmark's thesis and that there is no consensus on this very important issue which, as we shall see, is the opposite of what is communicated in the 4th IPCC Assessment.

We shall explore in the next chapter the hypothesis that if clouds respond to the amplitude of the drop in neutron flux (or some other parameter linked to the depth of the cycle) then a pulse of SW insolation to the surface will set up a pulse of warmed ocean surface waters. As we have seen from reviews of the latest oceanographic work, these pulses may be the timing mechanism for oscillatory phenomena and account for a large part of the north polar amplification of sea surface temperatures. These ocean cycles, particularly the Arctic Oscillation and the PDO, affect the temperature signal over several decades and account for the major dip between 1940 and 1980, and most likely also the time-lagged downturn after the solar cycle energies peaked in the mid-1990s.

The pattern of the last three or four decades of GCR flux comes *on top* of a clear centennial trend of declining GCR flux (as measured from proxy indicators such as the Be-10 and C-14 records). In the last 50 years there is a clear indication of a deepening trough in the GCR cycle (see data from the Oulu neutron monitor in Fig. 28).

This pattern of troughs needs to be seen with regard to arguments that the cosmic ray flux has not changed appreciably since 1950 and hence cannot account for any cloud change. On detailed examination the GCR

Fig. 28   *The neutron count (a proxy for galactic cosmic rays) at Oulu University,*
         Finland

pattern shows the deepest trough in 1991 – a full 25% below the peaks, compared to 15% in previous troughs.

The deepening troughs coincide with the rising peak of the solar cycle in the twentieth century. The latest peak of the 11-year cycle occurred in a double feature between 2000 and 2003 and the 2000–04 trough represents a 4–5% greater flux of cosmic rays than the previous 1990 trough and for a slightly longer period. This is precisely the period that the planetary albedo recorded by Pallé rose and 2001 marked a shift to cloudier conditions.

A simple arithmetic analysis of each trough shows that the 1990–91 trough carries 35% less neutrons over the total period of the trough than that of 1968–70, and that the fourth cycle with low points between 2000 and 2004 shows a 15% rise over the third cycle (i.e. *more* cosmic rays). These troughs occur at the high point of the solar cycles – the points of maximum irradiance in the visible spectrum. Thus, whilst the sun is at its brightest, clouds would be thinnest – amplifying the heat input. And when the sun is less bright, cloud cover would be highest, thus amplifying the cooling.

The use of arguments about 'trend' is therefore spurious. There is no 'trend' in the series because there is a break-point and the last cycle registers a phase change when the solar maximum was significantly lower than the previous peaks. Solar cycles 20, 21 and 22 show a deepening trough of reduced neutron counts, from 10% to 15% and then to 25%. Added to this, there is clear evidence of a decadal change from 1990 to 30% lower amplitude of the peak in 2000 thus contradicting the view that there was no significant change in relevant solar factors during the later period of global warming. In the latest data there is also a noted increase to an all-time high over the last two years, coincident with the recent downturn in global temperatures.

There have also been criticisms in the literature that Svensmark's correlations can only be maintained beyond cycle 22 by selective use of the satellite data; Svensmark has responded to these criticisms on the website of the Danish Space Centre. The literature on the subject is now quite extensive with several papers supporting the thesis, most particularly the latitudinal variability of the response, and others remaining sceptical (**Gray et al., 2005; Harrison & Stephenson, 2006; Kristjánsson & Kristiansen 2000; Kristjánsson 2001; Marsh & Svensmark 2003; Pallé Bago & Butler 2000, 2001; Tinsley & Yu, 2004; Udelhofen & Cess, 2001; Usoskin et al., 2004, 2005**).

There is one thing that is absolutely clear from surveying this literature and that is *there is no consensus* on an issue of great importance to climate science, with strong and accumulating evidence that solar activity of one form or another and by mechanisms as yet not entirely understood modulates cloud cover over decadal and by implication centennial time-scales.

And whatever the problems of selectivity with Svensmark's correlation data, all the relevant papers on global trends report a thinning of cloud over the crucial period since the 1980s and several report an increase after 2000 when the solar cycle began to show a lower amplitude and longer interval between cycles. Svensmark's theory may not account for the details of the phenomenon – there may be other factors at work that correlate with the solar flux. Furthermore, the *pattern* of cloud thinning and spatial distribution of cloud is as important as gross percentages and general trends.

It is worth alluding at this point to various 'attribution' studies that purport to analyse all of these factors through the ocean-atmosphere coupled computer models, as much store is placed upon them in academic circles and by the IPCC (see Fig. 1 and **Schellnhuber, 2006**). There was a major international conference and review of this work at Exeter, England in 2006, and all the modellers concluded that natural factors such as solar variability and internal variation such as cloud and oceanic oscillations *cannot* account for the late twentieth-century global temperature rise. Yet perusal of the reference lists of these numerous papers shows that none of these models incorporated either a reliable cloud model or an ability to replicate or predict ENSO and other oscillations, neither do they attempt to incorporate solar-cloud relationships. I have come to the view that the vast array of such attribution studies are fundamentally flawed for these reasons and constitute a grave and persistent failing in institutional oversight on the part of the world's science community.

In a review of the Schellnhuber proceedings (for the journal *ECOS*) I found no references at all to Svensmark's work. We are not talking here of some obscure 'contrarian' or fanciful alternative theory — Svensmark's work on cloud seeding was also published in *Proceedings of the Royal Society* in 2006. And as we shall see, there is a large body of evidence for solar cycle and climate connections that *requires* an *amplifying* factor for the variations of insolation in the visible light spectrum.

IPCC's treatment of the cosmic ray link to clouds and climate

As will be appreciated, there has been a great deal of work on the solar-cloud and cosmic ray relationship since the 3rd Assessment Report of IPCC in 2001, when the Panel dismissed the whole field as controversial and unsettled.

However, despite the wealth of evidence and number of research papers during the intervening period, in the 4th Assessment Report the Panel make only passing reference to the work on solar-cloud correlations, stating that it is 'still under ongoing debate within the Panel'. They reference Marsh and Svensmark's work and the critiques but at no point do they indicate the potential significance of this work, and there is no detailed treatment of Usoskin's important confirmation of the latitudinal effect. I quote their reasoning here in full:

> Many empirical associations have been reported between globally averaged low-level cloud cover and cosmic ray fluxes (e.g., **Marsh and Svensmark, 2000**). Hypothesized to result from changing ionization of the atmosphere from solar-modulated cosmic ray fluxes, an empirical association of cloud cover variations during 1984 to 1990 and the solar cycle remains controversial because of uncertainties about the reality of the decadal signal itself, the phasing or anti-phasing with solar activity, and its separate dependence for low, middle and high clouds. In particular, the cosmic ray time series does not correspond to global total cloud cover after 1991 or to global low-level cloud cover after 1994 (**Kristjánsson & Kristiansen, 2000**); {Sun and Bradley, 2002} without unproven de-trending (**Usoskin et al., 2004**). Furthermore, the corre-

lation is significant with low-level cloud cover based only on infrared (not visible) detection. Nor do multi-decadal (1952 to 1997) time series of cloud cover from ship synoptic reports exhibit a relationship to cosmic ray flux. However, there appears to be a small but statistically significant positive correlation between cloud over the UK and galactic cosmic ray flux during 1951 to 2000 (**Harrison & Stephenson, 2006**). Contrarily, cloud cover anomalies from 1900 to 1987 over the USA do have a signal at 11 years that is anti-phased with the galactic cosmic ray flux (**Udelhofen & Cess, 2001**). Because the mechanisms are uncertain, the apparent relationship between solar variability and cloud cover has been interpreted to result not only from changing cosmic ray fluxes modulated by solar activity in the heliosphere (**Usoskin et al., 2004**) and solar-induced changes in ozone (**Udelhofen & Cess, 2001**), but also from sea surface temperatures altered directly by changing total solar irradiance (**Kristjánsson et al., 2002**) and by internal variability due to the El Niño-Southern Oscillation (**Kernthaler et al., 1999**). In reality, different direct and indirect physical processes may operate simultaneously. The direct RF due to increase in solar irradiance is reduced from the TAR. The best estimate is +0.12 W m^2 (90% confidence interval: +0.06 to +0.30 W m^2). While there have been advances in the direct solar irradiance variation, there remain large uncertainties. The level of scientific understanding is elevated to low relative to TAR for solar forcing due to direct irradiance change, while declared as very low for cosmic ray influences.

The Panel thus effectively dismiss the whole field of research on the basis of criticisms, many of which have been answered by the scientists concerned but the answers are not referenced. It is evident to many of the climatologists working across disciplines that the signal of a solar-cloud relationship would be complicated by ocean cycles such as ENSO, and that the cloud effects would not necessarily be global or common to all cloud types. On the solar-cloud issue, the Panel are here downplaying a crucial area of breaking science in such a way that the importance of the issue to climatology is obscured and the significant *lack of consensus* in this important field fails to surface in the Summary for Policymakers.

This approach is not satisfactory. The Panel make the appearance of referencing the work, and hence dealing with an issue, but then make judgements which appear arbitrary and are not reported in the Summary for Policymakers, thus implying a consensus among experts which in actuality does not exist. The Panel thus select the work that suits their prior commitment and completely cover over the enormous challenge to that prior commitment that solar-cloud theory presents. As we saw in the previous chapters, cloud cover changes can explain virtually all of the warming signal and are strong evidence that the carbon dioxide model is flawed. There is also no reliable model for cloud feedback in the standard interpretation and one could just as easily ascribe the term controversial, unsettled or poorly constrained to this aspect of the standard model.

The Panel make the same response in the downgrading of the total solar irradiance. By referring to the level of scientific understanding of solar irradiance changes as 'low' (an elevation, apparently) and cosmic ray influences as 'very low' the Panel make the presumption that the science can be regarded as not important enough to disturb their computer-based prior assumptions. As we detail below, key scientists do not agree with the downgrading of their work. It is difficult for a non-specialist to assess the validity of the Panel's selection, but from the criticisms made of the IPCC process by its members Lindzen and Christy (see Chapters 11 and 12) one is led to be wary with regard to the way documents are finalized without the agreement or knowledge of key participants.

For example, the Panel refer to the work of Muscheler in 2007 to support their stand that solar activity has been just as strong during the recent past when temperatures were lower. This argument was picked up by *New Scientist* (19 May, 2007) and also by the 'RealClimate' blog which usually defends the IPCC consensus against scientific criticism. The *New Scientist* authors refer to two sources – Solanki's 'reconstructed sunspot number', and Muscheler's 'solar modulation parameter'. At that time I was unfamiliar with Muscheler's work, having seen only Solanki's model of solar magnetic activity, which is a more relevant parameter than sunspot number. Muscheler's paper appears to show (his figure is reproduced in Ethos A8.1) that the sun was *more* active in 1800 than now, when the globe was almost 1°C cooler, thus casting doubt not only on the relationship of solar magnetic activity and temperature but also on the recent claims that the sun was at a record high for the last 8000 years – as claimed by Solanki, Usoskin and others in the solar science literature.

I suspected this to be very likely an act of cherry-picking of the one paper that supports the IPCC viewpoint and that an internet search would clarify the matter. The 'RealClimate' blog showcased Muscheler and I suspect the *New Scientist* journalists used this source but failed to give any credit to the same website riposte by Solanki and Usoskin with whose analysis the

Muscheler work clashes. This riposte refers to a serious flaw in Muscheler's work which these authors published in the peer-reviewed literature (A8.2). According to these experts Muscheler uses an unpublished and unverified model and clearly not one that can claim a consensus among peers.

This riposte was posted on the *realclimate* website (August 2005) but only after 'being lost' and then later retrieved, so it is conceivable the journalists missed it and one cannot expect them to have read the original *Nature* correspondence. I have not had the resources to fully explore this debate among the experts themselves but it underscores the argument that IPCC are taking sides in a scientific debate, without full explanation, selecting that side which suits their previous orientation, and then failing to reflect this lack of consensus in their Summary document,

There is a major problem with the IPCC conclusions. Having downgraded the solar irradiance changes *and* the possibility of solar-cloud interactions, they provide no explanation for the historic pattern of temperature fluctuations – whether concerning the major cycles of ice ages, the minor cycles of 89, 400 and 2000 years, the Medieval Optimum or Maunder Minimum – nor for that matter the documented correlations of solar cycles to short-term oscillations in the atmospheric and oceanic climate.

If it were the case that solar irradiance changes are as limited as the Panel believe, then this section of the scientific community are running out of explanations for the longer-term environmental changes correlated to solar cycles and amply documented in observational data ranging from sea and lake sediments, stalagmites and tree-rings. In fact, the lower that irradiance is ranked, the greater the need for an amplifying mechanism such as advanced by solar-cloud theory.

It is not acceptable for IPCC to dismiss such lack of consensus with such phrases as (my emphasis in italics):

> For example, solar forcing appears to induce a significant lower stratospheric response (**Hood, 2003**), which may have a dynamical origin caused by changes in temperature affecting planetary wave propagation, *but it is not currently reproduced by models.*

> Whether solar wind fluctuations (**Boberg and Lundstedt, 2002**) or solar-induced heliospheric modulation of galactic cosmic rays (**Marsh and Svensmark, 2000**) also contribute indirect forcings *remains ambiguous.*

...of the difficulty in tracking the influence of one particular modification brought about by ions through the long chain of complex interacting processes, *quantitative estimates of galactic cosmic-ray induced changes in aerosol and cloud formation have not been reached.*

An initial effort reported exceptionally high levels of solar activity in the past 70 years, relative to the preceding 8,000 years (**Solanki et al., 2004**). In contrast, when differences among isotopes records are taken into account and the 14°C record corrected for fossil fuel burning, current levels of solar activity are found to be historically high, *but not exceptionally so* (**Muscheler et al., 2007**).

IPCC seriously at fault

In my view, the IPCC are seriously at fault in their treatment of the solar-cloud science. They have failed to adequately reference and analyse the recent reviews of satellite data, in particular the clear evidence of cloud thinning and increased solar flux to the planet's surface, all of which underscore the importance of the solar-cloud theory. It is not adequate to simply state that the issue is under review. The theoretical basis is sufficiently advanced and recent data indicative enough to justify a programme of incorporation of the theory into modelling work, and in my view IPPC should have called for this. These models need not utilize the apparently sophisticated GCMs. There are other ways of modelling atmospheric systems and estimating effects, and in my own experience simpler ways of working have produced results equal to and often better than GCM computer modelling which in many cases can give misleading impressions of accuracy.

If the missing solar-cloud factor is significant then all current models are inaccurate and misleading, and IPCC should highlight this implication and a significant lack of consensus. The appropriate *scientific* response of IPCC would have been to admit this lack of consensus among the world's experts and accord the solar-cloud theory sufficient status for it to be explored in detail, and to request that it is used in modelling exercises under a range of assumptions regarding the impact on cloud formations. In view of past failures in the history of environmental modelling (see Part Two), it is not acceptable for a UN body to reason that if a relation cannot be reproduced by modelling then that relationship is suspect. History shows the reverse is more likely — that the model is suspect. That IPCC do not take such a stance is entirely political and relates to the desire not to complicate the message to policy makers, and, one has to say, leaves them open to the charge of covering their own prior commitment to the greenhouse-gas attribution model.

My expectation is that a model which incorporated a solar-cloud relationship, even accepting crude estimates of variations over a solar cycle (and irrespective of the mechanism), would readily explain past fluctuations in temperature and better explain the late twentieth-century pattern. Further, if combined with oceanic heat storage, time lags and ENSO effects, a model could be derived that better explained the current regional distribution of warming patterns and the very recent cooling.

An IPCC-constructed 'consensus' that allows an editorial process to adjudicate on quite critical areas of dispute among experts in their respective fields cannot, in my view, be relied upon as a basis for policy decisions. It is clear to me that IPCC has made such a forthright commitment to the standard model that it has a biased attitude to new data that does not confirm that model. Such prior commitment problems have arisen before within the UN structure, as I have myself critiqued (and we will deal with this issue at greater depth in Part Two). But it would appear the lessons of the 1980s and 1990s have not been learned.

Extra-solar cycles: galactic intervention

Thus far in this analysis we have confined ourselves to solar and oceanic cycles. However, the longer-term proxy record for cosmogenic isotopes contains evidence of extra-solar events – where there has been a sudden influx of cosmic rays that appears to signal a galactic source of the cosmic ray variation rather than a variable ability of the sun to modulate the galactic flux. These events appear as sudden peaks in the record of cosmogenic beryllium-10, carbon-14 and chlorine-36 in the ice-core data.

There is also some evidence of an approximate 2500/5000-year double cycle of comet-derived dust debris which the astronomer S.V.M. Clube at the Department of Physics, Oxford University, regards as noticeable in the Bond ice-rafted data we reviewed earlier (pers. comm.). My own feeling is that galactic or extra-solar factors which have hitherto either been regarded as very low frequency (of the order of tens of million years, such as the passage of the sun through the galactic plane), or of irregular occurrence, such as radiation and debris waves from super-nova explosions, may well prove significant on shorter timescales and that much may yet be revealed with regard to the beginning and ending of ice ages, and perhaps also the 5000-, 2500- and 1500-year cycles that are recorded in the isotope records. Dust incursions into interplanetary space have the capacity to excite solar activity as well as to obscure sunlight or interfere with the solar wind.

In 2006, a small group of ex-NASA scientists reviewed the evidence of cosmic dust in sediments, animal remains and human artefacts as recent as 13,500 BP. They concluded that a massive incursion of cosmic dust particles (of interstellar origin) had penetrated the protective shield of the earth's atmosphere and caused the 80% loss of large mammal fauna hitherto assumed to have been perpetrated by human hunters. They note that human artefacts did not reappear on the North American continent for another 1000 years (**Firestone et al., 2006**). We would do well to bear in mind therefore that however much we may feel the boundaries of climatology have been delineated, there are still major cycles that lie beyond the confines of our solar system. At one time, the sun was regarded as a constant in the equations. Now we know that it has cyclic variability. As with the blindspots of a previous generation that assumed much of what happens in the variable climate of earth (and ocean) is *internal*, the same blindspot is now transferred to the solar system as it relates to the wider cosmos. My own personal view is that the sun's variability on the very long timescales of millennia and over ice ages will one day prove to have a galactic component.

9

New Theory

In the highly politicized realm of climate change, 'global warming' is a fact and not a theory. In the public mind, a 'theory' is something anyone on the street can have, almost as a passing conjecture, whereas a fact is beyond dispute. There is also just the slightest tinge of disdain for that which is *only* a theory. The public mind does not engage with hypothesis, still less with the concept of a null-hypothesis of the kind that has become central to environmental science. Yet, in policy terms, it is to the public mind that science must address itself on issues of major concern and find an effective language of translation.

Arriving at an effective language on issues of climate science has been the unenviable task of the IPCC's secretariat. The Panel cannot talk in terms of hypothesis and theory if they wish for effective action. The public mind demands facts and will certainly not consider major expenditure and restructuring on an economic level unless matters have gone beyond mere theory. Yet the IPCC consists, supposedly, of a body of expert scientists who, if they were to remain true to their science, cannot go beyond hypothesis and theory and all the caveats and uncertainties demanded of any science of the environment. For this reason, the IPCC necessarily consists of more than a body of specialists; it is headed by members with long experience of government advisory service, some with no practical connection with any of the disciplines of climate science, and supported by a secretariat of experienced functionaries and translators.

If it is to have any effect on the public mind and government policy, this body must develop a group-mind with a single, simple and certain message that has gone beyond theory. Global warming must become a fact.

Thus, in a recent pre-election speaking tour of the USA, the chair of IPCC, Rajendra Pauchari, a career economist, announced that the science of global warming was 'settled' and the world should now be concerned with effective action. Similarly, the head of the UK Met Office's Hadley Centre, Vicky Pope, recently declared that anyone who thought global warming has stopped was 'deluded'. Environmental journalists have been more forthright, with Steve Connor, science correspondent of the respected UK newspaper *The Independent*, arguing that we should not let uncertainties cloud the facts which were now beyond dispute and the world should get on with a policy of mitigation. The popular science magazine *New Scientist* has taken a similar line, regarding critical arguments as 'myths' and labelling critics as sceptics or diehard contrarians and global-warming 'deniers'.

We shall look further into this social and political background to the science in Part Two, noting at this stage that it will have a profound effect not just on the interpretation of the science but on the evolution of the body of science itself, because ultimately it is the public mind — as expressed by politicians in government — that determines to a very large extent what scientific avenues are explored. Further, working scientists are exceptionally sensitive to controversy and careful not to speak out against institutionalized authority.

Science and the popular image: the evolution of global warming from hypothesis to theory to fact

As ever, the popular image of science is far from the reality. It portrays an evolution of truth as a body of objective knowledge leading to the establishment of facts and the operation of immutable laws. It has little conception that science is also a social enterprise that is intimately bound to the political and economic world and further affected by the collective psychology of an era as well as institutional processes within its own arcane realm. The public view generally confuses science with law, speaking of established facts and certainty, and hence theory gets a disdainful caste, thereby downgrading decades of development in the philosophy of science — at least since Karl Popper clarified the methodology of modern science in the middle of the last century. Popper made clear that in order for science to stay true to itself all its laws must be regarded as provisional and theories are its lifeblood, their value lying not so much in what they establish but in where they may lead.

In this respect, modern science has spent considerable effort in clarifying the role of hypothesis and shifting the emphasis away from attempting to *confirm* a particular supposition to dealing only in those suppositions that could be *falsified*. In the public mind this would mean that we could never know that something was true for all time and under all circumstance, as the history of science amply demonstrates, but we could be more certain of what was not true under a particular set of circumstances. Henceforward, science embraced the value of a falsifiable hypothesis. If a hypothesis did not lend itself to being falsified, it could be of no value. And value itself lay more in the heuristic realm of whether emerging theories led to greater knowledge of the world, rather than greater certainty.

Environmental science thus came to embrace the use of 'null hypothesis' and the process of testing. Now, I would be the first to concede there are problems with this approach, but they are problems inherent in any method that seeks to explore the complexity of the planet's ecosystems. It became common practice to assume that there were no effects resultant on the varying of a particular parameter (such as the discharge of a toxic substance) and then to devote the investigation to testing that assumption by gathering data. In an ideal world, this is a sensible approach. However, the realm of science has its own institutional dynamics that affect the evolution of theory, observation and experiment: all manner of pressures relating to finance and politics affect the kind of hypotheses that are constructed and the resources devoted to falsifying them. In my own work in pollution control strategies and international legislation I have witnessed often enough how the field is laden with errors — most particularly, the failure to gather enough of the right kind of data when it comes to falsifying a very convenient null hypothesis in relation to the discharge of a potentially noxious substance.

In this real-world evolution of science a body of knowledge should never be regarded as settled and the consensus or majority view should have no more validity with regard to 'truth' than a dissenting minority voice. This does not mean that every dissenting voice leads to future evolution, but it does mean that no evolution occurs without that dissent being voiced. Science is therefore not about a body of absolute knowledge, rather it is a *process* of understanding with its value lying in the practical evolution of reliably predictive knowledge.

Once a science becomes 'settled' it is liable to stagnate. We will look more closely at the social processes that *want* the science to be settled. For the present we need to be clear that it is the duty of science to state clearly the boundary conditions of its knowledge and to draw attention to what currently lies upon the fringe — the place where breaking knowledge will inevitably appear and transform the current view. If science strays from this duty, it becomes a tool of the political or religious order of the day. It is with this preface that I now explore the theory of global warming.

The evidence base

In the case of global warming, we have a rather chequered history of the science, with many deviations from accepted and acceptable methods, together with a very high level of political interference, vested interest and emotional commitment such that 'theory' has been translated into 'fact'.

Yet, we have seen that 'global warming' is a mutable concept. Are we talking of the post-1810 steady rise in the global mean, a proportion of which is rebound from the Little Ice Age? In which case, the late twentieth century warming does little more than maintain the long-term trend.[1] Clearly, this is not what the public has in mind. Rather, we are referring to the unprecedented and apparently unnatural trend of the post-1950 period from which IPCC models have identified the human signature and concluded with rather complex linguistics that this late twentieth-century rise is 'very likely ... not due to known natural causes alone'.

'Very likely', as we discussed earlier, means a 90–95% confidence level which is *below* that required for a scientific hypothesis to be confirmed. IPCC could equally have stated that 'this leaves a 5–10% chance that natural processes acting alone' could be responsible for the signal. And this would lay emphasis upon an increased confidence level over their previous assessment which had left at the outside – using the same language – a 40% chance that natural processes dominate the signal.

Taking this convoluted language further, the 5–10% chance relating to natural processes acting alone, which would appear to be very unlikely, says little about the proportion of the driving force were the signal due to a combination of forces, as seems extremely likely, and this is the crucial issue. However, it is obviously not politic to say so.

To the discerning observer, IPPC's more recent statement leaves the door open for a future revision in relation to natural causes that are less well 'known' such as a potential solar-cloud mechanism. This more cautious science has become rather lost in the realm of media and environmental campaigning, but it is evident that much of the contradictory evidence that has been accumulating, even within the Working Groups of the IPCC, has had some effect.

Evidence

As I intimated in the chapter on the Uncertain Signal, there is an extensive controversy on the nature and reality of the global warming signal itself. I am going to set this aside as it is dealt with in other books and very widely in the world of climate blogs. I think the balance of probability is that the warming is real and unusual though not necessarily unprecedented.

The crucial question, as we stated before, relates to causation. Having identified an unusual signal with greater than 95% confidence (eventually) the IPCC *are far less confident* about attribution. And it is in communicating this that they get themselves into such twisted logic and language. The Working Group reports, as we have seen, are replete with uncertainty and contradictory data on natural processes. The Panel use terms such as 'poorly constrained' to describe the science in key areas of the oceanography of cycles, cloud thinning, radiation flux and ocean heat content, yet these are the critical areas of breaking science and central to any understanding of natural cycles and hence the identification of an unusual non-natural signal. The Panel cannot with one hand claim that this breaking science brings many unknowns, and with the other then claim that it has greater confidence in attributing the late twentieth-century rise to non-natural factors. This contradiction lies at the heart of IPPC's 4th Assessment Report.

For the chair of IPCC to then announce the science is 'settled' is clearly a political act aimed at convincing the public there need be no further questions. The science however is still asking fundamentally: what proportion of the warming is due to greenhouse gases emitted from human activity and what proportion is part of the natural long-term variability? In actuality, the public are encouraged to believe this is the kind of questions only 'deniers' and 'idiots' would ask.

We ought then to be very concerned as to how this political reality feeds back to the work of scientists. If those who pose the question are deemed to be deluded, deniers, idiots and heretics, what then the plight of the oceanographer, solar specialist or analyst of satellite data who seeks support and funding for investigating this crucial balance of natural and unnatural forces? Perhaps future sociologists and historians of science will have reason to explore the current social constraints on science research. We can but pose the question but are left in little doubt that such pressures exist.

Returning to what little we can be sure about, contrary to the media message the IPCC's high levels of confidence are rather limited. For example, it is often stated that temperatures have risen at an usually rapid *rate* even if in the recent past they had been as high, such as in the Medieval Warm Period. In fact IPCC restrict this statement to the rapid rise of CO_2, not temperature, and as we shall see, this is not necessarily very meaningful. With regard to temperature they state only:

Average Northern Hemisphere temperatures during the second half of the 20th century were *very likely* higher than during any other 50-year period in the last 500 years and *likely* the highest in at least the past 1300 years.

and,

Some recent studies indicate greater variability in Northern Hemisphere temperatures than suggested in the TAR, particularly finding that cooler periods existed in the 12th to 14th, 17th, and 19th centuries. Warmer periods prior to the 20th century are within the uncertainty range given in the TAR.

Declining levels of confidence

Any knowledge of the science immediately shows us that IPCC is not saying very much! The last 500 years includes the Little Ice Age and the long recovery from that low point from about 1810. When they refer to the last 1300 years they include the previously controversial Medieval Warm Period. The controversy arose not within palaeontology, but as a result of the IPCC espousing the statistical treatment by Michael Mann of the palaeontological record – the infamous 'hockey stick' from which they now retreat; but they state sheepishly that this warmer period was 'within the uncertainty range' given in the TAR. The likelihood that the current warm period is higher than the last is reduced to a 66% confidence level and hence rather unreliable as a scientific conclusion and no basis upon which to support a hypothesis.

Thus, the *pattern* of temperature rise provides no convincing evidence that the cause of such a rise is not predominantly natural. It is noteworthy that the Summary for Policymakers by restricting the time-frame to 1300 years avoids discussion of previous *warmer* periods (the Roman Warm Period at 2000 BP and the Holocene Climate Optimum at 8000 BP) and of cyclic phenomena with 400- or 1500-year periodicity, both of which would provide *contradictory* rather than supporting evidence to the claim that we are dealing with a non-natural cause.

Given that the science does not confirm the global warming hypothesis at the level of confidence required of science, almost everything else that

IPCC catalogues with varying degrees of certainty becomes irrelevant to the discussion. Glaciers' rates of change, ice-mass balance, sea ice loss, ocean heat content and sea-level rise are all *consequent* upon the acknowledged temperature rise and provide no further evidence in support of the theory. This ought to be the end of the story. IPCC should be admitting the failure of evidence to support the theoretical framework. There would then be a legitimate debate about the risks of taking action under the Kyoto Protocol and whatever may follow, as well as the risks of not taking action. But clearly that is not what IPCC are saying.

The Panel remain convinced that anthropogenic sources are responsible for the late twentieth-century rise. We must therefore look to what other evidence they call in support of this position. The only source is the computer simulations based upon concepts of 'radiative forcing' incorporated in equilibrium models of the earth's atmosphere and oceans. As we have noted, the concept of an equilibrium model is not universally accepted by some experienced climatologists (for example William Kininmonth) and has also recently been questioned by Katsouyiannis and colleagues. The planetary ecosystem is *never* in such an equilibrium state, being subject to many cycles of varying periodicity. Chief of these are very long-term solar-orbital cycles which at least partly determine ice ages and these will have long-term recovery components relating to the thermal inertia of the oceans. For example, palaeontologists favour a solar-orbital explanation for the drying out of the Sahara, a relatively recent 5000 BP event, and thus such processes can have time lags relating to ocean adjustments. There is not yet an explanation of clearly defined oceanic temperature cycles of 1500 years, nor is there a universally accepted explanation for the 400-year oscillations of the MWP and LIA. These shorter-term oscillations affect the amplitude of such globally important cycles as ENSO, which may have short periodicity (4–8 years) but be subject to longer variable *intensity*. Given the pivotal role of ENSO in amplifying the current global warming signal (in 1998 and 2002), we can see that the earth's real-world system is always very far from an equilibrium state.

There are many more detailed criticisms of the modelling process but we can certainly conclude that it is no scientific basis for *evidence* in support of a theory. This is not how science works. Models are predictive or heuristic tools under continual refinement as a result of feedback from the real world and, we shall see, with certain time lags of adjustment.

Radiative forcing and imaginary equilibrium

On closer examination, the supposedly scientific 'radiative forcing' factors or RFs are derived not from observation, but from atmospheric models that build in the equilibrium state. By definition, a radiative forcing is something *external* to the equilibrium state that forces that state out of its equilibrium. It can readily be shown that IPCC relies entirely upon this method to *attribute* global warming to carbon dioxide as a principle cause. It is important to hold in mind that this forcing effect upon surface temperatures is computed and is formulated in this way as a necessary input to the models with which future temperature increases will be predicted. Thus, these are not real-world data and it ought to be clear that they cannot be used as evidence to confirm the global warming hypothesis. The very most that can be said is that they may provide some basis for prediction. We shall assess this claim in the next chapter, after we have considered evidence that contradicts the model's expectations.

We may summarize the IPCC's conclusions from its attribution studies using models:

> Most of the observed increase in globally averaged temperatures since the mid-20th century is *very likely* due to the observed increase in anthropogenic greenhouse gas concentrations. This is an advance since the TAR's conclusion that 'most of the observed warming over the last 50 years is *likely* to have been due to the increase in greenhouse gas concentrations'. Discernible human influences now extend to other aspects of climate, including ocean warming, continental-average temperatures, temperature extremes and wind patterns.

Firstly, we should note the term 'most' at the beginning of the sentence; it is not very precise and could mean anything greater than 51%. As noted, *very likely* means greater than 90%, leaving a 10% chance that the temperature rise is natural and not enough to confirm a hypothesis. This is nevertheless claimed as an advance over the previous statement that left a 34% chance that natural factors are operating.

What evidence does IPCC advance for this leap in confidence levels? Only greater faith in the attribution derived from computer studies. The real world of science has actually moved the evidence in the *opposite* direction, as with the limitations acknowledged above in the understanding of past cycles such as the MWP and the LIA. Neither the rate of temperature increase nor the absolute temperature can be regarded as convincing evidence of a non-natural driver and all other indices are dependent upon temperature. There is therefore no basis for concluding that any of these indices provide further evidence of a 'discernible human influence', and most particularly not ocean warming.

The panel move further into less confident realms:

> It is *likely* that increases in greenhouse gas concentrations alone would have caused more warming than observed because volcanic and anthropogenic aerosols have offset some warming that would otherwise have taken place.

At a 66% confidence level, this evidence becomes little more than conjecture. Volcanic aerosols depress global temperatures for about two years, although they can be expected to influence the accumulated heat store of the oceans. There is considerable debate as to the role of anthropogenic aerosol, as we noted, with warming occurring despite the pollution in some regions and depressed temperatures occurring in regions not subject to pollution. As we have noted, this aspect of past modelling is now under revision as seriously flawed.

And finally the Panel conclude:

> The observed widespread warming of the atmosphere and ocean, together with ice mass loss, support the conclusion that it is *extremely unlikely* that global climate change of the past fifty years can be explained without external forcing, and *very likely* that it is not due to known natural causes alone.

'Extremely unlikely' claims a 5% chance of probability that natural forces acting alone could have created the 'global climate change' of the past 50 years (this is what is implied by the strange term 'external forcing' – a term that it must be said the public and policy makers would be hard pressed to define even after detailed reading of the report). Of course, no scientists (to my knowledge) would claim that anthropogenic greenhouse gases have *no* effect. But this IPCC statement has no scientific validity. It is not based upon real-world evidence, but upon computer simulation of RFs. Further, the supposed 95% confidence level is itself not a scientifically derived assessment drawn from the necessary statistical treatment of real-world data, but, as IPCC points out in a footnote, one of those assessments that is often based upon 'expert judgement'. The only source of such judgement comes from the same scientists who derived the RFs from the computer models.

When we finally come to the 'not due to *known* natural causes alone' the same expert judgement drops to 90% confidence and below the level necessary to confirm the hypothesis, which, in any case, is also not based on real-world data that can be subject to statistical treatment and normal scientific derivation of confidence levels. We have to conclude that there is no reliable scientific evidence to support the conclusions that anthropogenic greenhouse gases are even *partly* responsible for the recent warming. To do so, a null hypothesis 'that there is nothing unusual or unnatural in the recent temperature rise of the last fifty years' would have to be falsified at greater than a 95% level of confidence by real-world data. It has not been. Indeed, there is no evidence from the IPCC's work that a null hypothesis was actually constructed in the first place. And this would explain the lack of effort at gathering data that would test such a model. The actual path chosen has been via theoretical models based upon prior assumptions and which are not testable by traditional scientific methods. Moreover, as we have seen, the real-world data points to a greater role for natural causes than is attributed in the IPCC models.

The real world contradicts the virtual reality

It is now widely accepted that there is a natural long-term tendency for temperatures to rise and fall around the global mean. However, rather than become embroiled in analyses of past data, frequency, oscillation and causation, we have seen that the global warming signal ultimately comes down to the period of temperature rise between 1980 and 2000. There is no doubt in my mind that if temperatures had not risen suddenly after 1978 then James Hansen at NASA would not have been able to find any takers for his predictive models evolved between 1985 and 1988 and no Framework Convention on Climate Change would have been birthed at the Rio Summit in 1992. The phrase 'global warming' would never have been borne, despite the fact that temperatures were almost half a degree higher in 1945 than they were in 1900.

We have seen how predictive models have a very limited capacity to simulate past cycles. However, we would be sidetracked if we now devoted time to a critical analysis of the modelling. In my view, the real issue with models relates to their use as predictive and heuristic tools and we shall examine that issue in the next chapter.

As we have seen, in order to test what should be the null hypothesis, 'that there is no evidence to suggest factors other than natural variation have raised temperatures in the last century', there would needs be an exhaustive look at the natural world as well as any logging of greenhouse gas increases and examination of their atmospheric physics. There appears little problem in gathering data on carbon dioxide, though some of the sinks and pathways still contain areas of uncertainty, but it has been widely and consistently acknowledged in the science literature that considerable uncertainty exists with regard to natural cycles and radiation patterns at the earth's surface. Unless a similar effort were made to answer these uncertainties, *in advance* of any conclusion, then clearly, the conclusion would be biased towards carbon dioxide as the cause.

It seems clear from the evidence of the previous chapters that exactly this has happened. We have documented the limited understanding of natural variability with many unknowns and areas of breaking research. Moreover, it is my distinct impression that only very limited resources have been made available in crucial areas of this research – such as ocean cycles, ocean heat storage, solar-cloud relationships, atmospheric electric currents and examination of past environments in relation to cyclic phenomena.

The work on the PDO, for example, which is recently held responsible for the major downturn of global temperatures and the suppression of ENSO peaks, is limited to a handful of marine research laboratories, with the cycle only having been fully identified in 1995. There is ongoing research into its interconnections to other cycles. In the area of ocean heat storage, only a few teams have published analyses, with the main set of data suffering from instrument error, and only two or three breaking papers between 2006 and 2008 involved in the crucial reassessment of ocean heat storage (which is of critical importance in assessing the reliability of models). On solar-cloud relations, the leading Danish research team had their funding blocked between 2000 to 2006, before the European Space Agency intervened (and following publication of positive results from the Danes' limited basement-level experiments). I could track only two teams looking at cloud structure and decadal trends in albedo – each with conflicting results. Only one research laboratory in the USA and one in England appear interested in atmospheric electric currents, despite the obvious electrical nature of ionization and cloud formation. The field of palaeontology and the ecology of past environments fares better, with dozens of papers reporting correlations to solar magnetic cycles, but this does not register in the IPCC summary reports and the field is tiny compared to the massive investment in computer simulation and future prediction.

This situation is certainly disturbing from a scientific perspective, but it is also rather depressing on a personal level. I devoted some considerable effort to a critique of the UN system of groups, panels and committees on environmentally controversial issues during my work on marine pollution control. My critique sought to identify some of the reasons for the acute failures in the three decades to the Rio Summit in 1992 (**Taylor, 1993**). One of those failures was a reliance upon predictive modelling. If it failed with regard to heavy metals, persistent organic chemicals and radio-activity, despite huge investments in monitoring and simulation, what reason then to have faith in the same coupled ocean-atmosphere models operating to predict the impact of carbon dioxide?

Other and equally important aspects of that critique related to the failure of science institutions to examine contraindications to a hypothesis for which there were several decades of prior commitment (known as the principle of 'dilute and disperse'). Time and time again, those institutions falsely confirmed a convenient null hypothesis simply by not gathering enough data or failing to apply statistically robust treatment to the data they did gather. These institutions were compromised by their relation to government regulatory agencies, and hence to industrial strategy and the need for cheap disposal options.

In the case of the IPCC and the worldwide 'consensus' of science institutes, there has been a huge investment in collecting temperature data and constructing computer models very largely based on supposedly settled atmospheric physics, with neglect of those other areas of real-world data that might contradict the models. However, as the real world has started to impact the temperature profiles, IPCC have moved to adopt the traditional tactics of defence – selective use of the literature, carefully chosen experts and the maintenance of a false consensus. Such previous tactics at the UN with the group of experts on scientific aspects of marine pollution held up the adoption of the precautionary principle for almost a decade.

A reality test

As we have shown, there is very little evidence to suggest an unusual variation from natural cycles until after 1980 and even that change is not convincingly shown to be unnatural. There is clear evidence of a natural recovery from the low point of the Little Ice Age with a steady two-hundred-year trend and gain of about 0.5°C. Such variability around the long-term mean of about half a degree is quite normal over the last several thousand years. We noted also that despite the rapid rise in carbon dioxide levels from 1950 to 1980, global temperatures fell into a trough. There is no evidence that the current rise is unusually rapid, as evinced by the 1920–40 period worldwide and more particularly in the Arctic.

We can therefore choose the critical period of 1980–2008 because it is not only the period of unusual temperatures but also the period of advanced coverage by instrumental data on a global scale, in particular from satellite surveillance of key parameters. Whatever the outcome of the revisions and adjustments related to instrument calibration and the measurement of trends in solar radiation at the earth's surface, cloud cover, albedo and ocean heat storage, all of these sets of data show a clear pattern over these decades – first of warming, due to incoming extra sunlight at the surface, and later of rising albedo and cooling. In particular, something of major significance for the world's climate appeared to change between 1999 and 2002: in almost every region of the world, cloud cover changed significantly and there appears to have been a drop in ocean heat content shortly thereafter.

The fine detail of these decadal changes is not predicted by climate modellers. Nor is the recent levelling off of surface temperatures. But, in marginal defence of the models and Koutsoyiannis' critique notwithstanding, they do not pretend to replicate the fine detail (although they may now attempt to do so in the light of the recent downturn).

There is one very real test of the models. We can compare the real-world equivalent of the RFs as expressed in watts per square metre at the surface of the earth with the computed carbon dioxide signal (75% of the anthropogenic greenhouse gas effect). As noted in Chapters 3 and 4, the satellite surveillance of solar SW radiative energy shows that a decadal increase in the power of these natural forces dwarfs the computed RF signal from carbon dioxide by a factor of at least four. We could therefore attribute no more than 20% of that increased forcing to carbon dioxide.

IPCC might respond that its RFs are computed as additional forcing of a natural system (as in equilibrium) and that it is not appropriate to compare an RF with the actual wattage measured over this period. This is a weak defence. The fact is that five times the wattage required by the carbon dioxide model came in to heat the earth's surface at precisely the time the surface warmed. And the surface has cooled as that input has faded due to increased cloud clover and the movement of ocean cycles with which cloud cover is linked. The evidence points away from carbon dioxide in favour of natural cycles.

In addition to this real-world data, there have been two reassessments of past data trends, both of which severely compromise the carbon dioxide model and provide further evidence for the power of natural cycles.

The models are severely compromised in the re-analysis of past trends

There have been two areas in which the models have been regarded as successful. The first is with regard to replicating the major dip between 1946 and 1978, and the second relates to the steady build-up of heat in the upper oceans. In the first instance, the models used a simulation of the effects of sulphur emissions, residence times of the aerosol, and reflectance of sunlight to successfully replicate the period of global dimming. We now know that global dimming was not caused by sulphur pollution, which was a regional and largely land-based phenomenon, but by variations in cloud and natural atmospheric transparency. Thus the models replicated the pattern but not the mechanism, and this would make for a highly unreliable capacity to predict future patterns.

In the second instance, an acknowledged element of the predictive methodology of the models relates to ocean heat storage – the so-called long-term committed component of global warming. Much was made of the models having replicated the observed degree of upper ocean warming as recently as 2004. However, between 2006 and 2008 there were major reassessments, as we documented in Chapter 5, and the ocean heat storage factor had been overestimated by 200%. This is a very large factor considering the role of the oceans in future predictions. It further demonstrates the unreliability of the models, and currently the Hadley Centre has delayed the release of its eagerly awaited projections for use by the climate impacts and adaptation community (as of October 2008).

It follows from our previous analysis that the models have no predictive module with respect to cloud and SW radiation patterns and hence could not predict the pattern of the last 25 years. If the models cannot be validated by hind-casting and effectively failed to predict the key climatological changes on cloud and SW flux observed in the last 25 years, can they be relied upon to predict longer-term changes in terms of a general trend? According to Koutsoyiannis, there should have been enough time since the development of the models to test their reliability on a decadal timescale, and there is no evidence to suggest that any model that failed to predict an evolution over 30 years would be any more successful at 100-year timescales (**Koutsoyiannis, 2008**).

We are rather far removed, however, from such rational scientific critique of the models. In the current political climate, the models are used firstly to claim that *all* of the global warming (since 1980) is caused by human activity because, as we saw in Fig. 1, if left without the CO_2 input the models would show a slight cooling. It is upon this model alone that IPCC (and the world) bases its policy response.

Contrary to the IPCC's conclusion – and to similarly indulge in complex negatives – there is a strong and more *scientific* case for accepting the null-hypothesis that there is no evidence to suggest that recent warming is not a consequence of natural cycles and little-understood mechanisms. As we noted earlier, the IPCC case for carbon dioxide having the power attributed to it rests entirely upon one single *theoretical* assumption relating to the water vapour feedback. This underpins the model predictions that the extra gas has the power to override natural cycles and dominate the warming signal. This assumption is not supported by any of the real-world evidence from 1980 to 2008. Furthermore, such a hypothesis is not readily testable. An observed increase in water vapour – noted for some levels of the atmosphere – is, as IPCC note, consistent with the ability of warmer air to hold more water vapour. It cannot therefore be evidence in support of the initial cause.

As we noted earlier, more than one expert in meteorology within IPCC considered this theoretical amplification not only unproven but unlikely. We might well then ask what probability the IPCC ascribes to their decision on this issue being correct and how that probability was derived. The answer can have little to do with science and relates to their own judgement of experts rather than, as they prefer to put it, expert judgement.

Towards a testable alternative hypothesis

I have made a case for restricting the focus to the last 28 years of good instrumental records rather than seeking to establish patterns and causes over longer timescales, not because the latter are not relevant but because it is more difficult to construct testable hypotheses. In the case of the last three decades, we would have great difficulty in establishing what the natural world *ought* to have been doing because science has already admitted to not understanding the mechanisms of natural cycles whilst at the same time acknowledging that they are very real contenders given their previous role in affecting global changes in the pre-industrial era.

Thus I recommend we focus upon some basic parameters, such as the flux of radiation to the planet's surface (this can be measured and its heating capacity compared to the computed capacity of greenhouse gases). As we have seen, the real-world evidence relegates the radiative heating of the greenhouse gas effect to less than 20% of the observed warming potential. In my view, this data would be conclusive. However, there is a further check: as the variations in SW radiation are cloud-induced and likely to be a consequence of changes in the phase of ocean cycles, we should expect that when those ocean cycles change phase natural variability will then dominate the temperature signal. This is exactly what has happened in the period since 2001, when we noted a major phase change in all of the satellite data relating to radiation flux and cloud cover.

Thus, it is necessary to construct an alternative hypothesis. When I first considered this, I was of the opinion that the modellers should be instructed to factor in ocean cycles and a revised cloud model, as well as to explore solar-terrestrial correlations. This was based on my previous experience in the critique of atmospheric modelling rather than ocean modelling. In the course of my work I have been involved in critiques related to the outputs of both types of models – between 1978 and 1988 I was concerned very much with the output of ocean models, and between 1988 and 1998 with the output of atmospheric models. The former proved highly unreliable at tracking and predicting the behaviour of pollutants, whereas the latter were largely untested in the area in which I worked – that of the dispersal of radioactive pollutants, until the Chernobyl accident in 1986. Even then, there was little to be gained from questioning the accuracy of the consequence model compared to the initial inputs and conditions of an aerial release. In the latter case, therefore, my team successfully engaged in public 'model wars', which challenged the output by exploring supposedly improbable scenarios, and time was to prove us right (**Taylor, 1988b**).

In that very public game, we as independent scientists first gained access to the model software and then were able to program the models with alternative parameters and inputs, hence generating different outcomes. But I am far from convinced that the current GCMs replicate the real world and that they could generate reliable results. Also, I feel we should move away from the illusory world of models as predictive tools.

To be true to science, they need to be used for exploring environmental complexity rather than, as now, a soothsaying device offering illusory predictive power for the world of policy makers. To attempt construction of alternative models invites a descent from the principles elucidated by Karl Popper into a world more of Harry Potter and a battle of magic wands. My preferred approach is to use the methods of *synthesis* drawn from the wider world of ecological science and natural history. The human mind is a lot smarter than a computer model and the individual analyst a lot more likely to see relationships than the larger group mind that evolves around a modelling team grown blinkered by prior commitment. As I shall explore in Part Two, the use of computer models and *large* consensual groups of scientists has more to do with the political need for certainty and unquestionable authority than with the science of climate change. We can begin by assembling relevant observations – much of which has been marginalized in the desire to corroborate the models. Chief of these relates to ocean cycles.

The oceanic factor and prediction

It is evident from our analysis thus far that the planetary heat budget is subject to cycles of varying periodicity. Some are correlated with the 11-year, 22-year and 80-year cycles of solar output, others to fluctuations in the geomagnetic field and cosmic radiation, and there are very long-term cycles in orbital parameters (the relation of the earth to the sun). It is also clear that cloud patterns can change abruptly in phase changes and be subject to cycles and trends over periods of decades and possibly also much longer-term trends. These factors affect the input of solar radiation to the oceans, which are themselves subject to internal oscillations – some apparently driven by atmospheric circulation changes (e.g. the Arctic and North Atlantic Oscillation, ENSO) and others perhaps by factors relating to ocean basin currents and the global circulation of heat. Additionally, there is evidence of a feedback from ocean basin temperatures to global patterns of high-level winds such as the jet stream (see the work of Charles Perry, referred to earlier).

The oceanic cycles have all but escaped accurate prediction, in particular the ENSO, which has a varying 4–8-year periodicity and a greatly variable amplitude of temperature change. One analyst, however, Theodore Landscheidt, an independent expert on solar cycles and former advisor to NASA, acquired a reputation for predicting accurately the last three ENSO events using solar cycle mathematics. He predicted the 1998 record ENSO, and a much lower ENSO in 2002 which he said would mask the cooling response to the downturn in the solar cycle, which he also predicted (contrary to NASA's prediction).

Landcheidt then expected La Niña conditions would predominate. The Hadley Centre were expecting in February 2007 a major ENSO as temperatures seemed to be on the rise, but NOAA issued a note in mid-year that temperatures were staying in the normal range. Landscheidt expects no more El Niños for a decade or two, based on his calculation that there is an 85% chance that the sun will enter another Maunder Minimum by 2030 (**Landscheidt, 2003**). We shall return to these predictions.

This is where the controversy and debate leads to a vital crux: if the solar magnetic flux *is* the main driver of global temperature and the main cause of the recent rise, then prediction of future solar cycles is of paramount importance for planning and policy. However, the long-term trend that we saw in the rise from 1810 to 2000 is also subject to ocean cycling of the accumulated heat and it will not be an easy matter to ascribe percentage influences. But before we look at the issue of predictive power, we need to review what we know about the main driver of the changes we have seen in the 1950–2008 period of relatively good instrumental records.

These are the main elements of an alternative theory:

- Variations in the sun's output of visible light appear capable of producing significant trends over centennial timescales such as the rise from the LIA from 1800, and the post-1950 component of that steady rise could be extrapolated to give an approximation of what might be expected from this natural cycle.
- Variations in UV light are more difficult to extrapolate and have only more recently been shown to vary significantly over the 11-year cycle and to have implications for the polar vortex and jet stream. This larger amplitude would therefore imply a stronger 'pulsed' phenomenon.
- Variations in the electromagnetic part of the solar energy spectrum appear to have the capacity to affect cloud cover by a significant percentage and would act also as a pulsed phenomenon. The electromagnetic energy is also subject to longer-term cycles of relatively poorly understood frequency, but the high and low points correlate well with northern hemisphere temperatures, and may also have a global signature.

- The pulsed nature of solar energy can be seen to affect ocean surface temperatures over the 11-year cycle (**Camp & Tung**) and by inference, therefore, longer cycles of variation can be expected to create longer-term cycles of heating (and cooling) in the oceans.
- The oceans are subject to relatively well-documented oscillations, chief of which are ENSO (4–8 years with variable amplitude depending on the state of other cycles), the North Atlantic Oscillation with a decadal and multi-decadal signal (20 years and 70-year cycles), the Arctic Oscillation (60–70-year cycles), and the Pacific Decadal Oscillation which has an approximate 30-year period. All of these ocean basins teleconnect with their cycles, and when in phase they amplify and when out-of-phase they may lower the amplitude of the temperature signal. There is evidence for some of the cycles being timed by the solar cycle.
- The greater part of the sun's energy input is received and stored in the tropical and subtropical oceans but then moved by currents and atmospheric processes northward or southward to regions of permanent heat loss beyond the 60th parallel. The southern oceans have re-radiated more heat back to space than the northern oceans, where large gyres in the Pacific and the Atlantic have accumulated warm water at depth.
- During the course of 58 years of observation since 1950, the PDO in its negative (cold) phase coincided with the 1945–78 global cool period and in its positive 1978–2006 phase coincided with the period of global warming. At the end of the PDO cycle, the large Pacific gyre off Alaska lost its heat and the cycle turned negative, affecting the whole Pacific region and coinciding with the recent fall in the global index of surface temperature (2007–08).
- The Arctic rapid ice loss between 2000 and 2007 can be explained by the return of the Arctic oscillation peak warm phase and is comparable to the peak of 1940 as part of a 60–70-year cycle. The unusual bodies of warm water in the Pacific and North Atlantic have strong influences on Arctic climate, with the latter providing sources of warm water penetrating under the polar ice and both contributing to a decadal trend in increased cloud over the polar region.
- There are recent analyses which show that the warming on land is related to the transfer of heat by wind and precipitation from the ocean (the team concerned emphasized their analysis cannot identify a greenhouse gas signal).

- There is evidence that ocean temperature oscillations, monsoon conditions and ENSO variability are correlated to solar cycles and may be timed by them. The current drop in global temperatures, shift in the jet stream and loss of heat store in the oceans coincides with a 30% fall in solar magnetic energy from the peak in 1990 to the peak in 2000, and the lengthened interlude of very low magnetic energy from 2006 to 2008.
- There is strong evidence that the major part of the 1980–2000 warming was caused by cloud thinning and increased flux of visible light to the ocean and land surface; the cloud patterns show evidence of phase changes associated with ocean oscillations as well as the peaks and troughs of the solar cycle.

These are the observations and they are relatively undisputed. The mechanisms that connect them on a cause-and-effect level are likely to be complex and all of these factors will be involved in producing the observed pattern. Since the key mechanisms relating solar energy to cloud formation and the teleconnections of the ocean basins are not amenable to mathematical formulations, virtually none of these factors can be reliably built into computer models. However, we do not need to know mechanisms with any precision in order to work from correlations. Correlation, as we noted, is not cause, but a hypothesis and alternative theory can be constructed that makes sense of the data and makes testable predictions.

The combination of solar and ocean cycles

In the past, we can see that the additional radiation from the sun (as much as 0.3 watts per square metre since the industrial era according to IPCC) has increased the heat store of the oceans and that some as yet unexplained natural cycle kicked in to enhance the flux of SW radiation to the oceans through changes in cloud cover such that between 1980 and 2000 there were periods of years with an excess of 4 watts per square metre. Thus on top of the centennial rise in SW insolation a pulse of warming occurred in the late twentieth century. It would appear that the capacity of the oceans to integrate this heat over decades has been overestimated, perhaps because the southern oceans dissipate heat more than the northern, where a build-up occurs in two major gyres. The shorter cycles of 20–30 years have operated to amplify the centennial warming and produce most of the late twentieth-century 'signal'. The contribution of greenhouse gases is thus likely to be minor at less than 20% of this signal.

From the above pattern, we can expect that the large body of warm water that has been a homogenous feature of the North Atlantic for the past 20 years will begin to fragment and ultimately exhaust its supply of heat to western Europe. In September 2008 I visited oceanographers at the Hadley Centre to look for the first signs in their mapped sea surface data and there were small cold patches developing in mid-Atlantic and quite a large shift in the waters south-west of Britain and Ireland. This may mark the beginning of a negative phase of the NAO, in which case all three major northern oscillations will be in negative mode.

This part of an alternative hypothesis can be tested by a rigorous analysis of ocean heat content, ocean currents, time lags in surface temperature, prevailing winds and landward temperature variability. Given the power of the PDO and ENSO cycle to bring down global temperatures on their own, we should expect that with the additional effects of the NAO:

- the Arctic sea ice trend of summer losses will reverse (as appears to have happened in 2008);
- the cooler Pacific, Arctic and Atlantic waters will effect a shift in the jet stream to the south, creating a very wet and cool Britain and Ireland, a drier Scandinavia and wetter Baltic in the summer, with colder winters in western Europe generally; the Mediterranean may get warmer and have stormier conditions as African air moves northward (a consequence of the jet stream shift);
- the global downturn will last for about 30 years if the PDO maintains its normal frequency as a repeat of the 1945–78 pattern.

Thus we can anticipate on current evidence that ocean cycles will have the power to reverse the global warming signal. This should become clear within the next two or three years if there is no large ENSO event. We should not expect there to be a significant El Niño as powerful as that of 1998 because other conditions are not repeated, such as the positive phase of the PDO. If there is no such cooling, this will be evidence that carbon dioxide emissions have the power attributed to them by the models. ENSO notwithstanding, we will be able to monitor developments by watching the trend of summer ice loss in the Arctic. The reversal in 2008 should continue, with summer ice levels recovering to match those experienced in the 1970s before the trend in reduction began.

On the basis of past patterns, we might expect Antarctic temperatures to follow an opposite trend and for the continent to warm with significant reductions in sea ice, as the South Pole tends to do the opposite of what the North Pole does. Another complicating factor in the global picture would be the appearance of a large volcanic event that would enhance the cooling signal but make it more difficult to estimate the power of the oceanic cooling cycle.

Thus far we have not invoked solar influences other than in the documented flux of SW radiation. Yet there is evidence that the solar magnetic cycle affects sea surface temperatures as well as cloud percentage cover (which would provide a mechanism for the observed effect at the ocean surface). Again, we need not await discovery of mechanisms before formulating a hypothesis. The 24th documented solar cycle is about to begin after more than a two-year unexpected delay. At the time of writing, a single new-cycle sunspot has been recorded by NASA. In their previous predictions, NASA expect the 24th cycle to be higher than the previous cycle. If this is the case, according to past correlations this would thin cloud and provide a countervailing warming impetus to the expected effect of the negative ocean cycles. However, a number of solar scientists have argued that the current solar maximum will not be maintained, and they expect a lower 24th cycle. If this is the case, the solar-cloud effect would potentially amplify the cooling.

Thus we may conclude:

- if the solar cycle 24 is lower in amplitude than the 2000 peak – it will peak some time between 2012 and 2013 with a sunspot number between 45 and 75 – this may bring global temperatures down further (through mechanisms as yet unclear, but as a repeat of the 1810 Dalton solar minimum);
- if solar-terrestrial physics theory is correct then a low cycle 24 should increase the global percentage cover of low-lying optically thick cloud, unless the cycle effect is masked by a major ENSO event or obscured by a major volcanic eruption;
- if cycle 24 is low – and considering the status of the other ocean cycles any ENSO events can be expected to be of low amplitude – and there are signs of a pre-Christmas El Niño in the autumn of 2008 (so this will be a test case), global warming theory would predict a record ENSO above the 1998 peak.[2]

Virtually all of these expectations run counter to global warming theory where ocean temperatures should go on increasing, ENSO intensify and the Arctic sea ice maintain the momentum of losses. The next three years will provide a test-bed for both theories.

Longer-term cycles and prediction

In the next chapter we shall look at the possibility of longer-range prediction to the end of the twenty-first century. We have seen thus far that
ocean cycles have decadal frequencies of 10, 20 or 30 years in the Pacific
and North Atlantic, and 60–80 years in the Arctic and sub-polar regions of
the North Atlantic; and we have seen evidence for 400-year and 1500-year
periodicity in northern hemisphere and equatorial temperatures that have
a clear oceanic component. There is strong evidence that these cycles
correlate with the variable electromagnetic flux of the sun and also proxy
evidence from cosmogenic isotopes that these cycles have peaked in the
twentieth century and may now decline as part of this longer-term cycle.

In the case of the North Atlantic, where most of the accumulated heat
left from the previous decades of warming is now located, we have seen
that warming may have slowed down the 'conveyor' system and that this
can have global implications. However, the mechanisms remain unclear –
a combination of freshwater influx from the Arctic and changes in salinity,
density and temperature. There are recent signs that the freshening has
diminished. Further, recent modelling of the over-turning by the team that
has just recently incorporated oceanic cycles into the global circulation
models expects a return of the conveyor to its average speeds. It is not
clear how and on what timescales these changes might affect global
temperatures. There is no evidence, however, of any propensity for a
sudden shut-down that would interfere with normal cyclic patterns.

We will now look at two methods of prediction: (a) a synthetic method
based upon knowledge of the cycles; and (b) the use of the same mathematical models that we have seen are incapable of replicating the inter-
decadal trends of the past.

Notes

1. The recently retiring director of the International Arctic Research Center at the
   University of Fairbanks, Alaska, the internationally respected geophysicist
   Syun-Ichi Akasofu, has taken particular exception to the term global warming,
   issuing a recent note outlining the steady trend from 1800 to 2000 and stating
   his own view that there was no convincing evidence for the IPCC's view that
   recent trends could not be explained by natural cycles.
2. As it turned out temperatures in the Pacific reverted to neutral conditions by
   December 2008.

10

Prediction

In the previous chapter, I sought to develop a more synthetic approach to analysis and prediction than the use of models that were essentially not testable. We have to ask the question whether there is any value in prediction when the science is so uncertain. In my view, the answer is no. In fact, any pretence at prediction may proffer an unreliable knowledge upon which quite counter-productive policies could be based. I have given this advice on a number of occasions to UK government agencies as well as land management organizations in the voluntary sector – it is better to assume no knowledge of the future climate, but to examine current vulnerability to change in any direction. This is the concept of *resilience* or robustness that ecologists apply to ecosystems. We need to know what a robust human support system looks like. We certainly do not have one now, either regionally for developed countries or globally for developing ones.

But that is not to say we would be better abandoning all aspects of prediction. I need to address, however, certain failings in the current model-driven paradigm upon which a great deal of policy is based. There is a widespread popular belief (or perhaps yearning to believe) that it does not really matter what is driving climate change. We should be making the same changes anyway in the carbon footprint of our individual and corporate activities, thus not only lessening the risk to the climate but also in anticipation of fossil fuel production declining and oil becoming more expensive. I will deal with this issue when we come to policy in Part Two. In this section we need to examine both the carbon prophecies and the solar cycle in an attempt to delineate how severe things might become and to assess whether there are any so-called tipping points and undiscovered rapid feedbacks.

Estimates of climate sensitivity to CO_2

The term 'climate sensitivity' (to carbon dioxide) is used to express the expected warming in degrees Celsius from a doubling of CO_2 concentrations. Additionally, the amount of 2 degrees above the centennial mean has been chosen as the point beyond which climate change becomes unacceptably dangerous to humanity. The former provides a point reference in time related to the rate of carbon dioxide increase from human activities and the latter a reference point with which to guide a policy of emission reductions.

On average the concentration of carbon dioxide in the atmosphere has grown by 2 parts per million (ppm) per annum over the last decade. In the period 1800–2000 carbon dioxide levels have risen by about 100 ppm from 280 ppm (35%) and global average temperature by at most 1°C. If simple arithmetic were in order then we could expect that a further 180 ppm (doubling the pre-industrial level) would bring another 1.8 degrees and we would be over the danger limit. If the current rate were maintained that doubling point would be reached in 90 years and the danger level by 2050.

But simple arithmetic is not in order. Firstly, the relation of carbon dioxide concentration to radiative forcing is logarithmic – that is, the power to warm diminishes with each additional contribution to the atmosphere. And secondly, as we have seen, it is extremely unlikely that carbon dioxide can have accounted for the major part of the one degree rise of the last 150 years. The science literature would support at least a 50% contribution from solar irradiance to 1950 and 30% since that time (Solanki, Usoskin), with as little as 12% overall being a more controversial estimate (Muscheler). If we concede that 40% of the overall rise may be solar then the simple arithmetic approach falls to a 1.68 degree overall rise in 90 years and below the supposed threshold for humanitarian danger.

Given the logarithmic nature of the relationship, however, we would expect the effect to be less than this. But I can find no simple treatment of the carbon dioxide relationship and the arithmetic approach is likely to be no more reliable than we have found the computer models to be.

If we turn to the numbers evinced in the monitoring data from 1980 to 2000 when the 0.5 degree second-half rise was concentrated, then Svensmark's estimates of cloud forcing and the work of Pallé, Pinker, Wild and Wielicki indicate about 20% of the forcing at most could be down to carbon dioxide. This would leave 0.35 degrees for the first 100 ppm rise (after allocating a generous 50% carbon-effect for the pre-1950 rise), and an expected maximum of 0.98 degrees for the doubling in 90 years. These figures do not take account of the revisions in ocean heat content, which diminish the potential power of carbon dioxide to usher in a 'warming in the pipeline'.

If we were to take the evidence of the cloud-mediated solar flux and assume 80% of the late twentieth-century driving force was natural (and by inference) the rest of the centennial rise, then we have a figure of 0.2 degrees for the first 100 ppm and a final 0.56 for the 90-year endpoint which is close to Lindzen's figure for the non-amplified effects of doubling carbon concentrations. We thus have a likely range from 0.56–1.0°C for carbon dioxide doubling.

Jack Barratt reviews the 'saturation' issue from a pure physics standpoint in *Energy and the Environment* (**Barratt, 2005**), starting with the figure of 20% of the natural total greenhouse effect (all greenhouse gases including water vapour) allocated to carbon dioxide. He uses a figure of a 1% incremental effect derived from the basic atmospheric physics of saturation for the doubling of concentration. Thus, from a background effect of about 6 degrees for natural carbon dioxide levels, this produces a hardly significant rise of 0.06°C. As we noted, the climate models assume amplification processes, chief of which is the production of water vapour as ocean temperatures rise. But as Lindzen pointed out at the beginning of the IPCC exercise, these processes are not well understood and cannot be assumed to operate as positive feedbacks when cloud formation introduces a potential negative feedback loop.

Thus there is no sound scientific evidence that would militate *against* a climate sensitivity of about 0.5°C for carbon dioxide. The higher estimates are based upon computer simulations involving positive climate feedbacks which are theoretical and not validated by direct observation. However, other factors must be taken into consideration in the longer-term analysis:

i)   the rate of discharge of carbon dioxide from human activities may not remain constant;
ii) the rate of absorption and release of carbon dioxide by ecosystems may also change;
iii) it would be prudent to take a *precautionary* approach.

In the first of these, estimates of future carbon levels vary according to assumptions of the rate of economic growth and energy technologies deployed. In the second, non-linear changes may occur such that, for example, the oceanic, forest or soil carbon reservoirs either reach saturation point or become *sources* of greenhouse gases as the ecosystems warm. And in the third, due prudence also demands that we examine social and environmental costs of mitigation policies as well as the apparent benefits.

These issues cannot be ignored, yet the uncertainties and indeterminate non-linearities are also difficult, if not impossible, to factor in to computations. Of industrial and consumer emissions, it is likely that as oil becomes scarce coal use will increase in competition with renewable sources which are limited and relatively expensive. At the same time, there is little indication of any change in the global development model that both escalates energy demand and encourages least cost solutions to energy supply. There is every indication, however, that economies will be stressed by resource depletion (particularly oil), pollutant loads, food shortages and issues of social justice, such that economic growth patterns will not be stable and predictable and demand for fossil fuels will not increase to the level expected by many scenarios.

However, human ecosystems, particularly with regard to water and food supplies, are now extraordinarily vulnerable to even small regional changes in climate. A 0.5 degree *fall* in northern hemisphere temperatures might be a good deal more disruptive of these systems than another 0.5 degree rise or even a 2 degree rise.

It could also be argued that natural ecosystems functioned perfectly well in the last interglacial when the global average was about 2 degrees higher (and likely with no Arctic summer sea ice). Western European (Eemian) ecosystems were little different than today. For example there were extensive Atlantic oak forests with much the same mammal fauna but with additional northern forest elephant, rhino, hippo, lion and leopard, and there were also higher sea levels by several metres. Human life and the global economy would be severely compromised by such a sea-level rise.

The problem with this whole line of argument, as I will address in Part Two, is that even if two degrees were inevitable (which I do not think the evidence supports), it would take 50–100 years to reach. This is a huge span of time over which to make assumptions with regard to technology, economy, emissions and population and, most particularly, considering the inevitable impact of 'peak oil' on the current development model. There are many more threats to stability and on much shorter timescales than 90 years.

Future CO_2 emissions

I view all projections of future CO_2 emissions as so uncertain as to be beyond useful assessment. Most projections of the future impact of carbon dioxide assume that emissions *increase* as global industrialization and consumption gathers pace. A recent review of the scientific basis for mitigation strategies (**Metz & van Vuuren, 2006**), concluded that, even under the most optimistic mitigation scenario aimed at stabilization at 550 ppm equivalent greenhouse gases (CO_2 plus all other GHGs), projected global emissions rose 50% above 1990 levels by about 2020, before they would tail off and return to 1990 levels (by about 2040), and to 1970 levels by about 2050. Such a scenario is achieved largely by energy efficiency and demand reduction to 2030, with renewable supply options coming after that date. If a tougher target of 450 ppm equivalent is aimed for, which is about the same as current levels in carbon dioxide equivalent terms, supply options have to be developed much earlier in addition to the efficiency and demand reduction options.

The point here is that a huge effort is required just to get back to the current 'risk' level and avoid going much higher in the three or four decades ahead. I would add that from my understanding of renewable energy supply options such as the biofuel resource and its constraints, I find most of the energy policy options that would achieve this as not only fanciful but environmentally destructive on a scale that would match any of the predicted climate change impacts on the socially and economically vulnerable and upon biodiversity. I will return to this argument when I come to consider policy options in the face of such uncertainty.

Greenhouse gas concentration targets

There is at present a worldwide campaign to pressure governments into accepting a target of 450 ppm carbon equivalents as a precautionary measure because this leads to the best chance of keeping future temperatures below a 2 degree rise. The chance element relates to probability estimates based upon reviews of climate sensitivity for 550 ppm and 400 ppm (carbon dioxide equivalents) with the most recent review giving 8–57% (chance of keeping below the limit) for 550 ppm and 63–99% for 450 ppm (**Meinshausen, 2006**). These estimates are entirely based upon the flawed models that we have reviewed and hence should be regarded as an unreliable basis for policy. The problem with these targets is that limiting climate change to a 2 degree global average rise (even if that were assured) does not guarantee humanity against 'dangerous climate change'. The change we have seen of 0.6 degrees since 1950 has already brought 'dangerous climate change' to many communities.

The most recent scientific treatment of this issue, 'Observational constraints on climate sensitivity', by a team led by Oxford University's Department of Physics and including leading scientists from the Department of Atmospheric, Oceanic and Space Sciences at the University of Michigan, the UK Met Office, the University of East Anglia's Tyndall Centre, the Earth, Atmospheric and Planetary Sciences at MIT, Duke University and the National Center for Atmospheric Research in Boulder, Colorado (**Allen et al., 2006**) – and hence independent of the IPCC – concluded:

> ...it is premature to suggest that we can provide an objective assessment of the risks of future transient climate change associated with specific concentration pathways: the problem of the non-linear relationship between observable quantities and forecast responses applies specifically to stabilization scenarios.

This is good enough for me. The science is not adequate and in my view is not likely to be adequate. It is a question of the difference between *uncertainties*, which can be addressed and minimized by more observation, measurement and synthesis, and *indeterminants*. The sociologist Brian Wynne, Professor of Science Policy at Lancaster University, never tires in pointing out that the modelling community seldom really comes clean on the difference.

Thus, I would recommend abandoning the illusory 'science' of model-based predictions. There remains however the possibility that predicting solar cycles might enable us to better determine prospects for the twenty-first century. There is now growing evidence that past environmental changes are correlated with the cycle and a potential mechanism has been uncovered in the solar wind's relationship to cloud cover,

Predictions of future solar cycles and global temperatures: is a Little Ice Age imminent?

It came as somewhat of a surprise to find how little research there had been into the future cycles of the sun. Until the latest downgrading of the solar contribution, most climatologists would have attributed close to half the centennial rise to increases in solar irradiance and, as we have seen, there is a considerable body of palaeo-environmental knowledge relating solar cycles to climate change. If the sun were to repeat its pattern of decline into a Maunder-type minimum where average temperatures in

northern Europe dropped by about 1°C this would have serious implications for human society and biodiversity. Ironically, the carbon dioxide
excess, if indeed it has significant effects, would temporarily provide a
buffering for the worst of the cold. But there are very few papers that even
address this issue.

Solanki has put forward the view that on statistical grounds the sun is
unlikely to continue its current 'high' but makes no predictions as to how
low it might drop. The other solar-climate researchers appear to limit their
interests to the next single cycle – No 24 – using various methods of
prediction based upon the current behaviour of the solar dynamo. NASA,
for example, has had a number of teams looking at the issue of prediction
because it has great commercial relevance for satellite and various
ground-based electronic equipment sensitive to solar flares, which are
more numerous at times of heightened solar activity. In this respect,
NASA's teams have made a number of attempts without a great deal of
past success. They appeared more confident recently of their methodology, but their latest prediction expecting cycle 24 to be at least as high if
not higher than cycle 23 has been countered by a panel set up by NOAA in
the light of the delayed onset of cycle 24 (NOAA Space Environment
Center, April 2007 press release). The longer delay would correlate with a
lower cycle. However the panel was split:

> In the cycle forecast issued today, half of the panel predicts a moder-
> ately strong cycle of 140 sunspots, plus or minus 20, expected to peak
> in October of 2011. The other half predicts a moderately weak cycle of
> 90 sunspots, plus or minus 10, peaking in August of 2012. An average
> solar cycle ranges from 75 to 155 sunspots. The late decline of Cycle 23
> has helped shift the panel away from its earlier leaning toward a strong
> Cycle 24. Now the group is evenly split between strong and weak.

There is, however, one analyst who has maintained a long-standing
interest in solar cycle prediction – the independent specialist Theodore
Landscheidt (now deceased). An outspoken critic of the IPCC, Landscheidt had an impressive list of publications whilst working closely
with NASA and a track record of accurate predictions of ENSO as well
as the magnitude of cycle 23 which he predicted would be down on the
previous cycle at a time when NASA expected it to rise. Landscheidt's
work is interesting because his predictions are based upon a knowledge
of the mathematics of the cycles rather than an attempted understanding of the solar dynamo and magneto-hydrodynamics. He applied

a knowledge of the Fibonacci series to predict ENSO, and a similar mathematical approach to the movement of the centre of mass of the solar system around the solar disc — a phenomenon apparently eluci-dated by Newton. His analysis involves an assumption that the energy of angular momentum can be transferred to the sun's rotational momentum and torque in a mathematical relationship that derives from the periodic movement of the giant planets which determine the centre of mass of the solar system.

Having published several commentaries arguing for the enhanced role of the sun in global warming, he summarized his findings just before his death in 2004, in the journal *Energy & Environment* — 'New Little Ice Age instead of global warming?' (**Landscheidt, 2003**) — where he outlined the methodology for prediction of the amplitude of solar cycles. He showed how variable torque theory could account for the Gleissberg cycles and highlights what he calls 'phase changes' in these cycles relating to the Hale cycle of 22 years. He claims that certain low points in the Gleissberg cycle pre-date a shut-down of the electromagnetic field (no sunspots) in either short 'Dalton'-type minima, the name given to a period of two decades in the early 1800s when the cycle went dormant, or longer 'Maunder'-type minima which can last up to 100 years. He predicts that cycle 24 will be smaller and that it will presage either a Dalton or Maunder Minimum beginning in 2030.

He was a formidable critic of the IPCC's neglect of solar phenomena, and the paper cites 23 references since 1990 that demonstrate correla-tions of solar cycles with climate, claiming these were ignored by the IPCC. He refers to recent links between solar flare activity and UV radiation increases. He also updated the earlier Danish graph linking cycle length to temperatures to the year 2000 and argued that the recent continued increase in temperatures when the solar cycle was lengthening did not necessarily mean the correlation broke down as there would be a time lag of 4–8 years. He argued that it was the aa-index (a measure of the earth's field strength) that would be the best indicator. This had continued to rise to 1990 and, according to his estimated 8-year time lag, tempera-tures should peak 8 years later — which they did with the ENSO event that he also predicted.

He then argued that as the solar sunspot cycle is expected to decline so will temperatures. By analysing the past Gleissberg pattern and looking at how its peaks and troughs correlated with temperature, he estimated that cycle 26 peaking in 2030 will be the next minimum. This would follow on from the declining peaks he predicted for 1990, 2000 and then 2010 and 2020. He also expected that El Niño years, which peaked in 1998 and were followed by his predicted smaller peak in 2002, will be followed by La Niña years and he accords a 'high probability' of 85% that the minimum will be a Maunder-type 'grand minimum'.

Landscheidt's theory should be easy to verify by replication, firstly in relation to the sun's torque and the transfer of angular momentum – a simple astronomical fact that appears to correlate with the Gleissberg cycles of activity, but I can trace no papers that either verify or contradict this. Secondly, the mathematical patterns should be clear in the ice-core data and be easy to extrapolate over the next 100 years, yet no papers I have seen tackle this (this may be due to the low resolution of the strata).

Landscheidt died shortly after publishing this paper; it is easy to regard him as an independent maverick with a contrary attitude to the IPCC, but he had a remarkable record of practical predictions in relation to sunspot cycles and El Niño years and thus this paper should be taken seriously.

In 2006, David Archibald, an independent analyst, wrote a paper in the same journal: 'Solar cycles 24 and 25 and predicted climate response (**Archibald, 2006**). It is not a very clear paper, but at least it makes an attempt at prediction and gives some useful references indicating a correlation and suggested mechanism for solar cycles and temperature fluctuations – in particular, a recent update by Usoskin, Schuessler, Solanki and Mursula in the *Journal of Geophysical Research* (2005). The main aspect of the paper relates to predictions of the next two solar cycles and the author relies upon the work of Badalyan, Obreidko and Sykora (2001), and also refers to Svalgaard, Cliver and Kamide (2005) and Schatten and Tobiska (2003). I have abstracts of these papers but have not had sight of the text. It would appear that these works all expect a lower cycle for 24 and 25 with a maximum 50 to 75 sunspot number compared to 120 in cycle 23 and 150 in cycle 22. He quotes from Schatten and Tobiska who predict:

> ...a rapid decline in solar activity, starting with cycle 24. If this trend continues, we may see the sun heading towards a Maunder type of minimum – an extensive period of reduced levels of solar activity.

The 50–70 maximum sunspot number for these cycles predicted in these papers would produce a 'Dalton'-type minimum similar to cycles 5 and 6 from 1796 to 1820. Archibald then goes on to estimate the temperature consequences by looking at two data sets – one the annual average for three European stations, and the other for five USA rural locations. He uses a graph of correlation of solar cycle amplitude relative to temperature from 1705 to 2000 and also cycle length, and these crude methodologies show a correlation. A simple extrapolation is then used assuming recent temperatures reflect the strength of the solar cycle and he predicts a cooling of about 1.5°C.

The estimates and methodologies are crude but relevant. No other authors I have seen have explicitly attempted more sophisticated methods; there is clearly a relation between temperature and cycle length, and better methods could be used to derive the factor and apply it to the predictions. Surprisingly, the papers referred to on the predicted next cycle's strength all use precursors in the magnetic field and relationships divined over recent cycles when more sophisticated instruments have been able to measure differing solar variables relating to the magnetic flux. No papers use predictions based upon harmonic cycles that could be derived from the palaeo-environmental record or from Landscheidt's analysis of Gleissberg cycles.

In 'Prediction of solar activity for the next 100 years', Mark Clilverd speaking at the Lund Seminar on Solar Cycles in 2005 introduced a model of low-frequency solar oscillations, in the range 50–420 years, modulating the 11.1-year Schwabe cycle and used to predict the peak sunspot number of cycle 24 and those beyond. The peak sunspot prediction for cycle 24 is significantly smaller than cycle 23. Peak sunspot numbers are predicted to be ~45±27. The model also predicts a recovery during the middle of the century to more typical solar activity cycles with peak sunspot numbers of ~120. Eventually the superposition of the minimum phase of 105 and 420-year cycles just after 2100 leads to another period of significantly quieter solar conditions.

Research by Leif Svalgaard and Yohsuke Kamide of the Solar-Terrestrial Environment Laboratory, Nagoya University, Japan, and Edward W. Cliver of the Space Vehicles Directorate, Air Force Research Laboratory, Hanscom, USA, and published in *Geophysical Research Letters*, notes the peak amplitude of the sunspot cycle as a key goal of solar-terrestrial physics. The precursor method currently favoured for such predictions is based on the dynamo model in which large-scale polar fields on the decline of the 11-year solar cycle are converted to toroidal (sunspot) fields during the subsequent cycle. The strength of the polar fields during the decay of one cycle is assumed to be an indicator of peak sunspot activity for the following cycle. Polar fields reach their peak amplitude several years after sunspot maximum; the time of peak strength is signalled by the onset of a strong annual modulation of polar fields due to the tilt of the solar equator to the ecliptic plane. Using direct polar field measurements, now available for four solar cycles, this team predict that the approaching solar cycle 24 (2011 maximum) will have a peak smoothed monthly sunspot number of 75 ± 8, making it potentially the smallest cycle in the last 100 years (**Svalgaard, Cliver & Kamide, 2005**).

In addition, K.H. Schatten and W.K. Tobiska in a poster presentation to the American Astronomical Society 34th Meeting in June 2005 predicted that cycle 24 would also be smaller and that the sun could be heading for a Maunder Minimum.

Given that there is such serious consideration being given to the possibility and experts would say quite high probability of the sun heading for a Maunder Minimum, one would expect this issue to have received greater consideration by governments, the media and the science establishment, in particular the IPCC. That this is not the case leads to some speculation as to why the issue might be avoided.

Reasons are not hard to find. A Maunder-type minimum would have serious implications for world agriculture. Grain yields in the northern hemisphere would be at risk and at a time when world production has already been compromised by drought. Shortfalls in China, northern Europe and Canada would have serious impacts upon world grain prices and hence the supply of food to those countries currently in deficit. Furthermore, speculation in commodity markets in anticipation of poor harvests would be rife. In this respect, it is of interest that when downloading the Archibald paper from the net I was surprised to notice the parent website was *hm.treasury.gov.uk*! Economists have a longer-standing interest in solar cycles than is apparent among UN climatologists. (These paragraphs were written before the World Food Programme announced in the spring of 2008 that a global crisis was building, fuelled by speculation, poor harvests, biofuel demand and higher energy costs, and before the global economic crisis of October.)

Sea-level rise

This review would not be complete without some observations on the issue of rising sea levels. Predictions made by the IPCC based upon the models and future emission scenarios warn of a 1 m rise, and there is much media speculation of higher levels resulting from sudden ice loss either in Greenland or in the Antarctic. As we have noted, there is no convincing evidence for significant non-cyclic ice loss in either continent. About half of the noted sea-level rise of the last century is due to thermal expansion of the oceans, and most of the rest from ice and snow loss in mountain or coastal glaciers rather than the ice caps of Greenland or Antarctica. Given that there has been an indisputable temperature rise of 1 degree since 1800, there is no need to seek any other explanation for the sea-level rise other than that expected from the natural cycle of recovery from the Little Ice Age.

We can however address one pertinent issue. Almost all references to the past sea-level rise quote trends – there being a certain relatively small annual variability in the rate – and note that the rate increased over the last two decades in comparison to the longer-term rate. It is worth looking in closer detail at this rate of change because, given the slow-down in the global temperature and cooling of the oceans, we might expect a levelling off after 2000.

This is what we find:

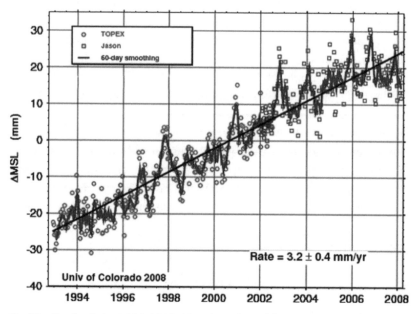

Fig. 29   *Sea-level rise 1992–2008. (Graphic adapted from University of Colorado at Boulder – http://sealevel.colorado.edu/)*

The trend from 1992 to 2008 is shown by the straight line as 3.2 mm/year, but it is clear from the data that there is a levelling off after 2006 – the year in which the Pacific began to cool significantly. If this level is maintained or sea level declines as the temperature of the globe continues to fall, this will cause a few surprises in the modelling camp because there has always been an assumption of much longer response times such that sea level would continue to rise even if carbon dioxide levels were held constant. If, as I understand, the oceans are the *primary* source of atmospheric warming, then as the oceans cool so will the atmosphere, and sea level will fall in line with ocean temperatures.

If we see further sea-level rise, it will be because natural cycles have not yet peaked. In the previous interglacial temperatures were 2 degrees warmer and sea level was several metres higher. Much therefore depends upon whether this current interglacial has run its course (models of glaciation based upon solar orbital cycles suggest we may only be half-way through). However, the descent *into* an ice age is a gradual process over several thousand years and some specialists believe that carbon dioxide levels might delay that process.

It follows that if, as I have argued, the period 2002–07 marks a turning point, then glaciers will begin to grow and ice-mass begin to accumulate again, thus levelling off the sea-level rises.

Ocean acidification

Another issue that correlates to global warming is the acidification of the surface waters of the ocean resultant upon an enhanced absorption of carbon dioxide. It is not an issue that I had considered important largely because of the relative amounts of carbon dioxide stored in the ocean, the atmospheric reservoir, the interchange between these compartments of the global system and the vast reservoirs of carbon in the deeper waters of the ocean and sediments. These interchanges dwarf the anthropogenic flux of carbon dioxide. I was also aware of the special pleading and alarmism that embellishes the press releases and publications of research teams eager for a slice of the expansive global warming budget. However, recently voiced concerns in the press by working scientists merited a brief look at the issue. It is well reviewed by a working group of the Royal Society in 2005, and my comments here are gleaned from their report (**Royal Society, 2005**).

My initial feelings about the fluxes are confirmed, but the Royal Society emphasize that on centennial timescales the carbon dioxide excess is trapped in the first 200 m of surface waters with deeper exchange only possible on very long timescales. Monitoring shows that since pre-industrial times the acidity of surface waters, as measured in terms of free hydrogen radicals, has increased by 30%.

This figure appears alarming at first, but the key questions relating to biological effects are (1) what is the natural variability regionally and over recent time? and (2) what is the longer-term geochemical variability? The first of these questions is required to elucidate what type of signal is being picked up and its causal relationship to atmospheric carbon dioxide, and the second relates to the natural adaptability of ocean organisms to acidification.

On the first of these questions, researchers reckon to have observed a 0.1 unit shift in global average PH values when the natural variability of the ocean ranges from 8.2 in the Arctic seas to 7.9 in zones of up-welling, such as the eastern Pacific and Arabian seas. The global average is $8.2+/-0.3$. The 30% shift is an estimate because good long-term data on natural trends are absent. On the second issue, proxy data show that during the Cretaceous CO_2 levels were nine times current levels but the oceans apparently managed to buffer this with additional releases of carbonate from the sediment store. The RS team argues that at this time changes were slow enough for the oceans to buffer, whereas the current build-up is too rapid for this to happen. I am not convinced by this argument, since the surface waters reach equilibrium with the atmosphere on much shorter timescales than deeper ocean mixing and this would therefore require a larger bicarbonate buffer concentration derived from the sediment stores and, as far as I can tell, there is no evidence for this.

The RS team predict that by 2100 pH could fall to an average of 7.9. Ordinarily this would be an unacceptable risk to impose upon a global ecosystem, but before considering policy responses we need to be aware that (a) that projection is based upon IPCC scenarios of future emissions, (b) the predicted average level is within the natural spatial variability, and (c) the apparently vulnerable ocean ecosystems consist of very old organisms from an evolutionary perspective – they have coped with far higher levels of carbon dioxide, and one would expect a certain resilience.

However, the team identify coral reefs as particularly at risk as they already occur in the more acid regions of the ocean. The risk may be real, but it is also a long-term issue (we shall return to the urgency question of policy response in Part Two). From my reading of the Royal Society review, I am not entirely convinced that the signal found is due to anthropogenic emissions – the team note that warming surface waters outgas CO_2 and this paradoxically *increases* acidity. The signal that has been observed will thus be compounded by natural warming as well as CO_2 absorption. Cooling oceans will absorb more CO_2 and alter this dynamic. The Royal Society team note that the science is very uncertain and requires much more work. I have no doubt that a doubling of carbon dioxide levels will alter surface water chemistry and that it is better not to risk perturbing the natural cycles and buffers, but there are also considerations of costs and timescales relating to the carbon economy and any policy requires a better understanding of the risks. There is no clear emergency that would effect the conclusions I draw in relation to a longer-term fossil fuel policy, which I outline in Part Two.

In conclusion

With regard to predictions, we have two possibilities that are worthy of scientific investigation and discourse: on the one hand, computer-based projections of a runaway global warming; and on the other, the prospect of imminent global cooling and a new Little Ice Age. In my view, the theory of runaway warming is not supported by the recent evidence of the global response to a 35% increase in carbon dioxide. On the other hand, recent observational evidence lends support to the thesis that ocean temperatures are driven by cloud changes which themselves are determined by the solar cycle. There is mounting evidence that the solar cycle is in decline and we can expect a cooler world. In Part Two we shall examine the policy implications of these two possible outcomes, along with why the greenhouse theory has taken such a dominant position despite the accumulating evidence of error.

Part Two

THE POLITICS

Introduction

'It is difficult to get a man to understand something when his salary depends on his not understanding it'
Upton Sinclair

In Part One we examined the nature of the false consensus portrayed by the IPCC. In several key areas of the science we saw how serious disagreements had been covered over during the progression from scientific discussion in working groups to finalization of the Summary for Policy-makers and entry into the political world. In Part Two we now look at the construction of this false consensus and its subsequent use in the realm of policy. We address issues of motivation, belief and bias on the part of those who have summarized the science, issues of collusion within the institutions of science, and the role of the media, lobbying groups and politicians. Only by exploring the full nature of the error can we expect to correct it.

The interaction of science and politics

In science, the interaction between established theory and its challengers is played out in the peer-reviewed literature, seminars and conferences. And although by no means free of heated emotion and cooler, often hidden, vested interests, the process is relatively open and honoured by time. All students of science are encouraged to think critically and examine the authority on which established views are based, and apart from radical shifts of perspective this evolutionary process is driven by the arising of new data that does not fit with old theory. Politics enters the game if the object of scientific enquiry seriously affects ordinary lives, consumer behaviour, business interests or government investment – with the latter extending beyond the financial to include the psychological elements of standing, status and face, all of which governments aim to conserve with some dedication.

There is, however, a new player in this theatre of operations in the form of the environmental campaign group. Such organizations now wield some considerable pressure on government as well as aiming to influence industry and consumer. They too have a psychological component to their motivation, often fired by a sense of mission to 'save the planet' and concerned with longer-term survival issues; they also have to keep their subscribing public engaged, and issues of standing, status and face apply as well.

I make these comments not out of cynicism. I have worked closely over the last 30 years with all of these major players, having advised government on thorny public issues ranging from nuclear waste disposal to renewable energy policy. I led Greenpeace campaigns as an advocate on the floor of the UN's conventions for the protection of the oceans; I also visited innumerable science labs, lectured in universities around the world and worked at grass-roots level for communities and local governments and often enough against their central government's interests. Thus, my own career has been steeped in this process of debate and involved in the interpretation and presentation of science at the level of policy in several fields closely related to the science issues we are dealing with. I have observed that when left to themselves scientists resolve their differences, theories come and go, and among their peers minority voices are given respect. Most scientists appreciate that science evolves very largely through this minority voice.

Once an issue becomes politicized, however, scientists are not left to themselves and I will slightly switch 'hats' from my ecological training and the critical review of environmental science, models and policy to that of social anthropologist, and reflect upon the social processes of science, policy and propaganda that underlie the climate debate. In this regard, the defenders of the orthodox view are as much in denial as those they accuse as climate-change deniers and where all reference to these underlying processes is dismissed as 'conspiracy theory'.

The stock response of those who cannot believe the IPCC could be in error is: 'How could so many thousands of well-meaning climate scientists have conspired together to provide such a false consensus?' Well, of course, they didn't. There never was a consensus among scientists. As we have seen, from the very first IPCC meeting there was disagreement even on the most fundamental aspect of the atmospheric model for carbon dioxide. Instead a smaller group of functionaries has presented a single voice.

When all of the current disagreements in climate science are added together, and the new data given sufficient weight, an alternative theory that downgrades the CO_2 model can provide a coherent explanation that better fits the current downturn in global temperatures. But this runs up against the massive global prior commitment that the IPCC, the media and the environmental groups have spawned – hence, instead of a rational discussion, we have a propaganda war.

The lack of consensus that has been hitherto hidden within the working groups of the IPCC has taken many years to emerge and influence the scientific community at large, and a number of petitions and declarations involving professionals in the climate field demonstrate that the current consensus claimed by IPCC is not as all-pervasive as it at first appears.[1] There is only consensus within the field of computer simulation. This community has large numbers of working specialists but centred in a few institutions worldwide. Another tier of modellers exists outside of the hands-on climate specialists and this set utilizes the climate models' output for other forms of prediction – such as flood defences, biodiversity responses, agricultural productivity, etc., and it is not common for any of this second tier to question the validity of the primary models. These two tiers make up the significant and powerful vested interest of the science community.

It is *this* agglomeration of interests that has developed around computer simulation that is presented as if it is a consensus of the world's scientists. It may well number a few thousand, but it ought to be immediately apparent that such numbers should count for very little – not because there is an obvious vested interest, but because the science of modelling has been so unreliable in the past. The problem on a political level is that very few people, even within the institutions of science, realize just how poor a record environmental modelling has.

In the following chapters I will show that the current climate debate has all the hallmarks of previous controversies – although, oddly, the environmentally 'good' and the environmentally 'bad' appear to have exchanged sides. In previous controversies such as acid rain, toxic chemicals and nuclear risks, the modellers colluded with well-developed industrial vested interests and their opponents represented the oppressed communities (constituting, at times, groups of nations as in the case of Pacific islanders fighting against nuclear testing).

In the current debate, the critics of the models are accused of being in league with big oil interests, or being adherents of liberal free-market economics and opposed to carbon taxes and other restrictions; and environmentalists occupy the same camp as governments as well as some apparently reformed industrial giants (such as AMEC, BP Solar, Siemens) and the science institutions. The media very largely support the environmentalists, except for the right-wing press which obviously sides with the apparent interests of the industrial free-market economy.

The necessary evolution of safeguards

In the following chapters I want to examine how the normal processes of peer-reviewed science can have become so subverted. I regard the current situation as quite unprecedented in the history of science and environmental politics and I want to offer some thoughts on potential safeguards. These comments parallel my experience of marine pollution issues in which I also developed a critique of the UN system and its failings, together with ideas for future safeguards, some of which are clearly still relevant.

I will also look at the question of urgency with respect to policy. There is great pressure from environmental groups upon governments to commit to mitigation policies that in my analysis will have no significant effect on climate over the crucial next five decades and probably not a great deal thereafter. Global cooling or indeed even small fluctuations in the natural cycles have serious implications for food supplies. The northern grain-belt currently subsidizes 67 countries in food deficit. Biofuel strategies aimed at mitigating temperatures in 50 years time will compromise the ability of the world to feed itself. In my view, there is no urgency with regard to energy policy. The urgency relates to creating systems resilient to climate change, especially in the supply of food.

Finally, I reflect upon the erroneous thinking and analysis that has created this debacle. I do so more as an anthropologist than ecologist. Science has become mired in a polarized, left-brain view of the world, which even within its own terms of reference has made it blind to the operation of cycles, irregular periodicity, the mathematics of spirals and elements of chaos. It would be trite to say that these are all elements of a feminine mode of thinking but nevertheless true. But the problem goes deeper than blindspots and gender issues in science. We are dealing with apocalyptic elements of religious intensity at a time of shifting paradigms and world-views. And in this we desperately need a different way of relating practically to natural cycles of change as much as we need a different way of thinking about them.

Note

1. The Oregon Institute of Science and Medicine *www.oism.org* in a petition led by Frederick Seitz, former President of the US National Academy of Sciences, now has over 30,000 signatures from scientists who object to the IPCC's summary of the evidence. The Manhattan Declaration is a petition to the UN by more than 100 concerned scientists and policy experts. The Leipzig Declaration is a petition by atmospheric scientists. Collectively there are thousands of scientists who disagree with the IPCC view, among them key members of the IPCC Working Groups, expert reviewers of IPCC material, climatologists and professional meteorologists. The most recent initiative involves 650 climate scientists in a petition to the US Congress (US Senate Committee on the Environment and Public Works: *Marc_Morano@EPW.Senate.gov*).

11

Virtual Realities

In the science section of this review we saw that despite the apparent worldwide evidence of climate change the proposition that this change was unusual, unprecedented and attributable to human agency cannot be validated at a level appropriate for scientific theory. The case for anthropogenic global warming has rested entirely upon computer simulations carried out by a specialist group of climatologists who hold that the observed warming pattern cannot be explained by natural factors acting alone and, as we saw in IPCC's key graphic (Fig. 1), the computer simulations predict that under entirely natural circumstances global temperatures would have fallen slightly over the period 1950–2000. Yet, in the body of its technical reports the IPCC documents the limitations and uncertainties in all of the key areas of natural climate fluctuation – in particular, oceanographic cycles, the variability of solar visible radiation, the potential for solar-cloud interactions and the cloud feedback in general. There is also accumulating evidence from the palaeoecological record of long-term cycles. Thus it was concluded that the claim by modellers of enhanced confidence in their ability to replicate the last 50, 100 or even 200 years of climate change is not supported by the real-world evidence. It follows that any attempts to predict the future based upon these models are equally unreliable.

We have seen that in the construction of what is in effect a *virtual* climate, the apparent success of the models in hind-casting (predicting the past) was illusory, with two key errors: in assuming that the global dip in world temperatures from 1945 to 1978 was due to atmospheric pollution (which led modellers to develop an erroneous sulphate aerosol model that simulated the pattern); and in over-estimating by 200% the accumulation of heat in the upper ocean – a factor which strongly affects future predictions.

Global cooling

Many of the criticisms of this virtual climate model have been borne out by the failure of models to predict the current downturn in global temperatures consequent upon a phase change in the Pacific Decadal Oscillation. There is, however, evidence that such phase changes are timed by solar variables, and at least one solar specialist, Theodore Landscheidt, successfully predicted this downturn.

As this review goes to press (in the early months of 2009), almost all past predictions of the solar cycle have been abandoned, despite the array of computing power and satellite monitoring available to NASA. Since 2006 there has been a 'quiet sun', with the solar wind losing power, the geomagnetic field declining and the cosmic ray flux rising. There is evidence of increased cloud cover and rising albedo. The PDO cold cycle is beginning to affect the western Arctic, with 30% more ice cover in October 2008 compared to the record low of 2007. Alaska is cooling rapidly and North America is experiencing a very cold winter. In western Europe, a shift in the jet stream has brought torrential flooding in the summers of 2007 and 2008, and the winter of 2008/09 has seen much colder weather in the Atlantic maritime region as higher pressure in the Arctic blocks the warm westerly air currents.

Meanwhile, the modellers at NASA and at the Hadley Centre, two of the main modelling centres, maintain faith in their assessments, regarding the current cooling as a 'blip' in what is otherwise a generally upward trend and the latest simulations apparently confirm the previous attribution studies despite the obvious lack of understanding of these natural cycles (**Rosenzweig et al., 2008**). We thus have a major disjunction between the realities of climate change, new data and new science, and the virtual realities of climate simulation upon which predictions of impending future disaster are based.

For many laypeople, this state of affairs is almost impossible to understand and it is inconceivable that the world's scientific community could have erred on such a scale. It is perhaps equally inconceivable that the vast array of climate commentators in the media and environmental watchdogs, political groups and various 'green' alliances could have failed to critically review the scientific basis. In this chapter I will show that such a scale of error is far from unprecedented – in fact, it is has happened on a number of occasions on just this scale on major environmental issues involving computer prediction. A great deal of collusion goes on to disguise the past history of environmental science, with errors covered over and policies defended in the face of conflicting new data.

The science of virtual reality

When faced with the enormous complexity of the natural environment and the need for prediction, science generally turns to 'models'. In some sense, science has always done this. It appears a powerful method to control the elements at play in a particular situation – for example, by holding a group of factors constant and exploring the effects of varying one of them. This works well in laboratory science, and experiments on this scale can generally be replicated by other laboratories. When, however, the global atmosphere and oceans are involved and computer models replace the physical reality with a virtual one all manner of problems beset the process of science.

Atmospheric models used by climatologists thus become virtual environments and work by simplifying the world into linked compartments or boxes; each box has a multiple of mathematical relationships with other boxes all devised to simulate material or energy flows through the complex system. In the modern generation of 'coupled' ocean-atmosphere general circulation models (GCM) predictive science conducts virtual 'what if' type experiments by manipulating the parameters of the model. The predictions of the IPCC are based upon comparisons of about a dozen complex models located at various institutions around the globe. This type of model approach has been widely used in the past to predict the consequences of pollutant discharge to rivers, oceans and atmosphere.

Whilst such complex models are of great value in attempting to understand the dynamics of complex systems, they have serious limitations and a poor history of success. These weaknesses became apparent in their past use to justify policies related to the atmospheric emission and coastal discharge and dumping of toxic substances. It is a history dogged by failure and provides some important lessons and perspective on the current climate models.

These weaknesses range from problems with the quality of input data, the constraints placed upon data assimilation by the requirements of mathematics, the choice of concentration and amplification factors (determining the rates of transfer from compartment to compartment), the strength of feedbacks, the choice of variables (and handling of dissent within the expert community and scientific literature when key policy decisions are at stake), and the collecting and analysis of data that may run counter to the model's predictions, particularly when policy and resources have been committed on the basis of previous prediction.

Lessons from the past history of environmental models and prediction

The now widespread use of such 'models' as a predictive tool in science was born of the mid-century electronic revolution and rapidly became the preferred approach to dealing with complex environmental phenomena such as energy flows, feedbacks, the tracking of pollutants and effects upon populations of plants and animals. The complexity of such models and their ability to mirror the myriad pathways in the web of an eco-system, paralleled also by the evolution of the computing power upon which they are dependent, appeared to offer an ability to simulate the global ecosystem itself.

It is important to bear in mind that this apparent capability was built upon the transference of a particular mindset from an entirely different discipline – that of electronic engineering and systems theory. A fashionable new branch of ecology developed using the electrical meta-phors of flow and the mechanical concepts of heat and stored energy, such that this virtual reality rapidly compartmentalized what then became the *eco-system*. This systems approach developed in concert with approaches and methods in the fields of cybernetics and owed a great deal to the insights and metaphors consequent upon electronic theory and technol-ogy. Equally, the advent of complex global models relied upon the rise of computing power and the capacity to model complex systems of flows and feedbacks. However, the limitations of this mathematical and digital creation of computerized 'model' realities are seldom explored, and most particularly by those scientists who build the models and proselytize on behalf of the discipline. For example, the physics of light and energy transfer, heat flow and storage is well developed, but that of solar-terrestrial magnetics, atmospheric electricity, ionization, cloud formation and cyclic phenomena is not, and this presents an immediate bias and limitation in modelling the real world.

This bias is inherent in the very basic principles of the coupled model and the way it uses *iteration* – many runs from the same initial starting point in order to gain a spread of results that reflects the operation of chance as well as complex variables and interactions. This means that a starting point is chosen for an equilibrium state that does not exist in nature. This method downplays (and may leave out altogether) the uncertain very long-term cycles that have accounted for past variability.

The discipline has not been quick to parade past or present limitations, but they stretch across the history of environmental assessment from the fate of organochlorine pesticides, radioactive isotopes and heavy metals in the marine environment to the origins and consequences of acid rain, CFC dispersal and ozone depletion. It was largely the *failure* of this predictive science that led to the revisions of international law relating to the burden of proof, the precautionary principle and the switch to clean production methods, rather than continue with dilute-and-disperse policies and the failures of monitoring and prediction (**Taylor, 1987; 1993a; 1993b; 1993c**).

Having been involved in the critique of modelling at the level of UN conventions and treaties and also in the running of computer modelling exercises (in the dispersion of atmospheric pollutants), I have an extensive range of experience of how models are constructed, how inputs are chosen, how mathematical functions essential to these virtual realities are selected and how they approximate to physical realities – as well as how such models can be 'tweaked' to perform and deliver outputs that may be more in line with ecological expectations and, not uncommonly, the expectations and interests of commercial and political 'interested' parties.

I am also in a unique position not to care too much about offending scientists within my own discipline. I have been fortunate to be able to lead a relatively independent existence running my own institutions and choosing my own research areas (and I should emphasize that, in addition to independent reports and advice, I have always published in the peer-reviewed literature). My independence is strengthened by currently working in the unrelated disciplines of restoration ecology and biodiversity. I make these points because they are important truths in the real world of politics, funding and research institutions, and they have a direct bearing on the 'science' of modelling and the operation of so-called consensus science.

I will relate a small example of the importance of an independent status from my early involvement in climate policy. I detail this particular experience because it illustrates some of the points I make about the limitations of modelling as well as the role of independent advisor. In 1996, long before I came to question the basic carbon model, I became involved in a review of the climate modellers' ability to generate reliable predictions with regard to the timing and onset of environmental effects. An earlier understanding of the science had led me to believe that it would take a doubling of carbon levels well into the next century before a human signature would be apparent, simply because of the amount of natural variability and 'noise' within the system. Yet, to the untrained eye, there were clear indications of climate change in the latter half of the twentieth century when carbon dioxide had only increased 35% above natural levels. It was clear to any ecologist that changes were happening much more rapidly than expected, and government agencies were turning to climate research units to answer questions regarding future policy.

In the previous year, the Countryside Commission, the UK government agency for rural affairs, had asked Professor Keith Clayton of East Anglia University (home to a specialist climate research unit) to review the issues of atmospheric pollution and climate change as they might effect countryside policies (**Countryside Commission, 1995**). His report leaned heavily upon the infant modelling techniques, but it was of limited use to the Commission because it could give little practical information relevant to their managerial responsibilities. Having a greater knowledge of the Commission's policies, I was asked to review the issues (**Taylor, 1996**).

As expected, Clayton relied upon simple models and predicted a continuing trend of warming for the British Isles such that by mid-century it might experience a climate similar to that then found in northern Portugal. As an ecologist with some experience of oceanography I was aware of the propensity of the Gulf Stream to weaken and move south on a periodic basis and that this was one of a number of potential cooling climate feedbacks in an overall warming world. Clayton made no reference to the phenomenon or to then recent Norwegian oceanographic reports that the circulation in the 'conveyor' belt (that part where the northern waters sink) was slowing down (by 20% over 50 years as we saw in Chapter 5).

This approach demonstrated the general rule that modellers try to avoid giving their clients' alternative outcomes that reflect major uncertainties as it tends to undermine the aura of sophistication regarding computer models and somewhat defeats the point with regard to massive computations aimed at greater accuracy. However, there was also no evidence that Clayton's team was even aware of the Atlantic conveyor's changing speed.

In terms of my own advice to the Commission, and not being constrained by such mathematical models or the need to talk up the potential future of climate modelling, I could simply advise the Commission that the future was uncertain and unpredictable − temperatures could go either way − and however sophisticated modern computer techniques were they were subject to human error, and it would be prudent to develop policies that were robust for a change in either direction, warming or cooling.

No-regrets policies and the concept of resilience

We then pursued many discussions within the Commission on 'no-regrets' policies and the building of climate resilience into agriculture, forestry, water, wildlife and recreational services, and indeed, this was also the direction I took for my subsequent work in biodiversity policy (Taylor, 2005).

In a later project with the Commission's successor, the Countryside Agency, my group at Ethos worked on visualizing landscape changes connected with climate and renewable energy (Taylor & Fraser, 2002). My team developed advanced virtual reality models that gave a visual perspective on landscape change and the impact upon rural communities. It was immediately apparent that the greatest degree of landscape, community and biodiversity changes would come from policies that sought to derive energy from the landscape itself and that these impacts would be far more extensive than the impacts of temperature or rainfall changes.

We assessed future energy demands as well as landscape change and in one scenario we looked at a potentially colder Britain. We did not say this was likely, merely possible with a high degree of uncertainty. The work was presented at a governmental Inter-Agency Meeting in 2003, when I demonstrated part of the visualization-of-change study that included a cold scenario where the conveyor slowed down. My presentation drew immediately dismissive comments from a representative of the Environment Agency (their head of climate policy) who assured me that a UN committee had reviewed the science and advised that such a shut-down could not occur within the next 200 years.

I asked the intervener if she had looked at the data on the slowing of the down-welling currents herself, or was relying upon the authority of the UN. She had not seen the data. In my earlier review of 1996 I had examined the graphs and noted that either linear or exponential extrapolations were consistent with the monitoring data – only time would tell which. I suggested that the Environment Agency might usefully carry out their own critical evaluation of this important issue and outlined why the UN might not have an interest in publicizing alternative analyses that would indicate a major impact upon regional economies and within the time horizon of modern investment and speculative commodity dealing strategies.

The Countryside Agency representative assured the meeting that they stood by my analysis (although at a higher managerial level I think they then decided that such work would not fall within their future remit!). My analysis was vindicated only weeks later when senior members of the US oceanographic community warned in a BBC *Horizon* programme that a slow-down was a very real possibility on timescales much earlier than had been considered. On my own reckoning, if the data concealed an exponential, shut-down was possible before 2020.

As I discussed in the Chapter 6, I followed this issue to more recent times, discussing new data with Professor Harry Bryden at Southampton University. If Bryden's data alone was linearly extrapolated then shut-down was as close as 2033; however, if the data from the previous survey was combined and extrapolated as an exponential function it reached a spread of end-points with the earliest at 2012, a timescale considered possible by the US specialists.

At that time, I was personally not convinced by Bryden's argument that the two data sets should not be combined — and clearly, senior US oceanographers were also not convinced. In the course of writing this book, however, I have realized that all attempts at extrapolation from short-term data are flawed and that major cycles are at work. Oceanographers admit that their understanding of how these cycles affect the conveyor is limited, but the issue raises important questions for science in relation to policy. Should oceanographers sound an alarm when data is so uncertain? And what policies could be developed based upon uncertainty? My own response has been to argue for the development of *resilient* systems, which I discuss in Chapter 15.

This issue also illustrates a recurrent historical theme concerning the high investment in models and computing power compared to the sparse investment in the gathering of basic real-world data. I am shocked at the paucity of monitoring and resultant data sets for what must easily be the most crucial climate feedback on the planet. This theme is repeated with regard to other data sets such as ocean heat content, cloud cover and satellite data on solar SW/LW flux.

This is an important issue. There is a massive worldwide investment in computer modelling estimated now at several *billion* dollars, having risen from a few tens of millions in 1990, yet the resources devoted to basic monitoring (and validation of the models) are tiny by comparison. This is a repeat of the problems encountered with pollution models. The models were very sophisticated in their construction, demanding huge amounts of computer time on the most powerful machines and employing ever-expanding ranks of programmers, students, supervisors and technicians, yet they ultimately proved to be incapable of effective prediction.

The limitations of modelling

The limitations of these models and failures of prediction revolved around inadequate monitoring strategies and the statistical robustness of the derived data. These areas of science are not fashionable and often subject to funding difficulties, yet they are vital to gathering data relevant to particular assumptions and hypotheses. There was considerable scope for selective monitoring and, in particular, *not* collecting data that might prove embarrassing and require the revision of established models and the practices licensed by their conclusions. If an effect was not predicted by the models, then nobody looked for it. Conversely, prediction of effects guided the monitoring towards finding those effects.

In the early decades of environmental protection when vast numbers of chemicals had been licensed for production and disposal, it was incumbent on the regulatory body or concerned citizen group to prove 'harm' by gathering such data and in many cases the regulatory body had already participated in the licensing process. Regulators often had quite friendly professional relations with chemists and engineers within the industries they regulated, and governments were not motivated to disturb the economic environment of key industries.

This cosy situation led to widespread problems as data began to accumulate on the toxic effects of discharged substances. The classic example is DDT and the thinning of eggshells in birds at the top of the food chain. Populations of eagles, falcons and hawks declined drastically throughout the northern hemisphere. Eventually this substance was banned, though with a let-out for southern hemisphere and tropical countries faced with malaria and unable, apparently, to afford its more expensive replacements.

Other chlorinated hydrocarbons had been released including the relatively inert and long-lasting PCBs. After its manufacture on an industrial scale and unconstrained release at disposal, data began to show that the substance was carcinogenic and its industrial use was subsequently banned. However, it continued to be released from uncontrolled landfills and circulate throughout the atmosphere as a volatile organic substance. By the 1980s it was suspected of causing immune deficiency in marine mammals and had spread to the point of serious contamination of human food supplies in hitherto pristine regions of the Arctic. Neither its eventual toxicity nor its propensity to be transported as a volatile substance from landfill in warm regions and to be later distilled into cold environments had been predicted by the models that justified its initial release.

A similar history involved the manufacture and unconstrained disposal of CFCs (chlorofluorocarbons), which were initially regarded as the least toxic substance known to chemists. Its future role in stripping the ozone layer was completely missed by those who licensed its production and dispersal. Concern is now arising with regard to the biological effects of compounds similar to CFCs that contain bromine instead of fluorine, commonly used as fire-retardants and similarly mass-manufactured and disposed without constraint. These chemicals are extremely long-lived contaminants and data is now accumulating on their 'gender-bending' properties in fish, amphibians and perhaps also mammals.

The problem of prediction was not confined to man-made substances. Many discharges were of naturally occurring heavy metals such as lead, cadmium, mercury and arsenic, which can be toxic at certain concentrations; and other substances such as iron or copper which could alter ecosystems if discharged in sufficient quantity. Nutrients such as nitrate and phosphate had the same capability. It is seldom appreciated that modern agricultural practice 'dresses' fields with nitrogen and phosphorous and that this causes an almost complete collapse in plant species diversity, with herbs and flowers disappearing to leave productive grasses. 'Pollution' is thus not readily defined, even when these nutrients 'enrich' river and estuarine systems, causing similar shifts in animal and plant populations and species diversity.

It was against this background that the precautionary principle was developed — a presumption against discharging or dispersing any suspected toxic substance even when there was no scientific proof of harm. This constituted a reversal of the 'burden of proof' which had aided industries in resisting rising doubts and indications of problems, as well as complex definitions of what constituted 'harm'.

As can be appreciated this was resisted by industry and most governments, and it took the engagement of environmental campaigners, particularly Greenpeace International, to push for change. I was heavily involved in this scientific, legal and diplomatic struggle in the lead up to the Rio Summit when the principle was adopted. At that time a small group of active scientists rewrote the global and regional ocean protection conventions; the process was eventually completed by 1993 (**Taylor &**

Jackson, 1991a; Stairs & Taylor, 1992; Jackson & Taylor, 1993; Taylor et al., 1993; Taylor, 1993).

A study of this historical process would not cast the scientific institutions in a good light. Although monitoring laboratories generated the conflicting data the public face of science opposed the new paradigm of precaution, with the greatest resistance centred within the professions involved in the regulatory process. They had been wedded for several decades to the old paradigm of 'dilute and disperse' and the emergent field of ecotoxicity was built around predicting the fate of substances that were released. Needless to say, a lot of this work arose from the study of substances that had *already* been released.

The problem with this whole approach was that environmental science simply could not cope with the complexity of the environment nor anticipate all of the negative effects. The problems were legion: toxicity studies missed effects such as suppression of the immune system and gender-change, carcinogenic potential was evident often only after populations had been exposed, and behavioural changes were almost impossible to identify or quantify. These were important issues for human exposure to mercury, lead, arsenic, cadmium and the organic chemicals. Additionally, ecosystem-wide effects were becoming evident from nitrate and sulphate deposition (acid rain) and of course there was the saga of CFCs and the ozone layer. Science could not predict toxicity and, even if it could, it would seldom predict the scale of change, its onset or rate of development.

The process of 'tweaking' the models

The panoply of institutional bias and defence against criticism meant that modellers could become seriously out of touch with progress in key scientific fields and miss vital components of the complex ecosystems they were modelling. On one example I dealt with, the first oceanic dispersal models for radioactive pollutants assumed that detritus falling as a constant rain upon the ocean floor simply joined the sedimentary layer, thus burying and isolating any dumped material from the rest of the ocean environment. For particle-reactive chemicals the model ocean was effectively self-cleaning. However, new oceanographic research identified a turbulent *mobile* layer above the sediments, such that re-suspended particles could travel in a continuous and long-lasting moving carpet that dispersed and deposited material many hundreds of miles throughout an ocean basin. The former inaccurate model had been used to license dumping operations, and when real-world data challenged the validity of those licenses it was resisted both by the science community that developed the model and the licensing authority that carried an overall responsibility to industries that had been allowed to dump (**Taylor, 1985a**).

The models could also be tweaked in more pro-active ways. In 1983 the UK Government had been pursuing a policy of low-level radioactive waste dumping in the North Atlantic south-west of Ireland and north of the Azores. My early work in reviewing models of dispersion had been supported by both Irish and Spanish Governments as well as Greenpeace, and as the political environment changed I was appointed to the UK government commission set up to investigate the science. In the course of that review I demanded sight of the International Atomic Energy Agency's draft environmental assessment that was being produced for the UN's London Dumping Convention. As the review period closed, the laboratory concerned (Ministry of Agriculture and Fisheries laboratory at Lowestoft) issued its final document prior to delivery at the UN. Having seen the earlier draft I noticed that the main impact, the point at which dispersed radioactivity would surface in active fisheries, had shifted from the politically problematic Norwegian Sea to the Antarctic Peninsula (**Taylor, 1985; Taylor & Jackson, 1991a; Taylor, 1993**).

This unpublicized change was later justified by the MAFF team as a result of a reassessment of the rates of diffusion at the North Atlantic conveyor site – leaving the offending toxic nuclides to reside in bottom currents and surface 500 years later in the South Atlantic. The shift had been made over a three-month period that happened also to coincide with the first statement of concern by Scandinavian environment ministers regarding UK dumping policy, which was due for review at the UN's convention on dumping at sea at which these countries were strongly represented. Delegates were not impressed when informed of MAFF's late tweaking of the model and dumping was banned until further research could be done. It remains banned, with all offending states that contributed to the UK-led dumping operation having found land-based storage sites.

I could enumerate dozens of such examples and have written extensively on this issue of models and their limitations. I have summarized these arguments as they affect the policy-making process in chapters in *The International Politics of the Environment* (the proceedings of Oxford University seminars and published by Clarendon Press) and in *Clean*

Production, edited by Tim Jackson (**Taylor, Dethlefsen & Jackson, 1993b**; **Stairs & Taylor, 1990**), in my critique of the UN's system of science committees and marine protection (**Taylor, 1993**), and in relation to monitoring and environmental discharges of radioactivity (**Taylor, 1988**).

However, the major point to note is that the institutions of science resisted the new paradigm. The old guard insisted the precautionary principle was 'political' and not 'scientific', failing entirely to acknowledge or understand their own hidden set of value judgements and the *limits* of their scientific methodology. These value judgements played out in the peer review process of publication. Contrary analysis was marked down as having been 'questioned' or 'criticized' or regarded as 'controversial', often with denigrating attitudes such that a fear-factor operated within research teams whenever they came to challenge the status quo. This led to the failure of monitoring strategies where data that might prove awkward were not collected, sometimes deliberately but more often because the scientists concerned did not expect to find any effects and hence saw no reason to waste resources looking (**Taylor, 1988**). The precautionary principle was also misrepresented and lampooned as unscientific even though it had been developed by senior fisheries scientists with extensive experience of pollution control working in German and Scandinavian ministries (**Dethlefsen & Jackson, 1993**).

The resistance of science institutions is not hard to understand, but a complex of factors are at work.

1) Many scientists had spent their working lives under an assumption that the work they were doing was of benefit to humanity (this especially applied to the work of nuclear scientists) but could not see that they were making all manner of non-scientific value judgements and trade-offs concerned with perceptions of the value of certain industrial activities (many did not share some of the values relating to 'clean' food and environments free of contamination prevalent among consumers). Few were appraised of risk assessment studies and the locus of benefit (i.e. the issue of who benefits from the industrial use of the substance and who is disadvantaged by the discharge) and thus there was a psychological barrier to accepting the new paradigm.

2) Some scientists in key positions considered that 'science' itself would suffer if a new paradigm involved political and social judgements. But they were incapable of seeing their own hidden agendas of political,

economic and social allegiances or – as I suspected in a few cases – were well aware of what was at stake but played to the gallery of scientific institutions.

3) There was an issue relating to jobs. Large numbers of scientists were employed in toxicity testing and monitoring programmes, as well as computerized models of prediction.

On this last issue, I dubbed the complaints that could be heard as the pendulum began to swing 'Goldberg's Lament' after Edward D. Goldberg, the oceanographer who specialized in tracing radioactive pollutants (or contaminants!) and who complained that as countries began to abandon ocean dumping programmes, or plans to do so, he could not recruit enough PhD students or fund the complex research and monitoring projects upon which the whole discipline of radioecology depended (at the time I was a member of the International Union of Radioecologists and on the editorial board of the Journal of Environmental Radioactivity). It was as if my fellow scientists needed a certain level of pollution in order to prosper!

In the decades that followed the adoption of the precautionary principle, monitoring and research programmes began to close, particularly in the field of atmospheric effects of sulphur. Industry saw little reason to do research because it could no longer use predictive science to justify discharges and was switching investment to 'clean production'. In the environmental movement our work shifted from negative criticism to the furthering of positive clean production. In my own work I helped set up the necessary infrastructure and networking of best-practice and could see how the new paradigm offered employment more to engineers and technologists than biologists – even those working as critic! I was happy to move on.

The belief in models rather than real-world data

In the current controversy we are witnessing not just resistance to changing an old paradigm of prediction but also resistance to new data that contradicts the models that underlie that paradigm. There are two historical examples that should alert us to the processes that operate whenever scientific institutions have made key commitments to a model that underpins wide-ranging political and economic interests.

As we noted early on, the first data indicating a loss of ozone over the Antarctic were generated by US satellite monitors. It was so unexpected (apparently, though, there had been warnings in environmental science) that the data were 'not believed' and assumed to be instrument error. It was the less sophisticated British Antarctic programme's ground monitoring that confirmed the losses and this programme had very nearly been shut down for lack of funds the previous year. This illustrates the propensity of the monitoring community to narrow their focus according to the expectations of the model (which may translate into inadequate or absent monitoring in areas that could be important) and to disbelieve data that contradict the expectation.

The other example relates to the X-raying of pregnant women. In the 1950s the radiological community, rising upon a wave of new technological applications in medicine, had developed a model for radiation effects and dose-limits. That model assumed there was a threshold for physical damage to cells and that low-level radiation was safe. The profession and its committees, including the UN's special committee on the effects of ionizing radiation and the UK Government's radiological protection service, with its members enjoying some status within the Royal Society and other learned academic bodies, showed no interest in following up the health of X-rayed women through epidemiological research. After all, the model showed no reason to do so.

The first indications of a problem arose when an epidemiologist, Alice Stewart at Oxford University, having been alerted by doctors to the rise in infant leukaemia cases, gained funding for an epidemiological survey and showed a direct link between the infant leukaemia and the X-ray dose. The data were not believed. Indeed the radiological establishment put up a long fight to discredit the work and Alice Stewart in person, though eventually bowing to the force of repeated surveys. The establishment's chief protagonist, Edward Pochin, was knighted for his pioneering work in comparative risk assessment, and Alice Stewart was shunned and regarded as an awkward maverick.[1]

The last example should provide a timely warning: it was not the learned members and knights of the Royal Society (key members opposed her election as FRS), nor the modellers, nor the UN committees and commissioners who pioneered the gathering of relevant data – their commitment was to the unreliable models. Further, there was considerable economic, institutional and psychological investment in those models. Even had they been motivated to do so, concerned scientists within those institutions would have had great difficulty gaining the resources to gather data and question those models.

Lessons for the current problem

The problem we have with carbon dioxide has some similarities. But firstly it is not so obviously a 'pollutant', being both natural and essential for most life on the planet. The oceans exhale and absorb vast tonnages in an annual cycle, about 90 gigatonnes, and land sources exchange 60 Gt. This compares to an anthropogenic source of about 7 Gt or 4% of the natural flux. However, it seems clear that the natural system cannot integrate this extra input and atmospheric levels are accumulating. There is some recent debate about how much of current carbon dioxide levels might be due to natural sources *following* a temperature cycle, or of long-term outgassing from the oceans. It is a complex area of science which I have not had time to review, but it is of interest that one commentator points out that the current consensus of a rise from 280 to 380 ppm since the eighteenth century relies upon (a) regarding past analytical techniques as unreliable and hence subject to much exclusion of data that does not meet the expectation, and (b) modern analytical techniques developed since 1950. There is therefore the potential to miss longer-term cycles of change (**Beck, 2007**).

However, the current model has the oceans absorbing about 1.9 GtC (gigatonnes of carbon) of the anthropogenic emission and land about the same, leaving about 3.3 GtC to accumulate annually in the atmosphere. Isotope studies appear to show that the majority of this accumulation is from fossil fuel burning (which releases about 5.4 GtC/annum) and this adds one half of one per cent annually to the atmospheric reservoir.

As with any biologically and chemically active substance (such as nitrate or sulphate), precaution demands that we try to limit the discharge. In fact when I was arguing at the UN for adoption of the precautionary principle I made a strong plea to include 'non-toxic' natural substances and to limit them according to the variability over time of the natural flux (**Taylor, 1991a; Jackson & Taylor, 1993**). As might be expected, the move to address natural substances and fluxes was resisted. The old school still believed they could do useful toxicity testing and the pragmatists in league with the economists argued that because of the high costs involved the flux targets were impractical.

Carbon dioxide had become suspect in the latter part of the 1980s and parties to the Rio Summit agreed that emissions should be reduced as a precautionary measure, with targets eventually set by the Kyoto Protocol. My very good friend and colleague Jackson Davis, then a marine biologist at the University of California at Santa Cruz, went on to become involved in the drafting of the Framework Convention on Climate Change and to represent the Pacific Island States. Another of my colleagues from that period, Tim Jackson, went on to play a key role in the development of clean production strategies and the Clean Development Mechanisms for the transfer of technologies. It is thus somewhat ironic that after 15 years I now find myself arguing for delay in programmes of limiting a pollutant!

The situation is rather similar to that which arose with CFCs − there was evidence of harm and the tap was turned off in what had proven an altogether very risky planetary experiment! The difference, however, is that economic stability and industrial strategy did not depend upon CFCs and the cost of replacements could be met. Fossil fuels cannot be cheaply replaced. Programmes of emission reduction have to be handled with great care in relation to the global economy and are complicated by the entrenched problem of industrial and economic competitiveness.

Whilst there can be no question that a significant effort needs to be put into reduction of emissions, it needs to happen steadily over several decades to take account of the practical economic imperatives. I argue this not to protect privilege but because there are social and environmental consequences attendant upon economic instability. I will address these issues of urgency and adaptation in more detail in Chapters 14 and 15.

In my review of the key science areas relating to the current IPCC model I detect similar problems with data, models and the political processes that surround the translation of scientific advice into policy. It is disturbing that I have seen no in-depth critical review of the models used by IPCC by any of the protagonists who, in the past, would have kept a wary eye on inter-governmental science and this despite ample evidence from within the science community that such models should be examined with a critical eye (**Koutsoyiannis et al., 2008**).

Such critical review requires motivation as well as funding. In this respect a recent issue of *New Scientist* (30 June 2007, p. 6) illustrates a point. Concerning the plight of polar bears − much hyped by the media and environmental campaign groups as at risk from global warming − Willie Soon at Harvard University's Smithsonian Center for Astrophysics published in the peer-reviewed journal *Ecological Complexity* a critical assessment of the apparent decline and vulnerability of this species, questioning the hype. *New Scientist* labelled Soon a 'climate sceptic' and pointed out that the research was part funded by Exxon-Mobil, as if that were reason enough to question its reliability.

There should be no stigma attached to 'scepticism' when it relates to climate models or any other construct of science. The very progress of science depends upon such scepticism. And we return once more to the issue of who is motivated to collect potentially contrary data. I make no judgement on the plight of the polar bear except that it is obvious that Arctic environments have warmed in the past. There was a Holocene optimum at about 8000 BP that was at least a degree warmer in the northern hemisphere than today and the polar bear recovered from that. Furthermore, it is evident from recent work on beach structures in northern Greenland that the Arctic Ocean was probably ice-free in the summers of this period (Astrid Lyså, Geological Survey of Norway).

The time factor and prior commitment to previous prediction

As we saw, considerable uncertainty exists in the data series regarding cloud changes over recent decades, and cloud is an integral part of the climate system such that even small percentage changes have major implications for climate change. Recent satellite data show that the whole of the late twentieth-century global warming can be explained by the apparent degree of cloud thinning recorded in most data sets between 1980 and 2000. This data was only available *after* the models were built at considerable expense, as well as political investment. The observed level of cloud change is not predicted by any of the models.

Yet this data is not presented by the IPCC in these terms. Instead the summaries repeat the past graphs and statements that the recent warming cannot be explained without invoking a human emission component from greenhouse gases. It would be more in line with the science and limits of the methodology to say that the recent rise in warming potential is not reconciled in the models, and the whole of the resultant rise in temperature cannot be reliably attributed to carbon dioxide because of these uncertainties and in particular the acknowledged uncertainties in the satellite-derived data for both cloud cover and radiation fluxes. However, this would constitute a world-shattering shift of confidence in the models and would bring into question the ability to predict future change.

There always has to be a balance between evaluating new data and incorporation of uncertainty into policy. In my view the IPCC is not well balanced in its approach and is resistant to changing the models. Scientists far more senior than I and even Working Group members of the IPCC have recently made similar criticisms. David Douglas of the Dept. of Physics and Astronomy at Rochester University, New York and John Christy at the Department of Atmospheric Science and Earth Systems Science, University of Alabama at Huntsville are about to publish a major critical paper on the IPCC's approach to modelling and prediction in *Energy and Environment*, arguing that natural ocean cycles can explain the greater part of the signal, and the residual effect of carbon dioxide is in line with a zero-feedback model. They state: 'These conclusions are contrary to the IPCC [2007] statement: "most of the observed increase in global average temperatures since the mid-20th century is very likely due to the observed increase in anthropogenic greenhouse gas concentrations".'

These problems are not best answered by delay and simply leaving out the offending factors (even when apparently justified by opposing or critical views among peers). The history of science and particularly environmental science shows that its entire evolution has been driven by voices dissenting from the consensus and this has been a particular feature of past UN committees on issues of pollution (**Taylor 1993a**). In the next chapter we shall look at this process of dissent in more detail.

Note

1. I have written about this issue and the period of controversy in my autobiography *Shiva's Rainbow*. I met and worked with Dr Stewart on several public inquiries in Britain (**Taylor, 2004**). A biography, *The Woman who Knew Too Much – Alice Stewart and the secrets of radiation*, by Gayle Greene (1999), is published by the University of Michigan Press. There is a good obituary in the *Independent* newspaper, 9 July 2002, which outlines her struggle against established interests. One of her chief critics at the Medical Research Council, John Reissland, met an untimely death on the eve of what Dr Stewart told me would have been a retraction of his criticisms.

12

Dissent and Consensus

'In my more than 60 years as a member of the US scientific community I have never witnessed a more disturbing corruption of the peer review process than the events that led to this IPCC report'
Professor Frederick Seitz, past President of the US National Academy of Sciences, on publication of the 3rd Assessment Report of the IPCC, 2001

Much is made in the current political world of an apparent scientific consensus; yet consensus is a tool of politics, not science, and in the political realm consensus politics is rare and seldom valued, that world being mostly populated by oppositions and the dialectical struggle for majority and power. In actuality, the institutions of science are equally political. Disagreement and heated debate are common and the history of science is replete with examples of the struggle of new perceptions against the guarded old. It is therefore rather odd that the current highly politicized climate debate lays such great store by an apparent consensus within the IPCC and, by extension, the world of science itself.

However, by continually *asserting* that consensus exists, public attention is drawn away from the disagreements we have noted in key areas of the satellite evidence for cloud changes, solar flux, the magnetic cycles of the sun, cloud seeding, UV light variability, ocean oscillations and upper ocean heat store calculations, as well as the poor record of computer simulation and analytical work and critical review of model parameters by leading specialists (**Camp & Tung, 2007; Compo & Sardeshmukh, 2008; Douglas & Christy, 2008; Friis-Christensen & Lassen, 1991; Hanna et al., 2005; Kininmonth, 2004; Koutsoyiannis et. al., 2008; Lindzen, 1991; Lindzen et. al., 2001; Landscheidt, 2003; Marsh & Svensmark, 2000; Pallé Bago & Butler, 2001; Perry, 2007; Pinker et al., 2005; Polayakov, 2003; Soon & Baliunas, 2003; Svensmark & Christensen, 1997**).

Thus, consensus acts as a political construct formed by the collusion of science institutions with the inter-governmental process. As such it is a potent element in the defence of the old paradigm. We shall look now at the potential motivation for this collusion and the means whereby what is quite simply a majority view based upon exclusion is perpetrated and maintained. We shall then consider what safeguards science has to adopt to prevent this happening in future.

Criticism and the treatment of controversial new data

In the science review of Part One, the following data emerged as among the most crucial scientific measurements with regard to the validation of any predictive power in the models relating to global warming:

i)   the flux of short-wave (SW) radiation at the surface of the earth and its trends; and the flux of long-wave (LW) radiation up from the surface, particularly over oceans;

ii) the percentage changes in global cloud coverage, its spatial distribution and changes in the global albedo index;

iii) the power of cosmic radiation to ionize and create cloud condensation nuclei;

iv) the complex interaction of variable UV radiation over the peaks and troughs of the 11-year solar cycle with the stratosphere/troposphere heat exchange system and the strength and spatial distribution of the jet stream;

v)   the rate of change of the Atlantic conveyor, particularly in the downwelling zones of the sub-Arctic;

vi) the past accumulation and recent rate of change of ocean heat content, particularly in the period post-2000, and major ocean basin oscillations such as the PDO;

vii) the complex interaction of ocean oscillations of 30–70-year timescales in the Arctic, Pacific and North Atlantic; and longer 400- and 1500-year low-frequency oscillation in either the solar magnetic or ocean conveyor cycles.

In each of these areas of measurement there have been recent publications of analyses that would bring the current global warming model and its predictions into serious difficulties and in *every* case these crucial data sets have been criticized, questioned, regarded as *controversial* and, in important instances, data and conclusions have been withdrawn after publication due to 'instrument' error or bias.

In some crucial areas I have been astounded at how *little* effort is devoted to important issues, for example: the six-year delay in setting up experiments to test Svensmark's theories relating to the ionizing potential of cosmic rays; in monitoring related to the rate of change of Atlantic conveyor belt currents; and in relation to ocean heat content studies.

These limited data sets are brought into question, at least where they would cause disagreement with the accepted models, for example as a result of instrument calibration issues or the lack of global coverage, whereas the basic assumptions of modelling are not. In my past work I have witnessed the tenacious resistance of modellers upholding their modelled 'reality' and its predictions in the face of new data from the real world. This new data is 'not believed' because it conflicts with the model and, vice versa, data that does not conflict with the model is not subject to the same critical appraisal.

The treatment of dissent within the IPCC

In the body of the IPCC Report, there is considerably more open admission of these limited and controversial data sets than is reflected in the Summary for Policymakers:

- With regard to the downward trend in reflected radiation, IPCC-4 (Chapter 9) comments (my emphasis in bold italics):

    These observations suggest an overall decrease in aerosols and/or clouds, while estimates of changes in cloudiness are uncertain. The model-predicted trends are also negative over this time period, but are smaller in most models than in the ERBS observations (which are considered more accurate than the ISCCP FD). Wielicki et al. [2002] explain the observed downward trend by decreases in cloudiness, **which are not well represented in the models on these decadal timescales** [Chen et al., 2002; Wielicki et al., 2002].

The Panel admit that:

    Nonetheless, there is evidence that climatic responses to forcing, together with **natural internal variability** of the climate system, produced several well-defined climatic events, such as *the cool conditions* **during the 17th century or relatively warm periods early in the millennium**.

Thus, whilst it is clear that cloud effects are of enormous importance in the climate response to any 'external' forcing agent, whether solar, volcanic, or anthropogenic, it is also the case that no models can adequately incorporate this response. Not only are there doubts about trends in data, there are admitted limitations in the models both with regard to clouds and also with regard to ocean heat storage. We note also that IPCC pre-judges the low frequency cycles of the LIA and MWP as 'natural internal variability' rather than forced by 'external' solar cycles.

- With regard to solar forcing, IPCC-4 (WG1 Chapter 9) comments (my emphasis in bold):

    Uncertainty also arises because the spatial response of surface temperature to solar forcing resembles that due to greenhouse gas forcing. Analyses that make use of differences in the temporal evolution of solar and volcanic forcings are better able to distinguish between the two. In such an analysis, solar forcing can only be detected and distinguished from the effect of volcanic and greenhouse gas forcing over some periods in some reconstructions [Hegerl et al., 2003, 2007], although the effect of solar forcing has been detected over parts of the 20th century in some time-space analyses and there are similarities between regressions of solar forcing on model simulations and several proxy reconstructions [Weber, 2005; see also Waple, 2002]. A model simulation [Shindell et al., 2003] suggests that solar forcing may play a substantial role in regional anomalies due to dynamical feedbacks. *These uncertainties in the contribution of different forcings to climatic events during the last millennium reflect substantial uncertainty in knowledge about past solar and volcanic forcing, as well as differences in the way these effects are taken into account in model simulations.*

Yet it is these inadequate models which are used to identify the 'very likely' anthropogenic signal with such confidence! Models which cannot yet simulate the PDO, ENSO or the North Atlantic and Arctic Oscillations, as well as the solar signals of 11-year or 22-year cycles detected in sediment studies and some surface temperature analyses, are regarded as accurate enough to determine natural events over a 50-year period. Furthermore, IPCC use this 'substantial uncertainty' in relation to past solar and volcanic forces almost as reason to place greater faith in the models.

Given these uncertainties, it is asking a great deal for non-specialists to accept the validity of these models in identifying a recent pronounced anthropogenic signal. Whilst there is some critique in professional circles of the reliability of these models (**Allen, 2006; Compo & Sardeshmukh, 2008; Gouretski & Koltermann, 2007; Koutsoyiannis et al., 2008; Lindzen et al., 2001; Polyakov, 2003; Stone, 1997**) this does not emerge onto the public face of the IPCC or in the pronouncements of the institutes that support its political call for action.

Somewhat earlier in these debates, the US National Academy of Science stood out as a source of public criticism. It was, of course, able to do so within the context of a society that was politically committed to a fossil fuel economy and a global business-as-usual strategy wherein lay great resistance to carbon taxation and global instruments of control. As we noted, Richard Lindzen, professor of meteorology at MIT and member of that first IPCC Working Group, had been critical of amplifying factors applied to the basic carbon dioxide model. As we saw, these factors related to the water vapour ecosystem response which he regarded as not predictable – in simple terms, if the water vapour condensed into cloud it would likely produce a cooling feedback to balance the warming. He was roundly critical of the committee and chair procedure that overruled or bypassed his concern, and as a member of IPCC he testified to the US Congress as part of a policy review by the US National Academy of Sciences, of which he is also a member as well as being one of eleven authors of a 2001 NAS report on the science of climate change.

In regard to the latter, he stated in evidence to Congress:[1]

The NAS never asks that all participants agree to all elements of a report, but rather that the report represent the span of views. This the full report did, making clear there is no consensus, unanimous or otherwise, about long-term climate trends and what causes them.

Our primary conclusion was that despite some knowledge and agreement, the science is by no means settled.

...we are not in a position to confidently attribute past climate change to carbon dioxide or to forecast what the climate will be in the future.

...distinguishing the small recent changes in global mean temperature from the natural variability, which is unknown, is not a trivial task.

He argued that contrary to media impressions (gained largely from simplified language in the NAS summary report) agreement with the three basic and undisputed elements of the science, (1) that the global mean is 0.5°C higher than a century ago, (2) that atmospheric carbon has risen over the past two centuries due to human emissions, (3) that as a greenhouse gas it is likely to warm the atmosphere (but it is one of many, with water vapour and clouds being the most important), 'tells us almost nothing relevant to policy decision'.

The NAS was asked by the US Congress to comment upon the Summary for Policymakers of the 2001 IPCC Report. As a member of the Panel, Lindzen wrote in the *Wall Street Journal*:

> The panel was asked to evaluate the work of the UN's Intergovernmental Panel ... and within the confines of professional courtesy, the NAS panel essentially concluded that the IPCC's Summary for Policy Makers does not provide suitable guidance for the US government.
>
> The resulting document has a strong tendency to disguise uncertainty, and conjures up some scary scenarios for which there is no evidence.

In 2003, John Christy, Professor of Atmospheric Sciences at the University of Alabama, and also a member of the IPCC's Working Group, testified to the US House of Representatives. He pointed out the tendency for models to dominate thinking, even when the real-world data did not corroborate important aspects of their predictions. For example, models predicted much higher rises in lower and upper tropospheric temperatures than were actually observed – 'this would suggest that at least some fundamental processes, for example heat transfer, are not adequately described by the models'.

These atmospheric temperature data sets had earlier appeared to show a cooling trend, but had been revised and did show a very small upward trend. Campaigners on both sides had used this data with 'sceptics' arguing that the cooling trend showed that global warming was not occurring, whilst the defenders of carbon dioxide models were quick to make use of the revised data to debunk the critics. In fact, as Christy testified:

> The inability of climate models to achieve consistency on this scale is a serious shortcoming and suggests predictions from such models be viewed with great scepticism.

He further argued that

> The conclusion in IPCC 2001 that human-induced global warming was clearly evident, was partly based on a depiction of the Northern Hemisphere temperature since 1000 AD. This depiction showed little

change until about 1850, then contains a sharp upward rise, suggesting that recent warming was dramatic and linked to human effects ... since IPCC 2001, two important papers have shown something else ... Using a wider range of information from new sources these studies now indicate large temperature swings were common in the past 1000 years, and that temperatures warmer than today's were common in 50-year periods about 1000 years ago. These studies suggest that the climate we see today is not unusual at all (quoted in Labohm, 2004).

We should note that there is more recent debate on these proxy temperature records with some analysts claiming past temperatures in the so-called Medieval Warm Period were not as high as today, and others claiming they were. There is evidence from the different data sets to substantiate either side of the argument, and this illustrates another tendency – for either camp to selectively quote that research which supports its prior position.

On Kyoto, Christy argues that none of the measures adopted can have anything more than a marginal effect upon climate change.

The evidence convinces me that none of these proposals would change, to a noticeable degree, whatever the climate is going to do.

The previous report of the IPCC in 1995 had also brought a caustic response from senior members of the US science establishment. Professor Frederick Seitz, president of Rockefeller University and a former president of the US National Academy of Sciences and the American Physical Society, took issue with the redrafting of texts that had been agreed by the scientific Working Groups, such that scepticism with regard to models had been suppressed. He wrote in the *Wall Street Journal* of 12 June 1996:

In my more than 60 years as a member of the US scientific community I have never witnessed a more disturbing corruption of the peer review process than the events that led to this IPCC report.

It is clear that the views of senior scientists, whose authority on these issues is recognized internationally and acknowledged within the IPCC's Working Group, are not then represented in the Summary for Policymakers.

I have witnessed this process of bias and exclusion by blinkered specialists first hand in my past work within the UN scientific regulatory process. I have also witnessed the enormous effort required to penetrate that process as well as the penalties working scientists pay in the process of dissent. This process begins with a paradigm of cause and effect which then biases data gathering and funding, and then any critique is mediated by the choice of chairpersons for the leading committees, and finally by the drafting and editing groups.

The charge of collusion in the IPCC with regard to the suppression of contradictory views and dissent has thus been made before and at the highest levels of science, but in the USA and not in Britain. The Royal Society represents British science and it has a history of unquestioning support of the IPCC. The only academic critique of IPCC procedures has come from the work of Sonja Boehmer-Christiansen, a social scientist at Hull University, reporting under the auspices of an Economic and Social Science Research Council project that the IPCC structure was dangerously narrow and self-selecting. Boehmer-Christiansen also concluded:

> ... an important insight gained from the research is that institutionalized science should be treated as a political 'actor' in its own right when considering its interaction with the policy making process.

Her views were published in *Global Environmental Change* in 1994 and led to an exchange of defensive letters, although with some agreement on her criticisms (**Boehmer-Christiansen, 1990, 1994; Moss, 1994**).

I make these points because it is my personal experience that the UN, the Royal Society and other professional institutions have all committed themselves to an uncritical acceptance of the climate models developed for the most part in the early 1990s, well before the crucial data that I have reviewed had been published and before alternative theories and mechanisms for global warming were postulated. Their leading members have made public statements urging governments to act and taken part in the development of mitigation policies dependent upon carbon dioxide computations and the General Circulation Model approach.

When a major authority such as the UN's IPCC aligns itself with the results of modelling, there is a widespread tendency for scientists who are in some way dependent on their governments and official funding bodies to accept such authority uncritically or to shy away from critical comments where they may have some concerns. This is an acknowledged sociological phenomenon related to authority. It does not require 'conspiracy theory' to appreciate this and I use the term *collusion* to describe the process.

I would also add that the climate theories and models developed in the late 1980s followed a period in which the reputation of environmental science and science generally had fallen to an all-time low and that pollution episodes and major errors were very largely to blame — from Chernobyl, Sellafield, acid rain, forest loss, dying seas, the loss of the ozone layer and nuclear weapons issues — and yet within a few years, environmental science was riding high again as the projected saviour of mankind. The effects within the institutions of science of the resultant prestige and the developing inclusion of science committees into international programmes that direct economic and social as well as environmental policy should not be underestimated. And nor should the scale of the *prior commitment* now entrenched in the global 'climate change' community of scientists, bankers, bureaucrats, journalists and industrialists that have created the mitigation edifice — in comparison with which those concerned with adaptation to inevitable change are virtually invisible.

Note

1. This material is presented in Labohm, Rozendaal and Thoenes' *Man-made Global Warming: Unravelling a Dogma*. Lindzen's comments were made in a letter to the *Wall Street Journal*, 11 June 2001. This book also documents previous dissent in relation to the late changes of chapters by functionaries in the IPCC when scientists in the working groups had thought the drafts were final — such changes removed elements of doubt or scepticism relating to the models. This was also aired in the *Wall Street Journal* (12 June 1996) when Frederick Seitz, president of Rockefeller University, wrote that he had never come across 'a more disturbing corruption of the peer-review process'. Seitz had been a past president of the National Academy of Sciences and the American Physical Society.

13

Collusion

'... readers of my column will know that I give contrarians, or sceptics, or deniers (call them what you will) short shrift, and as a close follower of the scientific debate on this subject I can state without doubt that there is no dispute whatsoever within the expert community as to the reality or causes of man-made global warming.'
Mark Lynas, Environment Editor, *New Statesman*

'You may have heard that there is disagreement among scientists ... well, not really.'
Al Gore, politician, in his documentary film *An Inconvenient Truth*

Thus far, we have looked at the delusions and collusions of science within its own institutions. Outside of those labs, corridors and drafting committee rooms, the conclusions of science are fed upon by a variety of hungry organizations in a heavily politicized debate not just about climate science but development policy: the role of the state and private sector; taxation, surveillance, monitoring and control; global inequity and human rights; biodiversity; the ultimate fate of the human species and even the survival of life on the planet itself. This is fertile ground for political commentators of all colours.

In the first of the above statements, Mark Lynas, an environmentalist who has in his own words 'devoted several years to the study of the science' and published a book entitled *Six Degrees* exploring the consequences of each additional degree of warming up to the most excessive predictions, is clearly convinced by what he has read. On the other hand, his work has apparently not acquainted him with dissenting opinion, such as that of Frederick Seitz, former president of the US National Academy of Sciences, nor any of the specialist criticism of modelling that I have documented.

Lynas may genuinely have researched the field without coming across major disagreements in the science. Al Gore, on the other hand, is well acquainted with the high level of disagreement, having been privy to US

Congressional hearings. In the former we have an example of narrow focus and selective use of material driven perhaps by a naive missionary zeal, whereas in the latter we are witness to a sophisticated, well-funded and very deliberate use of carefully selected science for the purposes of a political campaign. Of course, selecting facts to suit arguments is an essential tool of the politician, lawyer and propagandist, but the reception of those facts is subject to similarly biased filters and lay people can remain rather resilient in the face of campaigns of persuasion.

One thing I learned from involvement in the nuclear debates of the 1970s and 1980s was that information alone has little impact on the views that a person might hold. It was one of the very first conclusions of sociological analysis that apparently factual material would be evaluated differently according to some prior commitment that was more psychological than political and certainly had little to do with scientific training (**Taylor, 1977; 1986**).

A team of Dutch sociologists eventually described two mindsets: the technocratic, which evaluated all risk against an external sense of technical progress; and the communalist, which evaluated risk against a more internal perception of progress in terms of how human beings relate to each other. When presented with the same set of facts, for example, on the hazards of a nuclear reactor accident, each side came to completely different conclusions as to the acceptability of the risk.

I would add that each side was also motivated to investigate those risks in entirely different ways. On the one hand, the technocratic mindset tended to trust government and engineering institutions and felt in no way duty bound to look deeper into any ways in which potentially embarrassing information might be withheld, whereas the communally based mindset had a sharpened sense of what manner of deception might be afoot. And both sides indulged in emotive propaganda as they sought to influence the broader policy debate, the technocrats warned about blackouts and humanity being set back to the Stone Age, and the other warned about worldwide contamination, genetic damage, terrorist attack and nuclear proliferation.

I managed to talk to both camps. I could speak the language of science and visited laboratories and spoke in universities in many countries, finding dozens of scientists who were concerned at the risks and the cover-ups. The technocratic sense of progress is not confined to scientists and engineers, nor are scientists necessarily immune to the deeper values of community. Intelligent and well-meaning people could be found on either side of the psychological divide, and each had a capacity for delusion. In this, I mean the capacity to see only part of the picture and to zealously believe that it is the whole part and, moreover, the truth. Mark Lynas displays these characteristics, campaigning on behalf of humanity, global injustice and the diverse life-forms of the planet, all of which he sees endangered by global warming. The problem is that this zealotry is a matter of belief rather than of science and it blinds him to the genuine disagreement and level of uncertainty. Once such a person decides that there really is no disagreement, the blindspot develops, disagreement is not looked for, and hence none is found.

With Al Gore, however, we are dealing not with a personal delusion because Mr Gore is well aware of the very high-level disputes in the science, but with the artful presentation of an illusion. As a former Vice President of the USA, and with his contacts in the science community, Mr Gore was party to the briefings of the US administration by the US National Academy of Sciences, which as we can see from Professor Seitz's remarks was highly critical of the IPCC.[1]

Most commentators fall into the Lynas camp; they *believe* the science is unequivocal and, as believers, they denigrate what they then categorize as unbelievers. Labels of sceptic, heretic and denier are then used to imply that any such psychologically dysfunctional person should not be given a voice, lest, of course, they lead others astray. Fortunately, the editors of *New Statesman* did give the dissenters a voice. Lynas's comments arose after publication of an essay by Dr David Whitehouse, former astronomer and science writer for the BBC, a controversial opinion piece that maintained solar cycles and natural processes could account for global warming and in particular the recent cooling trend. Lynas opined on a *New Statesman* blog on the issue that though his magazine gave space to such views he did not support such liberalism. If he had his way, Whitehouse would not be given such space to mislead.

This is a common delusion that appears among many of the environment correspondents of the left and apparently liberal press — it is that any 'denier' has an underlying motive to lead others astray. This would appear a genuine belief rather than a deliberate attempt at defamation and distraction.

Al Gore's work is altogether more problematic. Gore presents himself as a world champion against ignorance and inertia and someone who has also studied the science and believes it unequivocal. Yet, in his film presentation he does not draw attention to the fact that the ice age record of periodic carbon dioxide increase *lags* the temperature fluctuations by 800 years. He also fails to point out that his 'mentor' Roger Revelle never agreed with his student's view that it would prove a problem. Nor did he publicize the fact that the disagreement became so acrimonious it led to a defamation lawsuit which Gore lost.[2]

There is no reference at all in the documentary to the dissenting views of IPCC member and National Academy of Sciences' expert nominee Richard Lindzen on the overplaying of feedbacks – a point Revelle would have agreed with – or to the NAS report to Congress that the 2001 Assessment Report should not be relied upon as a basis for policy.

In any other circumstances, it would be incumbent upon a politician with such a powerful message to declare any potential interests. In this case, most pertinent is Mr Gore's position as CEO of Generation Investment Management, a fund management unit set up by himself and drawing in several top investment bankers from the global corporation of Goldman Sachs. From published material, it is not transparent whether he or his clients would stand to gain from the USA signing up to a global carbon trading system.[3]

A key element to highlight here is the various desires to create a simple message – a unified voice, apparently based upon science that underpins the policy recommendations and business strategies. And that brings us to the IPCC and the science community itself. There is no doubt in my mind that pressures mounted upon the science community to speak with one voice when talking to policy makers and those pressures now continue with respect to the enormous investment strategies unfolding as a result of a belief that dangerous climate change can be mitigated by emission controls. I also have little doubt that these scientists were primarily motivated by a concern for the future health of planetary ecosystems and not their own prestige or standing. However, once committed to a single and simple message, processes are set in motion that militate against any revision, and these processes also become pro-active in attacking and 'debunking' any new science or theories.

An example of the debunking process: the BBC, *New Scientist* and the solar-cloud theory

I have watched as such important distributors of scientific debate such as the BBC and *New Scientist* have sided heavily with the 'consensus' and, on occasions, seriously distorted the science. Early in 2008, BBC Online ran a story that the solar-cloud theory had been 'debunked' by UK scientists. This report appeared in over 10,000 Google references and was widely picked up by the press. Richard Black, the BBC Environment Correspondent, stated:

> The research contradicts a favoured theory of climate 'sceptics', that changes in cosmic rays coming to Earth determine cloudiness and temperature... But UK scientists found there has been no significant link between cosmic rays and cloudiness in the last 20 years.

The research referred to a paper by Terry Sloan and Arnold Wolfendale, who presented their findings in the Institute of Physics journal, *Environmental Research Letters* (**Sloan & Wolfendale, 2008**). Black related how the University of Lancaster team of physicists explain that they used three different ways to search for a correlation, and found virtually none. And Sloan is quoted by the BBC website, 'The IPCC has got it right, so we had better carry on trying to cut carbon emissions.'

Black makes no attempt to talk to any of the scientists whose work this apparently contradicts, and thus cannot appreciate that this work is published not in the usual specialist journals that deal with this issue (see the work of Svensmark, Marsh and Usoskin) but in an organ of the Institute of Physics, of which the authors are members. When I eventually received the original paper I was startled to discover that far from contradicting the theory of solar-cloud effects the research *confirmed* the effect and merely pronounced on an upper limit of 23% for its role in the observed correlation of cloud in relation to the solar cycle, stating that other mechanisms should be sought to explain the remainder of the correlation. I quote:

> ... less than 23% of the dip comes from the solar modulation of the cosmic ray intensity, at the 95% confidence level. This implies that, if the dip represents a real correlation, more than 77% of it is caused by a source other than ionization and this source must be correlated with solar activity.

The work itself has been carried out by two retired physicists with no record in the field, using statistical techniques which at the very least needed to be exposed to criticism by specialist peers who do not share their conclusions. The BBC thus places itself in the position of arbitrator on science it is not in a position to judge and glosses over the fact that this is a still a major area where there is no consensus.

The *New Scientist* carried the story on their environment blog of 3 April, headed 'It has been said before, and it is being said again for good measure: there is no evidence of a link between cosmic rays and warming temperatures on Earth'. Catherine Brahic, the reporter, referred to the new research as a 'nail in the coffin' for solar theories of climate change. These two media organs are the chief sources of non-specialist information and display heavy bias. In the case of *New Scientist*, the blog was bannered by advertisements for a manufacturer of wind turbines.

The tide is turning however, and some senior scientists are speaking out against the prevailing orthodoxy. The recently retired director of the International Arctic Research Center at the University of Fairbanks, Alaska, Syun-Ichi Akasofu, a leading geophysicist who directed an international Japanese and American team of Arctic specialists for over ten years, began to speak out on the nature of the delusion. In an interview with *Executive Intelligence Review* of 11 May, 2007, Dr Akasofu stated:

> ... definitely climate change, or temperature, has been rising. Somehow the IPCC decided that the increase in the last 100 years is due to the greenhouse effect; however, a significant part of that would be just due to natural change. So, even if we spend lots of money on suppressing CO_2 release, it wouldn't do any good, because it's a natural change.
> And no one can explain the temperature rise from 1910 to 1940, or explain the decrease from 1940 to 1975. My point is, that until we understand the increase from 1910 to 1940, we just cannot say the increase from 1975 to the present is entirely from the greenhouse effect.
> ... IPCC state 'most of the present temperature increase during the last 100 years, from 1975, is due to a magnified greenhouse effect'. But there is no basis for them to say 'most', for they have not examined the natural component. So it's an assumption. Then, they say, computer models conform to that, but that's not true. What's happening is that computers try to simulate the present increase, but computers can't do that. So it's not confirming anything; their computers are just trying to simulate the initial assumption.
>
> Everybody's believing Al Gore's movie, which is nothing but science fiction.
>
> The top level, the very top-level climatologists or meteorologists, they don't join the IPCC, because the IPCC is too political. They stay away. So there's lots of – I don't know if it's the majority or not – but there's lots of silent people there. What I told you, that I wrote something on

that, people have to be careful, you could be assassinated. That's where we are now.

His team of researchers is well placed to assess the delusory nature of the science consensus. The institute collaborates closely in study of the natural and cyclic nature of the Arctic environment with the Russian Academy of Sciences and key Russian climate institutes. Just as the US National Academy of Sciences once had a strong cautionary note in its advice to the US administration, so too did the Russian equivalent, advising President Putin in 2004 not to sign the Kyoto Protocol – advice which he eventually declined to take, perhaps because Russia stood to gain financially from the operation of the Clean Development Mechanism and carbon trading.

In early 2007, the Russian Academy of Sciences had one of its members serving as Vice Chair of the IPCC – Professor Yuri Izrael, head of the Institute for Environmental Change and Global Ecology in Moscow. At the time of the release of IPCC's Summary document, on 14 April, Izrael was reported by RIA Novosti as saying:

I think the panic over global warming is totally unjustified. There is no serious threat to the climate. There is no need to dramatize the anthropogenic impact, because the climate has always been subject to change under nature's influence, even when humanity did not even exist.

However, this outspoken view proved short lived. Izrael later began to campaign on behalf of a Russian idea of seeding the stratosphere with sulphuric acid to cool down the planet – a project that his Institute had received funding to explore, courtesy of the Putin government.

The growing role of the environmental campaigners

In addition to this world of science institutions, governmental influence and media bias, there has also been a growing and powerful environmental lobby pressing for an unequivocal commitment from the scientists. NGOs well appreciated that governments will not move when there is major uncertainty, and a lobby has evolved out of a coalition of interests on the part of environmental campaigners and those industries standing to gain from a shift in policy. Naturally, there is also an opposing lobby from oil, gas and coal interests. The nuclear lobby has remained somewhat hidden, but has benefited enormously from the climate issue. Some campaign groups have allied directly to renewable energy interests, especially wind turbine manufacturers and solar collectors, whereas others have remained independent of commercial interests but used the projections of technology and capability to underpin their campaigns.

In addition to these straightforward political alliances, there has been a growing corporatization of the environmental sector. NGOs have grown from a few small back-street offices into a multimillion-dollar international organization – in the case of Greenpeace, with a fleet of ships, modern office suites, staff and pension funds. Such organization requires a steady income stream and does not have the option, as for example at the end of a successful ocean pollution campaign, to simply pack up and go home.

When an organization's ethos is essentially combative, it seeks out problems and threats – which is fine, as long as there really are serious threats that cannot be dealt with by trusted governments. But in my view, as a seasoned campaigner, the game changed significantly after Rio in 1992. The 'enemy' metamorphosed from being the dumpers and polluters, ably supported by a science-industry alliance (including the modellers), to a more subtle menace. As a result of the shift to the precautionary principle, industry and the regulators began to move in another direction – Clean Development Mechanisms were set in motion and large amounts of money shifted towards preventative strategies. This shift required a different type of environmental organization, and although the campaign groups made significant efforts to provide 'solutions', they were still ruled by the old ethos of campaign and combat. Whenever very large sums of money are being directed by public policy, corruption is rife and a different type of organization is required to see through the levels of deceit and 'greenwash'.

The problem with combat is that there always has to be a threat, an enemy, something to be fought against. And such was the demand that campaigners became professionals, with books and manuals drawing from the expertise of the advertising industry and the experience of political lobby groups. Image and simplicity together with achievable targets were essential tools of the trade. A corporate organization answers to the shareholders, but an environmental pressure group answers to subscribers who need to feel something is being achieved, otherwise they remove their subscription.

This dynamic is not talked of openly and any suggestion that it has played a role in the climate debate is met with hostility. But to discuss this issue is not to impugn the integrity of any organization; it is to draw attention to how a powerful and unacknowledged force can distort judgement. If you have published a best-selling book on global warming and achieved some status as an environmental correspondent, you are not motivated to seek out those scientists who disagree with the IPCC orthodoxy. Likewise, if you are a large multinational campaigning organization with a decade-long commitment to fighting climate change you are going to listen more to the views of Mark Lynas and Al Gore than to Dr Akasofu, Professor Christy and Professor Lindzen. And it helps that all of the world's science institutions also (now) speak with one voice. Thus, the environmental lobbyists become defenders of the orthodoxy, ably supported by all liberal-thinking, environmentally conscious laypeople and journalists. And that leaves only the conservative, business-as-usual economic optimists and free marketers to espouse the cause of the climate model's dissenters!

It is with great sadness that I now witness the level of collusion operating within environmental NGOs. They had finally begun making an impression on the international process of environmental protection in the lead-up to the Rio Summit in 1992 and helped produce the ground-breaking Agenda for the 21st Century that supported connectivity and interdependence – a true ecology that included all aspects of human well-being and, in particular, steps towards a global equity. Given the competitive nature of the world economy and the massive scale of inequity in wealth and economic power, these steps were never going to be easy. But I at least felt that environmental NGOs would be at the forefront of thinking.

Though this movement started out with great integrity of purpose, something has been lost. NGOs have embraced science to a greater extent but, in the climate debate at least, have come to rely upon and uncritically accept the authority of scientific institutions. In no other area have NGOs been so uncritical. There is a long history of former critical analysis: on the risks of low-level radiation; nuclear reactor hazards and waste disposal options; toxic discharges to the marine environment; incineration of toxic wastes at sea; the impact of acid rain; the deployment of GMO technology. In all these areas science institutions were part of the problem rather than the solution.

It was precisely the collusion between government, corporations and the science establishment that motivated NGOs to develop critical science expertise. The institutional science community has proven all too willing to accommodate government and business agendas and suppress or distort scientific assessments (there are many examples from pesticide studies, pharmaceutical trials, impact of GMOs, nuclear accident hazard analysis, the modelling of ocean dumping of nuclear waste and discharges of toxic chemicals such as PCBs). It was critical science, funded by the NGOs and supported by a very few progressive governments, that led to the crucial changes in the 'burden of proof' and subsequently to the 'precautionary principle' being written in to international conventions.

I played a role in that work, and had the privilege to work closely with some of the best environmental scientists of our day.[4] I hope therefore that the assessment I now make of the current situation will cause some reflection among my former allies. I have always held that we should work with the cutting edge of scientific truth, and that whatever the short-term goals or campaign advantages truth would ultimately serve our cause. I am concerned now that a 'corporate creep' has taken place whereby environmental NGOs have begun to behave like the large corporations we hitherto held to account. It is not hard to understand why this might happen – NGOs have grown in size and now command considerable resources. This requires a whole suite of corporate skills from professional personnel management to accounting and investment, the handling of press, media, publicity and public relations, as well as lobbying and strategic development of policies. This requires specialist training and hence recruitment of staff from business schools and organizational realms not known for being well attuned to the ethos of sustainability, nor for a commitment to scientific truths. Such specialists have to be represented at many levels of decision-making within the organization and it would be an act of naivety or denial to pretend this creeping effect cannot or does not now influence policy. This is not to say that anyone recruited from professional management and business circles lacks integrity, or feeling for the natural world, or concern for the overall well-being of humanity; rather it is a question of *how* these concerns are transformed into action and whether professional training can create blindspots, particularly with regard to the 'group-mind' that evolves within corporate entities.

Corporate 'creep' and the culture of targets

I would argue that this corporate creep affects the *kind* of targets chosen and the simplicity of the messages put out by the campaigners. Targets are selected that are *visibly* achievable and because they convey a simple message that can have effect in the 'market place' of parliaments and government policy. Complex issues are avoided.

One such complexity can be illustrated by renewable energy developments – almost all of which require industrial development in the countryside. These locations affect rural communities, national recreational resources, and are often in wildlife-rich and remote places. At present, almost all such operations are 'developer-led' – that is, the industrial developer selects the technology and the site. No alternatives or strategic assessments are available to the communities that must assess the proposal. This situation has changed little from the 1970s when the drive for more nuclear stations, chemical installations and motorways all benefited from the lack of strategic planning and the piecemeal approach – to the great disadvantage of communities that bore the impact. In this age, environmentalists should be embracing a paradigm that supports and empowers local communities, not the same old developer-biased 'trade-off' that operated in the past. Yet this would require campaigning for planning reforms and the empowerment of local communities – not an easy sell to the subscribers, nor welcomed by government agencies whom many modern NGOs are often intimately involved with if not in some way dependent upon.

In the latest proposals for 'planning reform' in the wake of the UK Government's most recent White Paper on energy policy there are disturbing announcements of the curtailment of individual and local democratic rights to both question government policy and, most disturbingly, the appropriateness of the sites chosen by developers. Thus the apparent urgency of tackling climate change is used as an argument to give developers greater power to select sites and technologies that reflect their reduced costs and profitability rather than considering the impact upon local communities.

Linguistic truth

The linguistic manipulation of 'climate change' has been an astute campaign move for any organization dependent upon public concern and support or government funding for research or implementation of solutions, because clearly anything the climate now does can be interpreted as human-induced. The issue of how much of the change is caused by greenhouse gas emissions hardly ever arises in any of the media treatments, nor in the alerts from the campaign groups. In the IPCC's 4th

Assessment Report, the authors even note that whereas in the 3rd Report 'climate change' meant human-induced, it should now be taken to mean *both* human induced and natural. The note, however, was barely visible.

As in all wars at whatever level of reality, truth is the first casualty and we are now encouraged to join the 'war on climate change'. It is not that I believe scientific truth can be totally pure, objective and unaffected by political worlds but that it should *strive* to be so. If that aim is not there, then it begins to sink into the world of propaganda and persuasion, a mere tool in a wholly other agenda.

A polarized political environment now surrounds modern climate science with even fellow scientists referring to their critics as *sceptics* rather than critical scientists engaged in one of the most fundamental aspects of the scientific method — that of questioning and testing prior assumptions. In this political world, it is easy to forget that global warming and 'climate change' are *inventions* of science. Such is the level of propaganda and the implicit meaning of terms, many people assume the issue is self-evident. It was a scientific group that set the alarm and it is science that is charged with identifying the changes, predicting the future, and advising on what to do. This places scientists in a position of some power and influence.

There is virtually no debate now within the NGOs on the social and environmental sustainability issues raised by the expansion of renewable energy supplies. The WWF, Friends of the Earth and Greenpeace have a website, for example, that urges factions who support wind turbines to write letters to the planning departments in areas where turbine clusters are proposed. The site does not discriminate and assumes all proposals should be supported. There is no comparable website urging community consciousness, responsibility and choice, and no data on the downside of applications in wild places — even sites where their organizations have balked at supporting proposals, such as the massive development that threatened the Hebrides.

There is an absence of the balance required by the principle of sustainability. This is an area in which I have first-hand experience. I wrote the first political assessment of renewable energy strategies with Ian Sanderson way back in 1980 for the Group of Independents in the European Parliament.[5] With my fellow scientists and analysts we gave educative seminars to the European Commission, which at that time was still wedded to an expanding nuclear future. In the 1990s I wrote much of the Countryside Commission's input to government's consultations on energy policy, and until recently sat on the joint Department of Trade and Industry/Countryside Agency 'Community Renewables Initiative'. The latter was a hard slog – to get the DTI to appreciate that a myriad of small-scale initiatives would lead to greater sustainability than would a smaller number of mega-technologies such as turbine 'farms', tidal barrages and carbon capture from power stations.

It was readily apparent in those corridors of power that the industrial progressives had little comprehension of the 1970s 'ethos' of community, scale and appropriate technology. Nor had they any working knowledge of the 'clean production' and 'waste minimization' methodologies developed in the 1980s and implemented eventually in the 1990s. These were the models that *worked*: auditing and awareness, community responsibility (e.g. river-catchment based 'identity' for collective action to reduce waste-water, discharges and resource use) and all under what could be called 'good housekeeping'. When the UN set up its Clean Production offices in Paris, the lead staff were female, and many of us reflected that the male ego was ill-suited to the task. For men there was not enough glamour, not enough 'power', not enough kudos for whatever ambitions they might have other than to clean up the mess.

Whilst our efforts to clean up rivers, estuaries, and coastal discharges have proven successful in the longer term and nuclear expansion has been held at bay, the renewable energy front seems to have lost all sense of balance and purpose. I have heard representatives of Greenpeace, Friends of the Earth and the Centre for Alternative Technology argue that the division of local communities, the overriding of local democracy, the degradation of wild land and beauty *should* all be sacrificed in the cause of renewable energy *targets*. When pressed, reasoning for this sacrifice is that we must save the 'Bangladeshis' of this world, those more vulnerable to climate change than ourselves, and then ultimately to prevent 'dangerous climate change' – meaning the prospect of a runaway global warming driven by our excessive emissions of carbon dioxide.[6]

Thus, in the name of 'preventing climate change' it now possible for large corporations to move in on Iceland's pristine river-wilderness to harness the 'renewable' power for aluminium production with metallic ores being shipped from Australia, or for hydro-schemes to press into SE Asia's remote and highly biodiverse cloud-forest uplands which also are a refuge of indigenous tribal peoples. In Britain, the Hebrides, islands of ethereal beauty and home to endangered and recovering populations of sea eagles, are being mooted as the 'renewable energy' powerhouse of

Europe with 700 MW of wind turbines – far in excess of local needs and requiring long-distance transmission to Scotland's industrial belt. Similarly, the Greek island of Skyros, fabled for its wild beauty, indigenous ponies, sea eagles and Eleonora's falcons, is threatened with a 300 MW turbine installation.

Further, the immediate imposition of 'targets' for biofuels in transport will hasten the conversion of natural forest areas into agricultural production and compete with land for food (and raise food prices). The NGOs have belatedly awoken to the consequences of the policies they have advocated, and are now taking out full-page press adverts saying biofuel development should take place only with appropriate safeguards. Unfortunately, the supply policies are in place well in advance of such safeguards. There seems little chance that governments will impose ecological standards on biofuel production and every likelihood that multinational energy and agribusiness corporations will seek out the cheapest and most profitable sources of production, threatening wildlife-rich marginal land and forests in South America, Eastern Europe and SE Asia. In a recent review by the *Guardian* newspaper (22 November 2008), it was reported that sovereign funds in the Arab Gulf States, South Korea, China and Japan were purchasing vast acreages in Madagascar (1.3 million acres in this country alone), Brazil, Laos and Cambodia – primarily to secure food production and biofuels for their own populations.

There are two aspects of this situation that cause me increasing concern. The first is that environmentalists are now using the same language and reasoning as the 'developers' of the past. Beauty, feeling, continuity and community are dismissed as 'subjective' emotional responses and locality must be sacrificed for the wider good, if not as a national target then as a global responsibility. And the second is that they have begun *acting* as corporate players, seeking alliances and moving into markets that generate cash flow for their organization.

On the question of language, we have already witnessed the agenda of sustainability become hijacked by the development lobby. 'Sustainable economic growth' has been added to the criteria by which any development is to be evaluated, despite the fact that no one had ever defined what that meant or come up with any examples of recent economic growth that were not ultimately dependent upon the exploitation of a cheap and finite source of highly polluting fossil fuel. Such developers were able to use 'sustainable economic growth' as a trump card over all other aspects of sustainability.

The modern corporate NGO does not specifically ally itself to sustainable economic growth, but neither does it pursue rigorously a critique of economic growth in relation to consumption. Instead, the large corporate NGO, mindful of its cash flow and client base, now tailors its campaigns to sets of targets that exist *within* the current political economy. This is illustrated most clearly in the uncritical support for onshore wind-farms. The *targets* and *context* are accepted, and criticism – if it exists at all – is restricted to the choice only of siting. It is as if all of the lessons of the past two decades have been ignored. If the driving force of development and the absence of strategic planning are not tackled, no amount of critique on specific proposals will prevent the inevitable trade-off mentality inherent in corporate and governmental thinking. This is now evident with the proposals for large-scale wind-farms in the Hebrides, tidal barrages on the Severn estuary and a tranche of new nuclear power stations.

In the latter case, the RSPB is now in active opposition to the Hebridean development, only to find its voice competing with powerful corporations offering local economic progress and in the name of global equity and national targets. The RSPB had hitherto rather uncritically supported wind power and even allied itself through corporate finance to renewable energy companies. How far can corporate NGOs now pursue a strategic critique, especially when (as in some cases) these groups have forged active business links with wind developers? The 'green' bank Triodos now holds substantial investments in wind energy and backs the developer Ecotricity who applied to build a large turbine in the Mendip Hills of England. This was opposed by the local community and voted against by the democratic planning authority on the grounds of damage to local quality of life – only to be appealed by the developer. The local community's values were overruled by central government (such decisions being strongly influenced by the attitude of the DTI) in the name of government policy and national targets. It would appear that local democracy and community feeling – an essential element of sustainability – can be sacrificed in the name of an apparently ecological policy. Given that offshore turbine sites have far greater potential than onshore sites but are more expensive, the key point of debate should be the differential cost which is also the cost of safeguarding the landscape; but in a time-pressured, target-led, least-cost approach with no strategic planning safeguards this point is missed entirely.

It is not just the corporate creep of attitudes to community and development but that other aspect of business mentality, the focus upon goals and targets, that is driving policy. These are arbitrary figures set to a timetable whose purpose has less to do with a rational assessment of the environmental dynamic (such as timescale of causes and effects, lead-times of technology, and business cycles of capital renewal) than with the need of the corporate entity (and governments work in this paradigm) to have purpose and goals by which to assess *itself*. This applies in particular to any corporation that has *growth* targets, but is applicable to all such entities that must have some measure of their *performance*.

There was a time when such phenomena were restricted to govern-ments in the old enemy of the 'East' where there were five-year plans and the whole state was run as a business corporation under a sham of democratic decision-making and a huge shadow of corrupt practice. In the 'West' there evolved an apparent disengagement of government from the corporate world with the sale of state enterprises and assets in energy, water, transport and forestry (the sale of nature conservation assets was blocked by popular opposition). Privatized concerns then developed a corporate set of targets and goals that is far removed from their previous 'public service' mentality: efficiency and price dominate above any wider public goods such as community, biodiversity and sustainability.

Needless to say, western privatization has had its own share of corrupt practices, but in this new millennium we are now all 'westerners' and environmentalists must become aware of the nature of the necessary collusions they have made with this expanded corporate reality.

This leads me to the second of my concerns. Environmentalists are not only adopting the language of corporate power, they have become *players* in that corporate world. Organizations have grown enormously in the last 15 years such that they have large central, in some cases international offices and staff, all of which require professional management. These professionals are recruited either from schools of corporate management or from the ranks of experienced practitioners. The organization develops a corporate identity and presence. It is not uncommon for Chief Executive Officers to have little grass-roots connection to the environmental move-ment, little scientific knowledge, little feeling for community develop-ment, but a lot of experience in large corporate management. The problem is that the very size of the organization and its formidable management tasks *demands* this professionalism. For the more overtly campaigning groups, the CEO may be parachuted in from the ranks of professional lobbyists, political advisors or public relations, rather than coming up from the ranks of the organization itself.

I am concerned that such corporate entities then develop their own internal agenda that has as much to do with their position in the 'market place' or the 'political world' as it does with the issues they were set up to tackle. The danger is that these internal processes begin to direct the thinking and mentality of the organization and it loses contact with its origins. It does not, however, lose sight of its 'mission' because in the position in which it operates that would obviously diminish its power, but it now occupies the dangerous political and corporate world where appearances are often as important as substance. It is a world of marketing and image. And it is in this new world that the 'mission' is redeveloped, packaged and marketed.

Thus 'climate change' becomes a simple slogan – apparently meaningful, but in scientific terms quite vacuous. By absorbing the complexity of anthropogenic causation and its less than certain signal into a single phrase that now implies all such change is man-made, the propagandist pre-empts debate on uncertainties. The mission is taken further when any such critics are branded as 'deniers' of climate change itself – an apparently obvious reality, on a par with 'flat-earthers'. I see little sign that these processes are recognized or understood. On the contrary, NGOs are in knee-jerk reaction to any hint of collusion or criticism of their targets and general approach.

Deception: hidden agendas and marketing strategy

In this century we have become much more aware of how the political world has embraced shallow image and marketing to the point of blatant deceit. Many countries, including my own, have embarked upon a worldwide game of power and control of resources under a smokescreen of 'the war on terror' with such infamously doctored analyses as the 'dodgy dossier' on weapons of mass destruction. Many people had reservations about the document drawn up before the war by UK intelligence agencies. It alleged that Iraq had biological weapons capability. Yet this dossier went through the whole UN process without being directly impugned. Considerable effort was devoted to protecting the way in which the drafts of expert analysts had been doctored to match the political aim of providing a context for invasion. This one single and unstoppable act of collusion with an obvious corporate agenda related to oil, armaments and reconstruction contracts has sequestered vast resources that could have been used effectively to secure clean water, sustainable communities, protected forests and wildlife across the globe.

The 'war on terror' that ensued is widely criticized for creating more terror, as if its protagonists would not have been aware of this propensity at the outset. They were indeed so advised and chose not to listen and to fabricate and doctor evidence in a global drama that was disturbing in terms of how few intelligent politicians, civil servants and scientists raised issues concerning the truth. It was a classic example of 'the emperor's new clothes', but with the further step that there *were* plenty of voices crying 'he is naked' only to be ignored by the masses of courtiers.

One could conduct similar analyses of collusion and intention in the 'war on drugs' and the 'war on poverty'. In the former case, despite massive investment in military-style operations involving navy, air force, army and police around the world, drug supply increases and prices fall. And a lot of uniforms, guns, helicopters and consultant contracts get sold and suitably conservative voting people empowered by the uniformed salaries. In the war on poverty, policies are pursued on the basis of a 'trickle-down' theory. In fact, the rich get richer and poverty increases even in those countries where the overall wealth is still growing. The 'war on climate change' should be seen in the context of all 'war on' mentalities – the paradigm is military, and the military mind is not the best solution to any of these problems.

Firstly, the military mode of thought focuses on combat to deal with a symptom, not a cause. It creates for itself unquestioned and simplistic goals and targets relating to control. It does not question the *context*. In particular, it does not question the nature of its alliance with the corporate business world that benefits from its 'solutions' to whatever problem it is directed, whether in relation to security, armaments or reconstruction. Such vested interests have no obvious commitment to outcomes that would lessen their business, but the military mind accepts the goals that are in large part derived from business agendas. Governments thus readily adopt the military mentality with its language of war and simple targets and excuse themselves any deeper analysis of causes and any real commitment to solutions. The real agendas, which relate to economic power, security and control, are not addressed.

There is now no question in my mind that environmental organizations have engaged in this global war game of markets, interests, power and propaganda with some considerable purpose, but not a lot of awareness or experience. They have sought to ally with other players – government, inter-governmental and corporate. What I do not see is any kind of self-reflection and analysis regarding their role, their goals and their successes. What is far more visible is their adoption of corporate goals, the mentality of the marketplace and, most disturbingly, the uncritical acceptance of political and economic realities. I call this collusion where others might use the term conspiracy. It is not a conspiracy because that implies conscious intent and manipulation. It is collusion from self-interest, lack of self-awareness, ignorance and naivety. This is not to gainsay a genuine caring and desire to change the world in a better way.

This lack of self-reflection has something in common with the process of 'denial' as understood in personal psychology. Self-reflection (and hence the healing process) is blocked because the persona *knows* that something very painful lies beneath the surface of enquiry. Perhaps it is the death of a dream — some knowledge that what is now lived is a betrayal of the very creative purpose for which they are born. In order to maintain the persona in what may well be an aggrandized and powerful position, the body of denial develops an irrational certainty of purpose. If this is challenged, and in particular if that challenge hits the nerve of that original purpose, the response is irrational and quick to point the finger.

That finger can never address the issue but instead focuses upon the persona of the challenger, inevitably projecting upon that screen an image of their own denied constellations of purpose. In George Bush's War on Terror, the terrorists 'hate our freedom' and there is no admission that the oppressive actions of the warmongers created fertile ground for the terrorist.

Terrorism and global politics may seem a long way from the concerns of the Royal Society for the Protection of Birds and the WWF, less so Friends of the Earth and Greenpeace. But all four have entered into the collusion, if unconsciously, in the 'War on Climate'. None pursue any kind of radical critique of the corporate world. Yet this world stands to gain enormously from the war and military mentality: vast sums will be invested in aerospace turbines, tidal barrages, and biofuel plantations; there is a growing market in traded permits and carbon credits. Above all, these big corporations are seizing the opportunity to *look* as if they are part of the solution with ethanol, green diesel and hybrid motors; but this technology is not sustainable, nor is it affordable by the majority of people in developed countries let alone the developing world.

These corporate solutions *depend* upon an ever buoyant consumer economy, if not for their existence then for their growth and ability to remain a player. That constraint now applies to environmental NGOs. The critique of development, or indeed the restructuring of the *developed* world, is hardly pursued at all. At most there are calls for restraint, for efficiency or for safeguards, but all within the politically acceptable model of development.

A simple analysis of carbon emissions and current policies demonstrates the deceptive nature of the 'prevent dangerous climate change' mission. If current developed world policies to reduce emissions by 60% or even 80% on 1990 levels by 2050 are successful global emissions will still rise 55% by 2030 and carbon dioxide will reach double natural or pre-industrial levels by 2050. Annual CO_2 emissions are currently around 7000 million tonnes where less than half of 1% is currently being traded or offset. The Global Environment Fund set up to finance ecological development spent $1.8 billion between 1991 and 2004 on climate change projects (almost entirely mitigation) compared to the total official and private general 'development' aid of $2,700 billion (i.e. 0.07%).

A sizeable $400 million of this GEF money went to China, which currently runs a $200 billion annual surplus on trade and had over $1000 billion accumulated as sovereign funds by 2007. China is busy trying to buy up profitable fossil fuel energy companies in the US and Europe! The International Energy Agency estimated that the global investment required in electrical power production to 2030 is $20 trillion; by 2030 83% of this will be fossil fuel based, with half the increase occurring in China.[7]

It is clear that a complex global game is being played whereby Kyoto signatories reduce their emissions (at present only marginally and at least-cost) whilst exporting capital and production to China (and latterly also India) where emissions are set to rise considerably. China's emissions grew by 67% to 1000 MtC from 1990 to 2004 and are rising at 5% per year. Against this context, the whole of the UK savings to 2050, which would amount to 100 MtC/annum if 60% reduction was achieved, are eclipsed by the projected worldwide growth of emissions by 2030 of 3000 MtC per annum (from the current 7000 MtC to 10,000 MtC/year).[8]

It could be argued that NGOs have little choice but to enter this world in order to stand any chance of being effective. But this requires players who can enter the game powerfully and consciously and environmental NGOs currently show no sign of taking on any major critique of the game. Better, I would argue, to focus upon those issues they understand and have experience of, and to convert that experience into effective policy on the ground. These are the areas of biodiversity, sustainable habitation and community well-being that need to be *resilient* to inevitable changes.

The organizations that should be my natural allies now show all the signs of corporate creep, collusion and denial. They are not willing to look at the way their internal organization has changed, their embrace of the corporate ethos and how this affects their goals and policies, or their collusion with government and alliance with other corporate entities, or their assessment of climate science. Moreover, they are adopting the same irrational response to criticism that has marked government and corporations throughout the decades of environmental campaigning.

In my view, all of the major environmental campaign groups need to radically assess their campaign agendas and ways of working and bring themselves back to a more grass-roots orientation to the benefit not only of the world's poor but also to those people living in their own rural back-yards. The climate campaigns are symptomatic of entirely other aspirations influenced by their growing focus upon world agendas, lobbying in high places, and the need to draw subscriptions from a largely urban based, wealthy and basically fearful clientele.

Media collusion

One of the strongest safeguards against institutional bias and any form of collusion lies with the global media – in particular science journalists. Another candidate would be the take up of scientific dissent by environmental campaigning groups who have the capacity to fund data gathering, analysis and critical review. Occasionally, TV companies have made major contributions. In the case of low-level radiation hazards, my Oxford research group was funded by Greenpeace to gather epidemiological data in areas of nuclear contamination and our work was taken up, expanded and publicized by Yorkshire TV. This led to a major government inquiry on nuclear risks. Greenpeace International funded a large proportion of the analysis that underpinned the emergence of the precautionary principle and TV played a major role in highlighting acid rain damage and the death of seals in the North Sea. Science journalists brought much of the contrary data of specialist reviews to the attention of other scientists, where the institutions of science itself were generally laggard.

In my experience over the three years I have been compiling data and voicing my concerns, I have come to realize that these historic safeguards have considerably diminished. There are no environmental campaign groups with any interest in critiquing the IPCC model. Science journalists almost without exception support the consensus. The only break in this sea of approval has come recently from a TV programme on the UK

Channel 4 called *The Great Global Warming Swindle*. The programme was presented by the respected science journalist and one-time editor of *New Scientist*, Nigel Calder, and included interviews with both Lindzen and Christy who made forthright criticisms of the IPCC. However, the programme also mishandled some graphic data and included some poorly edited material from the oceanographer Carl Wunsch, who made a vociferous complaint that reached many UK newspapers. Science journalists (for example, Steve Connor of the *Independent* newspaper) reported extensively on the failings of the programme, but completely ignored the arguments of Lindzen and Christy, as did all reports of the controversial programme in *New Scientist*.

The programme had 'extended' the timescale of a graph showing the 1800- to 2000-year correlation of solar cycle length with the northern hemisphere temperature record. In fact there were no readily available data for the post-1990 cycles and the cycles and temperature are filtered by 11-year running means. The relationship is held by some to 'break down' after 1990 because temperatures continue to rise whereas the solar cycle lengthens. But the C4 programme gave no hint of this recent criticism, simply showing the old data and extending the timescale such that it looked as if the relationship still held. It was either a mistake or a conscious avoidance, and again was roundly criticized. But, as we shall see, the point is a minor one and criticism draws attention away from the real issue of the longer-term correlation, which is dramatic and requires explanation. In addition, since that time temperatures have fallen following the lengthening of the solar cycle and the graph's relationship would be confirmed.

The media and the message in the 'war on climate'

At the present time, the environmental campaign groups are riding a wave of apparent success on climate change. It has been placed firmly upon the political agenda. It is in everybody's minds in the community, in the media, and even in the business world. To even raise a critique of the global scientific monitoring community and the modellers who work with that data is to be consigned to the basket of 'conspiracy' theorists — as if the sociology of science had never been born. In my own case, in former work on ocean pollution and the collusion of UN agencies the standard technique was to ignore peer-reviewed papers and we had to rely upon the media to publicize these omissions. There were, however, some senior scientists in editorial positions who made invitations to publish what were in effect damning critiques of the way the UN operated and allowed scientific vested interests in research, monitoring and predictive models, corporate interests (for example, in the development of new chemicals and cheap forms of disposal), and governmental collusion in licensing discharges and dumping operations.[9] In these earlier years, the media were engaged in a profound critique of the old paradigm – yet now they seem to have fallen under a spell of naivety or vested self-interest in the scary-climate story.

These sorts of questions have become common ground when analysing the 'defence' industry, pharmaceuticals, pesticides, agrochemicals, nuclear and GMO industries, but the 'global warming' fraternity and its UN operation is placed beyond question. To even suggest that similar forces may be at work is to be denigrated and dismissed without argument. These attitudes are known as 'group-think' in the corporate world and, interestingly, some reflection emerged in the 14 April 2007 edition of *New Scientist*, where the editorial drew attention to a main feature article on the way 'groupthink', in particular the operation of authority, detracts from good decision making within the corporate world (and also in government and science). The editors then comment upon the latest IPCC report which they concluded leaves 'nowhere to turn for climate change deniers...' and then state apparently on reflection, 'assuming there is no "groupthink" here, or a global scientific conspiracy, the only other occasionally voiced argument is that the IPCC scientists have staked so much on the greenhouse gases that they are unwilling to brook any alternatives. This notion runs so completely counter to what science is about that it is as likely as a global conspiracy.'

New Scientist, hitherto a journal noted for a radical perspective, particularly with regard to corporate realities, shows itself completely incapable of even allowing the possibility of group-think not only within the IPCC but within science *generally*, despite decades of evidence to the contrary (often expressed within its own pages). The most disturbing element in this editorial is that no *evidence* is required to support this final end-of-argument assertion that 'not brooking any alternative' would be as unlikely as conspiracy. Scientific conspiracy may indeed be very rare as it involves the deliberate fabrication or suppression of data and a measure of collusion with other powers, either corporate or governmental. About the only current realm in which commentators might vote an exception with regard to the rarity of the phenomenon would be pharmaceutical drug trials or pesticide toxicity testing in which governments collude on the 'privacy' of data.

On the issue of this type of collusion, I wrote an opinion piece in *Nature* in 1978 on how opponents of nuclear power were denied access to probabilistic accident analyses on the potential for nuclear reactor cores to melt down. It turned out that government had knowingly colluded with the industry's science and engineering analysts in limiting the availability of data on large-scale accident consequences as well as the methodologies of risk assessment. Had such information been available at the outset of the nuclear programmes, it is doubtful any such power stations would have been licensable. The same could be said for reprocessing and certain high-level waste storage technology.[10]

At least *New Scientist* allows some critical comment in the Letters page of the same issue in which it features 'groupthink': Roger James writes,

> In Karl Popper's philosophy of science, a hypothesis can never be proved, only disproved ... given this, the scientists' task is to be able to show that there are no established facts that are not compatible with a theory. The challenge for those who hold carbon dioxide responsible for climate change, therefore is to ask what theoretically conceivable or possible observation would convince them that they are wrong.

The problem is that within the 'global warming' paradigm there is no room for 'we may be wrong, but...' Such doubt is not entertained. Hence everything that comes under the umbrella term of global warming divorces itself from science (and does so, paradoxically, in the name of science!). The IPCC is *the* authority, and supposedly a scientific body. The deeper problem, as with the NGOs, is that it is a *corporate* body and as such, in order to function, it must have *one* mind (the group-mind speaks with one voice). This is referred to by the appealing term of *consensus*, as if having so many scientists agree on something lends it sufficient authority to be beyond question — except by those with a questionably contrarian disposition!

The requirement of a certainty, one-minded, unquestioned authority and a simplified message that exhorts action has historical antecedents in every culture, and it would be as well as be aware that in those circumstances what is focused upon with such certainty usually diverts attention away from something the propagandists do not want anyone to look at. In the case of governments, a focus upon emissions, efficiency and supply technology draws attention away from any critique of the development model and its political economy. In the case of NGOs they draw attention, including any painful self-reflection, away from their own impotent lack of such a critique and their dependence upon that political economy. It is, of course, the political economy in its structure and value systems that drives the source of emissions ever upward. In the case of the media, simple messages make good campaign material, which also sells product. None of these players has the remotest self-interest in gathering potentially conflicting evidence, and that was always the flaw in Popperian science.

In all these campaigns we are asked to focus upon carbon emissions – usually our own – and upon a simplistic interpretation of climate science with the setting of arbitrary targets and timescales for emission reductions. These actions bear no relation to effective responses to climate change itself. Realistic emission reductions in Britain, Europe or America will not *reduce* current atmospheric carbon levels in the short or even medium term and not even in the longer term if developing nations industrialize as they currently plan to. It is questionable whether even a massive programme of worldwide emission reductions would prevent a *doubling* of atmospheric carbon levels. Yet in the simplistic messages of exhortation, these facts are not highlighted – clearly, to do so would be counter-productive.

Thus the *actual* relationship of carbon levels to the dynamic of climate change is ignored. This is precisely the realm of greatest uncertainty in the science. This uncertainty, widely expressed and admitted by *working* scientists, gets suppressed in the process that translates the working science into a policy message. This is the difference between the IPCC's 'Summary for Policymakers' and the body of the report. The writers of the Summary reflect a 'group-mind' that filters out dissent.[11]

Where *New Scientist* errs so profoundly is to deny that this group-mind can have the propensity to feed back to the working of science itself. There is a large body of sociological evidence that this feedback exists both in the corporate world, governmental realms and inter-governmental realms.[12] Indeed, paradigms exist within the institution of science itself which shift according to the social milieu of the political, economic and even religious world in which it has to operate.[13] The fact that science strives for an objective body of knowledge should not blind its institutions to their historical propensity to oppose this process. But it invariably does, with leading scientists in the fields of physics, chemistry, engineering and biosciences proving woefully ignorant of any sociological analysis of the relationship of their assumptions, theories and knowledge to social, political or psychological factors, or of their institutions to the broader world of political and economic forces.[14]

As scientists involved in the policy process, we need to reflect upon the structure of our institutions, academies, funding bodies and government advisory councils. If we are to learn the lessons from what may one day prove to be the single most damning failure of predictive environmental science, we need to carefully examine the way science colludes with political agendas, how it represents itself in the policy process and how it communicates to the public. In the British system, Civil Servants control and direct departmental policy and are adept at playing scientists involved in the processes of decision making. Advisory committees recruit from established figures with a preponderance of university professors. Likewise, research councils allocate funding with very little outside input and follow the interests of those institutes seeking funds. Government departments place contracts for outside consultant's analysis but seldom seek to challenge old paradigms and do not reflect upon an evolved culture of consultants whose basic dictum is to give the customer what they want to hear and make sure they stay in the league of safe choices.

In my 30 years of experience, there are a few individuals who stand out – in particular Lord Flowers, who led the Royal Commission on Environmental Pollution on the issue of nuclear power in 1976, and Sir Richard Southwood, who headed the same Commission's review of energy and climate issues in 2000. Professor Sir Fred Holliday led an exhaustive review of radioactive waste dumping. But it is the exception that academics can find the time and resources to appraise themselves adequately, and the academic climate has changed and become far less favourable to effective involvement; government quickly learns which people to avoid. In the case of climate issues, it has surrounded itself with figures with limited experience of the science and no propensity to challenge the political correctness. The present chief scientific advisor to the UK Department of Environment and Rural Affairs used to be an IPCC official. In the USA, the incoming President has appointed John Holdren, an advocate who regards any dissent from the IPCC view as a form of denial. The British press referred to Holdren as one of the world's leading climate scientists; yet he has no such hands-on experience, being a policy analyst with a physics background in energy technology. With the national science institutions, advisory committees and advisory posts thus selectively arranged, government insulates itself from the difficult task of constructing an effective policy in the face of major uncertainty. It could come seriously unstuck if the climate continues to cool. In the next chapter we look at what an effective policy might look like.

Meanwhile, my concluding recommendation is that governments review their safeguards with regard to controversial science-based policy areas. There are effective review procedures with various audit commissions and parliamentary select committees, and in my view it is the task of parliaments to ensure that these bodies are adequately funded and effectively reported.

Notes

1. I should point out that the US National Academy of Sciences somewhat revised its stance only four years later in 2005, when it joined with science academics worldwide in calling for action on carbon dioxide emissions. No new science was referred to in that time.

2. This story is told by Lawrence Solomon in the *Canadian National Post/Financial Post*, 28 April 2007, *tp://www.canada.com/nationalpost/financialpost/story.html?id=58e0c50c-1631-46ca-8719-78c0973526e*. The story is also told by Professor Fred Singer, a fellow target of the lawsuit, *The Revelle-Gore Story: Attempted Political Suppression of Science,* Hoover Press: Gough/Alchemy, DP0 HGOUAP1100 rev1 p. 283.

3. The company does not disclose its holdings, but an interview with *FastCompany.com* by Ellen McGirt provides some insight on Gore's extensive business and media interests and how to get the message across on long-term sustainability issues. *www.fastcompany.com/magazine/117/features-gore-an-inconvenient-portfolio.html*

4. In the run up to the Rio Environment summit I worked closely with a small team of scientist-activists that included Klaus Sperling and Volkert Dethlefsen, two fisheries ecologists working within the German government, as well as Professor Brian Wynne, a sociologist of science policy at Lancaster University, and my colleague Tim Jackson (now Professor of Sustainable Development at Surrey University) who worked closely with the Stockholm Environment Institute (see Taylor, 1990, 1991, 1993; and Jackson & Taylor, 1993); see also my chapter 'The Precautionary Principle' with Jackson and Dethlefsen, in *Clean Production Strategies* (Jackson, 1993); and my chapter 'Non-Governmental Organizations and International Treaties: the role of Greenpeace' in the Seminar series *International Politics of the Environment*, Oxford University, and 'Non-governmental organizations and the legal protection of the oceans' (with Kevin Stairs), in *International Politics and the Environment*, ed. Hurrell & Kingsbury (Stairs & Taylor, 1992).

5. In 1982, collaborating with Ian Sanderson, a transport specialist, I wrote *A Critical Review of Alternative Energy Strategies in the EEC*, a report for the Group of Independents in the European Parliament. It critically reviewed all of the low-energy and alternative-energy strategies then current in the EEC countries, providing a summary of their technical basis, political, social and economic feasibility, and a review of their impacts upon the environment. The four-volume 'Taylor-Sanderson Report' was translated into all the languages of the EEC and seminars were given to the Parliament and the Commission.

6. Peter Harper of the National Centre for Alternative Technology in Wales, writing in a letter to *Resurgence* magazine, advocated sacrifice of our aesthetic need of beautiful mountain country on behalf of the Bangladeshis of this world who are threatened by rising sea levels (a letter by me in response was published on their website).

7. See the Worldwatch Institute's *State of the World in 2006* and my discussion in 'Climate Watch' in *ECOS*, July 2007.

8. See the Worldwatch Institute's *State of the World in 2006*.

9. This work led to: 'The State of the Marine Environment: a critique of the work and the role of the Joint Group of Experts on Scientific Aspects of Marine Pollution (GESAMP)', published in *Marine Pollution Bulletin* 26, 3: 120–127 (**Taylor, 1993**), and was developed during work for the International Maritime Organization and presented at two symposiums; 'Environmental Capacity and the Limits of Predictive Science: the precautionary principle in the control of hazardous substances', a paper presented to the Joint International Symposium on Hazardous Waste Management, and published in *Proc. Symp*, Swedish Environmental Protection Agency & CEC, Stockholm (**Taylor, 1990**); and 'The Precautionary Principle and the Prevention of Ocean Pollution', a joint paper with Tim Jackson to 1st International Ocean Pollution Symposium, Puerto Rico 1991 and published in *Chemistry and Ecology* (**Jackson & Taylor, 1993**). The groundwork had been done with earlier critiques of the discharge of radionuclides to the marine environment in: 'The Interpretation of Monitoring Results' in *Radiation and Health*, ed. Southwood & Russell-Jones (**Taylor, 1987**); and 'Radionuclides in Cumbria: the international context', in *Pollution In Cumbria*, Institute of Terrestrial Ecology (**Taylor, 1985**); as well as 'Environmental risk analysis in Great Britain' (in Italian) in *Annuario Europeo Del'Ambiente*, DOCTER, Milan (**Taylor, 1984**).

10. I reviewed this issue in debate with the nuclear industry, 'Environmental Issues in Nuclear Risk Assessment' in *Nuclear Technology International* (**Taylor, 1988a**), at a major CEC conference, 'Large Consequence Low Probability Accidents' at the Standing Conference on Health & Safety in the Nuclear Age, CEC, Radiation Protection, Report EUR 11608 EN (**Taylor, 1988c**), and 'Nuclear Power in Central Europe' in *The Ecologist* (**Taylor, 1977**). But this

work began in the late 1970s when scientist-activists first began to penetrate the wall of secrecy surrounding nuclear 'safety' studies and licensing procedures. The secrecy allowed the industry and government experts uncritical access to policy makers and licensing committees – see 'Nuclear Energy: how the odds are stacked against the opponents' in *Nature* (**Taylor, 1979**).

11. All commissions and committees that work with large numbers of scientists in areas where disagreement exists have to handle dissenting voices. How this disagreement is handled is crucial. The more 'political' the situation (i.e. with policies that have large-scale consequences for investment, environmental impact and past political decisions) the stronger are the forces that operate to suppress dissent and uncertainty and also to revise past assessments in the light of any new science. Richard Lindzen, a lead author in the working groups of the 3rd Assessment Report of the IPCC in 2001 has been openly critical of the process, in particular the language adopted in the Summary for Policymakers and the changes in already agreed drafts at final stages. Additionally, Sonja Boehmer-Christiansen at Hull University studied the formulation and impact of scientific advice in global climate change as part of an Economic and Social Science Research Council project, concluding that these pressures operated not only to suppress 'plurality' of opinion within the scientific groups, but also to direct the research agendas of climate science itself, which could not be regarded as a neutral player – the science institutions were becoming political 'actors' on this international stage.

12. See Barry Barnes, *Interests and the Growth of Knowledge*, for insights into how political and social worlds impinge upon even the most apparently pristine fields of science. Barnes is now Professor of Sociology of Science at Exeter University and author with David Bloor and John Henry of *Scientific Knowledge: a sociological analysis*.

13. The classic work of Thomas Kuhn *Structure of Scientific Revolutions* also lays bare the paradigm shifts of science that always take place within the context of a social and political environment.

14. In this respect, a new book by Mike Hulme, the founding director of the Tyndall Centre for Climate Research at the University of East Anglia, *Why We Disagree About Climate Change* (Cambridge University Press), stands as an honourable exception, and sadly published too late for my own work to benefit from its cogent, reflective and scholarly analysis. It is rare for someone in Hulme's position to step back and take a sociological perspective on the way scientific knowledge is generated and used, and he has risked the criticism of his peers by speaking out on the use of alarmist terms such as 'climate chaos' and questioning the range of apparent 'fixes' for an intractable problem.

14

Urgency and Error

As I tackled the final editing of this chapter, Britain was experiencing the first heavy snow of approaching winter, the USA was in the grip of record snowfalls, the Arctic sea ice was 30% up in October compared to last year and central Siberia was experiencing such unusually warm weather that migratory ducks, geese and swans had yet to take their winter vacation on my local estuaries. The data centres recorded a continual drop in the global mean temperature and the trend lines showed no overall warming since 1995. Despite what the instruments say, politicians and media pundits, environmental campaigners and the chairman of the IPCC warned that global temperatures were rising faster than expected. There were whole-page adverts in the press purporting to represent scientific opinion that the planet is fast approaching a 'tipping point' when runaway warming will be unstoppable – and the UK Parliament voted to increase the country's commitment to carbon emission policies by setting a long-term 2050 target of 80% reduction. The EU has already tied all member states into a 30% target by 2020, with a 5% target for biofuels in transport. The new administration in the USA is expected to throw its weight behind a target-led approach to carbon emissions, much as several cities and the state of California have already done.

Yet a perusal of the real-world scientific data that I have reviewed provides no clear evidence for such tipping points, no convincing evidence that the climate change seen over the past 25 years is driven by the rise in greenhouse gases, and very little evidence that future climate change can be predicted. Further, the simulations that do exist, point to significant effects of greenhouse gas build-up only in the latter half of the century and, as importantly, no effects until then of mitigatory policies pursued so urgently in this next decade. Yet the world's science institutions back the IPCC in its call for urgent action to tackle climate change, and its representatives, the media and environmental groups continue to assert that temperatures are rising. This repetitive and false assertion bears comparison to the endlessly repeated claims regarding weapons of mass destruction in the lead-up and aftermath of the invasion of Iraq despite the evidence to the contrary. It is chilling in its Orwellian undertones, as if by repeating something often enough, irrespective of the truth, has become a learned technique to convince the populace.

I have colleagues who agree with my analysis of the climate science but say, 'What does it matter? At least the money will be spent on things that will ultimately prove useful.' Unfortunately, the great majority of the resources now devoted to fighting climate change are directed towards energy supply options, many of which are extremely damaging and of limited relevance to the real risks of a changing climate. The perceived urgency now directs government to ways and means of supplying energy without burning carbon. This path appears more reliable in terms of meeting targets than that of reducing the demand for that energy by restructuring the economy. This is how government has always operated, turning to supply options to meet rising demand, rather than attempting to reduce that demand. Renewable energy is thus treated in the same way. This path has never actually delivered in the past, as I will show; a much more sophisticated approach is going to be necessary, particularly during times of economic recession.

Taking time to get the policy right

My original purpose in examining the science was to assess whether there really was the urgency that is being pressed upon government, businesses and consumers. My motive stemmed from my review in 2002 of the Royal Commission's proposed energy policy for the UK in order to achieve a 50% reduction in emissions of greenhouse gases by 2050. That policy, enacted without stringent safeguards, would lay waste to the British landscape. That which I love, and which sustains me, would be turned into a technoscape of turbines, barrages, plantations and powerlines. And at the same time, energy demand and carbon emissions in the UK are projected to rise for several decades and certainly much more on a global level.

With care, however, these changes could be managed and integrated into a response to climate change that honoured all of the criteria of sustainability relating to rural communities, landscape, farming, forestry, recreation and wildlife. And to this end I worked for several years, first bringing together a team of sustainable energy experts and planners, ecologists and landscape designers to begin a visualization process that could be developed at local and regional level that would educate communities with regard to the technology available and the impacts generated by each choice. In this regard, I sat for three years on the National Advisory Group for the government's Community Renewables Initiative, setting up ten regional centres of expertise and information. This work is showcased on the Ethos-UK website.

We are now eight years on. Very slowly, the UK has begun to take loft insulation seriously. Future housing standards, it is promised, will adopt a 'zero' carbon standard. Meanwhile, land-based wind turbines have grown fivefold from about 300 MW to 1.5 GW and there are 10 GW 'held up in the planning process'. An 8 GW Severn barrage scheme has been resurrected for further study (it was recommended by the Royal Commission), a £200 million plant is being built on Humberside to turn grain into bio-ethanol to meet the EU directive and the 4 GW Drax coal station is scheduled to burn 10% biomass by 2010. Plans are well developed for several large arrays of wind turbines offshore with talk of 20–30 GW capacity, but also of massive logistical blocks relating to the availability of construction barges as well as turbines.

Meanwhile, the global economy faces recession, the inflated oil price has fallen from $140 maximum to less than $50, and although the economies of the Middle East, China, India and Brazil are still registering above 8% growth they too are faltering. Just in the last few months, the political and economic climate has changed considerably. We might call this a 'phase-shift', somewhat time-delayed from the climatic phase shift of 2001, but entirely coincident with the depressed solar cycle.

This economic shift has enormous significance for the 'climate wars' response. The Green Party in the European Parliament is calling for a 'Carbon Army' to go into battle – fighting for efficiency and with ever better carbon-freeing technology. All that is missing are the uniforms! Britain's special Cabinet Minister for Energy and Climate is talking of part-nationalizing the electricity industry.

Beyond this economic world, the sun is still very quiet. Its electro-magnetic field flickered into life last month (October), and curiously the markets rallied briefly on the same day – but a mere flicker and two years behind NASA's computerized schedule.[1] More solar scientists are now talking of the possibility of a Maunder Minimum, but not yet the politicians, media pundits and environmental campaigners – and of course not yet the IPCC! It is against this background of many changing climates that I turn now to an effective policy response.

The illusions of mitigation

It follows from my analysis of the science that mitigation is an illusion. Firstly, even accepting the orthodox model, politically and economically feasible emission reductions, such as the most efficient and socially conscious scenarios of the IPCC for example, that could achieve a 50% reduction by 2050 would not mitigate the effects of the next 50 years of climate change and very questionably affect the 50 years following that. Given that the IPCC scenarios are based upon maintaining economic growth within the standard economic model (which ignores major cycles and phase changes until they happen!) *and* have overestimated the carbon dioxide effect as well as failed to incorporate climate cycles and phase changes, then we can see that mitigation is a dangerous and expensive illusion. The most that could be said for it would be that it might make for a more stable world in the twenty-second century, but it would be at the price of distracting attention from necessary adaptation over the next few decades.

If I am right about human-induced emissions of greenhouse gases contributing less than 20% of the driving force of the global warming and carbon dioxide is responsible for 75% of that (the other greenhouse gases such as methane make up 25%) and, further, that the carbon dioxide component is made up of 75% industrial/consumer emissions and 25% from land-use changes, then a simple calculation shows that a 50% reduction of global industrial and household emissions (including all transport) will address only 6% of the driving force and an 80% reduction only 9%.

There is, therefore, no urgency in relation to climate mitigation policies. We can therefore take time to look carefully at the technology, timescales, finance and environmental impact of an energy policy that would still need to take account of the future limitations of oil and gas supply, as well as the major atmospheric pollution of coal burning. Nuclear power also has its constraints of resource supply as well as wide-ranging social and environmental impacts.

There is however another source of urgency. There is a very real possibility of a major global cooling episode. This possibility is not presently admitted by the experts at NASA and Hadley, nor talked of openly by governments and environmental groups. But it is nevertheless real, and if this next solar cycle falls further back, as predicted by Landscheidt, then within three years we could hear governments and science institutions discussing the possibility of a Little Ice Age and a shift in policy.

Global cooling and famine

Cooling is not good news, even for a supposedly over-heated planet. The past 30-year warm phase has seen a doubling of the global population, which now stands close to 7 billion, and agriculture has kept pace. Another billion will be added to this by 2020. Most of this increase will be in populations already stressed by poverty and diminished natural resources. In 2008, there were 67 countries in food deficit and for the first time in its history the World Food Programme ran out of money to purchase food on the inflated world markets. Agriculture does not respond well to falling temperatures, especially in the northern hemisphere. In this respect global cooling is worse than global warming.

I warned about the possibility of global cooling and pressure on food supplies two years ago.[2] I predicted then that food would become the overriding issue of urgency within three to five years. When the World Food Programme issued its warning in summer 2008, there was still a world food surplus; it was just expensive due to the pressures of some poor harvests in Australia and China, the high cost of oil, commodity speculation and the burgeoning demand for biofuels. If we add to that a drop in yields from the northern grain belt – the main surplus – then at some point within the next three years the world could enter food deficit.

I could say that we are not prepared for this global humanitarian emergency. But that would be only partly true. Some countries are well prepared and protected and have seen this crisis approaching for some time. In the last year alone, as I noted previously, sovereign funds in the Gulf States, South Korea and China have been buying millions of acres of productive land in some of the world's poorest regions of subsistence farming – Laos, Cambodia, Mongolia, Brazil and Madagascar. Large international corporations have also been purchasing or leasing land for dual purpose food-biofuel crops such as soya and sugar cane in South America and Africa.

Much of the motive behind trade 'liberalization' has been to free the movements of capital and open up land ownership in advance of this state of affairs. Of course, in the standard model this is not just to improve production and 'feed the world' (with genetically modified crops) but also to reduce food prices on the world markets. The negotiations have stalled because the 'developing' world's food producers want something in return – not just access to 'western' markets, but also the curtailment of government subsidies that protect western producers and distort the world market.

This system is not set up to cope with a world food deficit, at least not for the majority who live in relative poverty. But it is well adapted for the richer countries to buy their way out of a famine. Lest anyone should doubt the potential failure of humanity in this regard, at the height of the British Empire and the opening of the first great global market for commodities three million landless peasants emigrated or starved to death on Britain's doorstep in the colony that was Ireland. Their subsistence-level potato harvest had failed and they could not turn to alternative foods because English landlords exported the grain harvest onto the 'global' market. Of course, the peasants could not afford to buy the grain.

We could perhaps note some antecedents. Ireland had been colonized by Britain in 1801 and 60% of its democratically elected representatives to the UK Parliament were landowners. There was 30% unemployment and widespread poverty in the years leading to the failure of the potato crop in 1845 due to the fungal disease blight. Most were landless labourers. One English Lord, for example, owned 240 square kilometres, and those employed on the land raised crops and livestock for export and profit for the largely absentee landlords.[3]

Learning from history: the Irish famine 1845–52

In 1841, Ireland was part of the British Empire, with a population of 8 million people, two-thirds of which were engaged in agriculture on some of the most fertile land in Europe. John Mitchell comments on the famine of 1845–52 (**Mitchell, 2005**):

> That an island which is said to be an integral part of the richest empire on the globe ... should in five years lose two and a half millions of its people (more than one fourth) by hunger, and the consequence of hunger, and flight beyond sea to escape from hunger ...

The island population suffered an estimated 1 to 1.5 million deaths, with another million emigrating. During the six years numerous bungled attempts were made by the British administrators to retrieve the situation, but they were hampered by the constraints of the market and the lobby of landlords to protect prices. At the time, an offer of foreign financial aid from the Ottoman Empire was curtailed because it made the British aid look paltry in comparison. As hundreds of thousands faced starvation, the peasant landholders received aid only on condition that they gave up title to their lands. In the midst of this time an Irish rebellion with the goal of repealing the Corn Laws and banning exports was put down by the British military (some commentators believe Irish exports of meat products actually rose during the famine years).

Ironically, a small financial aid package of about one hundred dollars was delivered on behalf of the Choctaw nation living on a tribal reservation after the infamous Trail of Tears, when they had been forcibly removed from their lands and marched to new lands in Oklahoma. This event was recently commemorated by the Irish President.

There are a number of lessons from this history that are relevant today – social, political, economic and ecological:

- subsistence farmers with age-old systems adapted to their climate were by force of law and economic circumstance driven into wage labour on very low incomes for the production of exported food;
- the population became reliant upon alien crops and small plots of land, and condemned to poverty by low wages and taxes that kept them in the global economy; housing was poor and health provision inadequate;
- the crops were blighted by a fungal disease imported by accident in fertilizer from South America;
- the system collapsed very rapidly and there were 'market' constraints operating against retrieving the situation – social unrest ensued, which involved military action and punitive courts.

If lessons are not learned, history does repeat. A recent Channel Four documentary on Paraguay and the plight of its Amerindian peasant farmers – whilst not at the famine stage – shows the same process of dispossession at work as more and more land is bought up by international consortia engaged in soya production. The main difference in the modern global economy is that the diaspora created by the current 'development' model leads millions into either *favellas* on the edge of mega-cities such as Sao Paulo or into the high-rise apartment blocks of the Gulf States, Bombay and Shanghai, where they service the global economy. None of this is remotely sustainable on an ecological level and, given the economic crisis of 2008, probably not socially or politically sustainable either. Climate change and especially global cooling will bring this system into a major crisis where millions of people face famine, and the deciding factor of who lives and who dies will be who has access to the global market.

Against this rather predictable consequence of climate change, the urgency with regard to energy supplies appears blind and morally bankrupt. Economic policies have been concerned with 'wealth protection' and hence the control of inflation in the so-called developed economies. Yet the wealth protection of the global warming years has only been achieved by export of capital and services (and hence emissions) to regions of cheap labour and resources such as China, India, Indonesia and the Gulf States. These countries have maintained an 8–10% growth rate (and a similar or higher return on capital invested), and their return export of cheaper services and goods has helped the western economies avoid economic stagnation. These countries have also accumulated vast currency surpluses which for some time they reinvested in western securities and derivatives. The latter term is an obfuscation for the selling on of mortgages and loans by western banks eager to funnel more money into the ailing western economies. Having sold on the mortgages, the banks used this money to lend more cash into the bubble, thus fuelling the housing market price rise and promoting release of equity and second mortgages. This enabled the richer to carry on purchasing luxury goods and the poorer to buy houses they could not afford. It only took a small alarm in the US mortgage market to cause the market for mortgage-backed derivatives to dry up and the banks to find themselves with no more cash to lend.

Having recently been 'bailed out' by governments to the tune of hundreds of billions, these banks are still reluctant to lend into ailing economies and would rather purchase equities at the prices they helped to knock down, or invest in the still growing economies of Brazil, Russia, the Gulf States, India and China. I detail this state of affairs because it is very relevant to the current decade and anything I might recommend with regard to a climate policy, which must necessarily be paid for and take account of current market conditions. But there are other parallels – as with the peaking of ocean and solar cycles, the economic cycle is unique in human history. It builds upon everything that has gone before. And everything that is vulnerable to the climate changing is also vulnerable to the economic climate. There are feedbacks.

If the uniqueness of the current situation is not recognized, then policies will fail and the consequences will be severe. If old remedies are sought, with the same kind of thinking and systems that produced the crisis, then they will almost certainly fail. Almost everything that is proposed for climate mitigation is not only old thinking, it is counter-productive – unless, that is, we are to abandon humanitarian concern and our imperatives to safeguard sustainable lifestyles and biodiversity.

It is against this backdrop that the arguments relating to transforming the global economy from fossil fuel dependency to a low carbon future should be seen. We require a different mode of thinking with regard to energy and economy, yet the main protagonists are talking of little more than trying to power the current system from an alternative source. In my opinion and with years of experience of looking at the technologies available, their costs and impacts, it is not feasible. Indeed, it can be viewed as a major deception to rank with the 'war on terror', the 'war on poverty' and the 'war on drugs', which is to say that the architects of war know perfectly well whom the war serves.

I feel it necessary to add to this matter of 'deception' as it is a hallmark of our times. As the banking crisis unfurled it became apparent that chief executives of the twelve leading banks had paid themselves close to a billion dollars in bonuses. The pattern repeated down through the levels of management. The true state of the banks' finances was apparently hidden from depositors. In one case, a senior banker managed to embezzle over $50 billion, so lax was the oversight. This is deception on a grand scale. But only partly deception. A strong case could be made for mass collusion. As long as the profits were rolling in, however unsustainable, many people who knew exactly what was going on had no reason to raise an alarm. The bankers and economic pundits now act as if nobody knew or understood what was happening, and government colludes in this further deception and then channels further billions into the failed institutions, many with their old management still in place.

There is no reason to suppose that government is any more intelligent with regard to climate science and energy policy, or that any lesser degree of deception pertains. This realization is particularly important considering the equally vast sums that will be involved with carbon trading. We will return to this issue.

The dangers of targets and timescales

Currently, the main focus of policy is upon carbon emissions from industry and households. Environmental campaign groups are producing figures justifying the most substantial cuts, some as high as 80% by 2030,[4] whereas the UK Royal Commission on Environmental Pollution argued for a 60% cut by 2050 and many governments have adopted this longer-range figure (**Royal Commission, 2000**). As we noted, these cuts are predicated upon the models and projections of the IPCC and a working assumption that global temperature rises must not exceed 2°C if we are to avoid 'dangerous climate change' (**Schellnhuber, 2006**).

In order to stay below this threshold the IPCC have reported on a number of scenarios and projections to the end of the century or beyond based upon varying levels of economic growth, models of development (such as energy intensity and environmental protection) and population growth. Even the most optimistic of scenarios, involving a major shift towards environmentally friendly technology and lifestyle, indicates a continuing rise in greenhouse gases to the equivalent of doubling CO_2 levels by 2050, and the Panel provide a range of forecasts for the resultant global average temperature rise. These projections vary according to the feedback assumptions of the models and simulation runs, which display differing 'sensitivity' (the term used for the assumed temperature response to a doubling of CO_2 concentrations). High sensitivity when coupled to high assumptions of population growth, economic growth and business-as-usual scenarios can generate simulated temperature rises of 6°C.

In view of the scientific uncertainty evident in the models there should be little confidence in the sensitivity factor, yet this is not questioned in the debates around mitigation timescales. We have already documented key uncertainties relating to the water vapour, cloud feedback and ocean heat storage, with recent temperature patterns showing that natural forces are now dominating the greenhouse effect. In addition to these uncertainties, there can also be little confidence in the future pattern and level of economic growth. It is not at all clear, therefore, that this huge effort at scenario building has a great deal of scientific value. Predictions vary between 1.5 and 6 degrees and any probability assessment has virtually no reference to the real world, simply reflecting 'expert' judgement and the degrees of statistical uncertainty in the models. As we saw, some specialists have made much simpler calculations based upon the temperature increases that have been seen so far and have advanced a sensitivity factor as low as 0.5 degrees.

Whatever the sensitivity factor, it seems to me that a doubling of CO_2 levels by 2050 is likely. Further, as we have noted, there are a number of solar scientists who expect the sun's magnetic activity to decline. These solar lows have coincided with low temperatures in the northern hemisphere with historic lows of as much as one degree and lower falls in some key regions. When commenting upon recent attempts by solar scientists to bring this information to public knowledge, *New Scientist* talked of potentially 'buying time' to respond to global warming ('Saved by the Sun', *New Scientist*, 16 September 2006), but no mention was made of the current vulnerability of modern populations and the world food situation.

It could be argued that longer timescales of effects should also be considered – for example, a possible trebling of carbon levels to 2100 and beyond – but the impending social and economic changes make such long-term prediction of emission trends of little value. The impending peak of world oil production coupled with the peaking of world population and the stresses this will cause upon the global economy and ecology should make all such forecasts an illusory exercise. There is also the question of technology cycles related to energy and transport, which will have a greater and unpredictable impact in the later decades of this century.

The future is not predictable

I see little benefit in running sophisticated multi-scenario computer simulations of dubious accuracy in the light of these major uncertainties. The immediate future is only relatively predictable. Analysts can only work with known technologies, the implementation of which largely depends upon economic factors; and these economic factors, despite what economists would like to think, are likely to prove very unpredictable (I wrote these lines months before the current global crisis). Oil analysts do not expect prices to remain at their current low of $40 per barrel (having peaked at $140). Meanwhile, there is an artificially stimulated price for biofuels. In a recently reported bankruptcy, a biofuels company failed to deliver on a major contract because the price of palm oil had doubled from $350 to $700 per tonne within a year. The much heralded Stern Report by the UK Government, on the projected costs of mitigation compared to the projected costs of climate damage, operated on a maintained oil price of $57 per barrel for many decades. Clearly the future is uncertain with a difficult climate for long-term investment and many are calling for a 'new world order' to direct funds appropriately.

Despite such modelling attempts to cost both the investment required to reduce emissions and the environmental damage consequent upon failure to do so, I place little faith in such projections. Economists maintain their projections based upon past trends and hence the future is always of perpetual economic growth, and do not entertain slumps or crashes until they happen. It is always assumed that growth will resume. The prospect of serious structural change brought about by population, resource pressures and social unrest is hardly ever considered.

When this last sentence was first drafted, the current economic crisis was not incorporated into any scenarios; almost all benign development models assume a transfer of wealth from developed to developing countries together with functioning clean development mechanisms for technology transfer and a carbon credit system. In many ways, the wealth of the industrial north is exposed as rather fragile and a period of reduced growth, recession, higher taxation and several state bankruptcies is likely over the crucial decades ahead. Further, the large accumulated foreign currency reserves of export-led-growth countries such as China, the Middle East, Russia and India are proving equally fragile, with those states lending their dollar reserves back to the USA and Europe in order to keep those economies from recession and hence their own export markets afloat.

When we come to consider adaptation strategies in relation to water, forests and agriculture, in particular the globally vulnerable states, the required investments are measured in tens of billions of dollars (for example one estimate for securing global water supplies in both hygiene and availability is of the order of $20 billion, and similar figures exist for forest protection and biodiversity), yet such sums are being paid to aid a few banks in the developed world, and the global banking crisis is estimated at nearly $3000 billion. China, India and Russia hold approximately $2000 billion in their trading accounts.

Carbon credits and capital transfers

I will not here review the complex flows of capital in relation to carbon credit systems; at present they measure only a few billion, mostly within the European Union, and a similar order from the Global Environment Facility of the IMF. The disbursement of these funds has attracted severe criticism with corporations benefiting but with minimal impact on carbon emissions. In one case, customers signing up to a Barclay's Bank credit scheme paid for carbon reductions using hydroelectric and wind turbine schemes in China. These schemes could hardly be described as meeting the criteria of 'additionality' (i.e. the schemes that would not have happened without the funding) when cash-rich China is already expanding its renewable energy supplies (though tiny compared to the expansion of coal production!).

I will give these modelled projections and mitigation schemes little further attention. The imminent economic and ecological future upon which most policy depends is far more important and more readily predictable. This is the main point of my argument and the central reason for writing this book. Humanity is heading for what John Mason, the Australian diplomat who has campaigned so strongly on this issue, called the 2030 'spike' – by which time all of the current system's ecologically headless chickens come home to roost (**Mason, 2003**). Emission controls can have no effect on climate during this timescale. We are committed to climate change, whatever the uncertainties of the climate science. And even if the climate stabilized, the changes coming are no less severe; I think they will come much earlier than John Mason envisaged.

This issue of the 'spike' is what is missing from the ongoing climate debate with its focus upon emission targets, energy supply technologies, and technical 'fixes' generally. Certainly the debate broadens out into issues of international development and equity, particularly where contraction and convergence is concerned and the notion that a planetary just carbon allowance could be agreed,[5] but the debate is remarkably naive when it comes to models of development and the entrenched power of competitive industrial economies and their goals. Contraction and convergence could not come about without a significant shift in economic power and privilege. There are no signs that this is going to happen, at least as a result of a concerted effort at redistribution. And contraction and convergence does not address the historic investment in technology and production, the purchase and distribution of goods, the migration and movement of labour and the fundamentally different nature of developed and developing economies.

Policy in a real world

After ten years of advising UK government agencies and working on inputs to government consultations on energy policy, I have a good idea of how government in this new century works, what is feasible and why. There have been some significant changes in recent years and I am not convinced that environmental campaigners have taken these changes on board. Governments are well aware of the popular concern for the environment, and indeed most politicians share that concern. However, their job is to balance this with their commitments to global economic growth and social stability. In former times, government could dismiss environmental issues as peripheral or at most a question of cleaner technology and better regulation; but climate issues, whether or not the science is sound, bring into question the sustainability of the whole enterprise and hence cannot be ignored.

The problem for government is that the force inherent in the commitment to economic growth runs counter to the orthodox climate imperatives to which they publicly subscribe. Actions that would be required to effectively address the climate issue run up against the economic imperatives of controlling inflation and maintaining full employment – and soon perhaps also recovering from recession. Thus, (most) governments play the game of agreeing with the climate analysis, thus diffusing any criticism, and then pretending to do something about it. The favoured means is to enact policies at a minimal but highly symbolic level, thus appearing to respond, but not actually, and thus not compromising the economic goals or raising any question of restructuring towards real sustainabiity. And, of course, it is in everybody's short-term economic interests to collude in this – even the environmental campaigners.

As I argued, collusion and deception is intrinsic to the climate game, which is, after all, a war game. The preference for simple targets that government can sign up to, with the UK commitment to 80% reduction by 2050 as the main example, is underlain by the most simplistic analyses of how such a transformation can be achieved and at what cost. But in terms of the game, campaigners have won and hit their (meaningless) target. Furthermore, because it is a game, no comparative environmental impact assessments have been performed. As I will argue, the supposed remedy is likely far worse than the sickness, and the lack of impact studies indicates the lack of real ecological and caring consciousness. I recall that when the dumpers were arguing that the ocean was the best option, they too failed to provide impact assessments for other options despite international regulations advising them to do so. The EU also has regulations for the environmental (and social) impact assessment of major policy decisions, but these are also not being implemented because there is no lobby for them.

This simplistic approach is most evident in the popular baiting of the US administration for failing to join Kyoto and embrace emission reduction targets. I have never seen any mention alongside the US's 25% of global emissions figure that the US also produces 25% of tradeable goods and thus drives the world economy from which virtually all other nations benefit. Much as there is a huge potential for efficiency within the US economy, cheap fuel has maintained its economic growth and powered the global economy. American investment and trade also powers the Chinese economy (*vide* $200 billion in their annual trade surplus), which then feeds back into the global economy from which all states benefit.

I am no apologist for this global growth. The measurement of GDP might as well be an index of planetary destruction, and the competitiveness of the US economy has come at a high price in lack of social investment, widespread inequity, the seizure of land from indigenous peoples and the prosecution of global warfare such that the US accounts for half of the world's annual military expenditure (or as much as the rest of the world combined!) at $500 billion.

It is important to understand this dynamic, if effective policies are to be developed. 'China' is now an illusory nation state. It is part of the global (US-led) economy, and even talk of the balance of power shifting is relatively meaningless when China is dependent upon that economy. What is at stake is the relative wealth of sub-populations within each of the apparently independent nation states.

The defence of wealth

One of the key imperatives relating to economic growth is the control of inflation, which in a world divided into industrialized states and commodity producers means protecting the wealth of the industrialized population. All industrialized economies are structured to prevent a rise in inflation as an uppermost goal. What this innocuous word actually means in the modern context is to hold onto the privileges and power that has already been accumulated and to increase it. This is not bluntly stated because that would upset global sensibilities in an era when the rich are supposed to have the eradication of world poverty as one of their key humanitarian goals. In fact, rich nations operate policies that resist price increases in commodities, which is all that poor countries usually have to contribute to the world economy.

It is important to note that none of the environmental groups that espouse the IPCC approach question these imperatives. They have, of course, become corporate entities reliant upon financially buoyant consumers and have thus no motives to provide any real-world analysis of the problems of vulnerability and inequity. And with regard to policy advice by groups of government-appointed scientists, there is of course even less criticism of the basic development model.

A recent joint policy initiative by FOE, the Cooperative Bank (which has a commitment to ethical and environmental standards) and the academic Tyndall Centre for Climate Studies at Manchester University illustrates the point. The parameters for an energy policy scenario were set to allow projections of economic growth and continued privileges and freedoms that are totally aligned to government policy, including, remarkably, an increased measure of transport mobility, (**Tyndall Centre, 2007**).

Energy, society and security

Development and energy supply are synonymous. Industrial society is founded upon supplies of cheap fossil fuels. As soon as anyone talks of 'sustainable' development, the issue of sustainable supplies of energy comes to the fore. Indeed, despite the obvious inadequacy of this world view, almost all environmental campaigners fall into the same paradigm with only a nuance changing: sustainable development has been taken over by climate issues and climate issues are reduced first to carbon dioxide, then to emissions and then to energy policy.

All modern protagonists in the energy debate view nature (and humans) mechanistically. Energy metabolism has become more than a metaphor transferred from animals to human society; it is regarded as a scientific reality (if not the reality). Thus emerges a language of energy efficiency, energy flows, consumption and waste. This scientific and technical linguistic view then translates into a focus upon technology – mostly of supply, but also of increased efficiency. The scientific world view does not dwell on 'purpose', leaving that to theologians, nor does it often turn its attention to economic and social structures and vested interests. And as we noted in our brief discussion of the nuclear debate, the modern environmentalist has more in common with the technocratic mind of nuclear protagonists than the communally oriented mind of the original green visionaries and writers of Rio's Agenda for the 21st Century.

Furthermore, it has always surprised me how few scientists are aware of sociological enquiry into their own behaviour as players in the political and economic world that feeds them and accords them status. Science maintains a self-image of objectivity and sees itself as the central enterprise that underpins technological progress. It eschews the issues of purpose and meaning and pretends to an independence of the political and economic world.

Thus the scientific view of sustainable development is that carbon dioxide emissions are a consequence of a particular technology, and hence in the face of climate change that technology must be replaced. But on a political level the technology of energy use is fundamentally a consequence of the development model – the demand for energy at a cheap price. It is an issue of consumption and consumer choice. This is often referred to as 'lifestyle' choice, but for the vast majority of people (and this includes people in the 'developed' world) there is very little choice involved.

Any political and social analysis of the situation in the developed, rich world would show that a large proportion of the population lives with very small margins of freedom, with significant numbers in fuel poverty (spending more than 10% of their income on keeping warm) and unable to cope with inflation in this sector. The capacity of government to ameliorate effects on the scale required is dependent on the very small margins of maintained economic growth – of the order of 1–2% – and any stagnation or recession of even 0.5% coupled with price inflation is likely to bring widespread social distress and the threat of disorder.

The shadow of development

This real-world pattern of energy use in modern society arises from the historic development model, and it is this model that is currently exported to the developing economies:

- Capital and enterprise creates 'industry' by replacing handwork with machines and fossil energy accelerates this process because it is so much cheaper than human labour.
- Few people would choose to spend five days a week looking after or operating a machine on a production line unless they have no alternative means of providing for a basic livelihood – and few see such work as a potential road to such wealth as would liberate them from these basic concerns.
- The managers of industry, their financiers and other service sector personnel, having more access to finance and schooling, do strive to liberate themselves and some do so successfully, with very high bonuses and investments that more than secure their basic and future requirements.

- This model strengthens itself by creating basic insecurities sufficient to propel significant numbers of people to work in boring, sometimes soul-destroying production lines but to reward them sufficiently that the small amount of time they have free makes up for the drudgery. In later developmental stages, entertainments, illusory opportunities and lotteries keep the dream of personal freedom alive.

- At an early stage in the process it is necessary to enact policies that uproot people from self-sufficient, subsistence levels of occupancy of the land into cities, introduce property rights and real estate markets and to concentrate ownership of the land, as well as the production facilities, into a few hands – the land then becomes another 'industry' which can then feed the cities as well as the profits of banks, corporate land-owners and wealthy individuals.

- Once this movement from land to city is complete (as in a 'developed' industrial economy) it is then imperative to stave off unrest and disorder resulting from poor living conditions, crime, drugs, gang warfare, prostitution and mental disease, and in this more enlightened late-industrial age this is largely accomplished by efficient government programmes of health, education, housing, sanitation, full employment, high wages where possible, and eventually shareholdings and ownership of valuable real estate. (Note: as China, now a vital part of the global development process, emerges from 'developing' status, it has reached the point where large numbers are ensconced in city high-rise and appropriately named 'appartments' but not yet secure in the anticipated developed-world benefits, and many commentators report widespread social unrest).

- As the developed economy gets richer it begins to stratify further and sucks in immigrants from poorer countries willing to work for low wages and do the menial tasks. Thus where normally a developed economy would peak in population and even begin to decline, it continues to grow in its resource demands for space, housing, infrastructure and energy use (a classic example is Spain where carbon emissions have increased by 50% above 1990 levels and several million foreign workers from Africa and the former Spanish colonies support the economy).

- Further factors arise as each developed economy or trading area competes with others in the more developed industries and technologies. The chief factor is price, but important also is skill and inventiveness in manufactures, and the economic power to secure its resource demands.

- Eventually a small number of states become financial centres, with that sector dominating the economy, and servicing the development and trade structures of the rest of the globe (for this reason, the UK, as a global financial centre, suffers more from a downturn).

- In the later stages of the developed economy another factor emerges as a result of the historic exploitation of the resources through various phases of empire (at first military and then economic): the political backlash of economic and cultural colonialism has to be managed and in the modern world this is served by the transporting of 'democratic' institutions of government and a limited amount of trade benefit largely to elites trained in the colonial capitals' business and law schools. However, the system has lately broken down in relation to the Islamic world and the appearance of 'terrorism' and the expansion of 'coalition' forces onto foreign soil (at least those countries with important resources of oil, gas and strategic minerals).

Terrorism, drugs, crime, mental ill-health, prostitution, trafficking, obesity and violence against women are now pandemic features of the development model's underbelly – and shadow. Indeed al Qaeda means 'the base' – a reflection of the inherent insecurity upon which the developed nations' social systems depend, and which is then psychologically projected onto an external enemy. That enemy also espouses a puritanical zeal in the face of such a parade of corruption. As with the climate, to think such 'opposition' can be beaten is to buy into a convenient illusion. Such opposition will fade once the shadow is embraced and the underlying corruption and injustice dealt with.

With respect to the rising tide of obesity, in a recent issue of *New Scientist* Ian Roberts, professor of public health at the London School of Hygiene and Tropical Medicine, argued that there was a clear link between the obesity epidemic and global warming (he called 'pandemic obesity' an 'energy vortex' that was spiralling out of control). An obese person eats about 40% more in calories and this feeds back through the rich and fatty food choices to excess agricultural production, food processing and transport – all sectors that are significant emitters of greenhouse gases. Obese people also increase their energy consumption in the home and with motor cars. And in the US where they are a large percentage of the population, they have begun to campaign against the social stigma that Roberts regards as the only thing slowing the epidemic. He makes the point 'we live in an environment that serves primarily the financial interests of the corporations that sell food, cars and petroleum', but he does not point out that this is perhaps the final stage of the global development model.

As an example of what can happen when a rich nation adopts a particular food source, consider the fashion for farmed salmon. It is farmed in Norway and shipped frozen to a processing centre at a port in China where it is filleted and packaged for the world's supermarkets. Container ships then bring the packaged salmon back to a port in Britain for distribution. The fuel costs, however great, are outweighed by the labour-cost savings in China and the whole enterprise is 'economic', contributes to economic growth and has made salmon economically accessible to larger numbers of people.

World trade rules and economic values support ecological madness

The global development model allows such ecological madness. Indeed, there are WTO laws that not only enable this but work against any one country imposing environmental regulations that would inhibit trade – particularly in food products.

Unless the shadow side of development is acknowledged as the driving force of carbon emissions (and all other environmental degradation) not a lot is going to change, and this returns me to the issue of deception. It is not productive for environmental campaign groups, liberal journalists, government chief scientists, the Royal Society or the IPCC to venture into these political waters. So they do not. Instead they pretend that these issues do not exist, even when they are intellectually well aware that social forces are the fundamental driving forces of technological choices. This leads them to policies that advocate technical fixes or small 'lifestyle' changes that can be made within the paradigm, and then to pretend that this approach has the capacity to deliver the longer-term targets. Thus, commentators in Britain and other EU countries can enumerate all manner of such changes without fully disclosing their absolutely minimal impact on global carbon projections which are entirely dependent on the development model and the expansion of the world economy.

This naivety is not so prevalent in the USA, where critics of the global warming orthodoxy also tend to be anti-socialist, anti-taxation, rather gung-ho supporters of the standard business-as-usual development model and see 'global warmers' as a quasi-religious socialist cult intent upon global control and a world government! The best that can be said is that they are not willing to collude with what they see as a scientific deception.

The myth of the low carbon economy

Even a cursory glance at the policy proposals for mitigation is enough to see this deception (and self-deception) at work. A book published in 2007 by the environmentally conscious and academically well-regarded house of Earthscan illustrates this. Chris Goodall, in *How to Live a Low-carbon Life*, deals with the issue of high carbon emissions from the average British and European household. These emissions amount to about six tonnes of carbon dioxide in direct emissions, and six tonnes indirect (downstream in the economy) per person. If each person or household does the following:

- installs a new and efficient boiler and better insulation and bears lower winter temperatures, they could save 0.3 out of 1.2 tonnes/annum;
- with shorter showers and better shower technology, another 0.1 saved;
- replace the fridge, no tumble drier, smaller TV and no stand-by, 0.2 saved;
- smaller car, fewer miles, another 0.3;
- no air travel, 1.8 saved;
- other emission changes not specified, another 0.3.

Then household demand is reduced by 50%. But if contraction and convergence is aimed for at 3 tonnes/person/annum in total, the whole of the external demand must be reduced or offset. This is achieved through:

- changing food buying habits – 1.3 tonnes;
- investing in commercial renewable energy companies – 1 tonne;
- buying domestic green electricity – 0.8 tonne;
- installing a wind turbine – 1 tonne;
- getting a new car after 15 years use instead of 10 – 0.1 tonne;
- buying less heavy goods – 0.2 tonne;
- working from home – 0.8 tonne;
- purchasing offsets for the remaining 1.3 tonnes.

Thus, our intrepid green consumer reduces carbon emissions down to 25% of their former level towards an ethical standard for the planet. Other authors, such as George Monbiot in his book *Heat*, take similar approaches, as does the Tyndall Centre study.

On the face of it, these changes are hardly lifestyle threatening, but a little closer examination reveals just how naive they are in the real world. Firstly, in most advanced industrial states there is a great deal of poverty. In the UK at least 30% of households would be too poor to consider replacing boilers and fridges, let alone investing in wind turbines (which offer very poor economic returns). Very few of the legions who work in factories could 'work from home'. And it is the nature of those factory jobs that leads to the recreational needs for air travel and larger TVs. Likewise, changing food habits to organic and local produce increases food bills by 20%. Furthermore, from work that my former research group pioneered, I doubt that even a reduced and healthy average UK diet could be sourced locally and organically considering the systemic elements of organic supplies for feed grain, such as fertilizer and land area. However, this is not to say that I think the overall direction is not appropriate. Lowering demand for fuel and resources is an essential step towards future resilience, but there is a crucial issue in relation to the knock-on effects on the economy (employment in particular) as well as how to deal with the question of fuel poverty.

In relation to this strategy, 'green' electricity is currently heavily subsidized and occupies a tiny percentage of demand. For this to make an impact it would have to rise substantially beyond the ability of governments to subsidize and hence the cost of this option would also be prohibitive to large sections of the population, as would the installation of a wind turbine or solar panel for the average household. In my own work as an energy policy advisor to one of the UK government agencies more than ten years ago, I put forward a proposal to make solar panels and decentralized micro-generation a legal requirement for all new housing. It would have added 10–20% to the cost of a new house and this was considered economically unsound. In the course of that ten years, British house prices rose by about 300% without any environmental benefit and to contribute to the current considerable economic distress. This is a salutary lesson and I firmly contend that only such approaches through regulation and long-term finance can significant progress be made.

The voluntary consumer revolution is not going to happen and all of the recent social surveys point to this. People express their concern but do little — not because they would not like to, but because they are trapped by the demands of urban life and economic realities. The low-carbon economy is a convenient myth. Its convenience lies in the ability of commentators to parade a solution, pretend that such a solution exists, and blame an ignorant public if they do not enact the proposed remedy. Goodall has awoken, at least, to the implications of his 'buy green' answer to the externals, commenting in an *Independent on Sunday* article that supply of 5% of UK transport needs for ethanol would consume 70% of the UK wheat crop. Few other commentators have considered the consequences of a massive investment – of the order of £100 billion, in barrages, wind turbines, hydro-schemes and biomass burning.

As we noted, one UK electricity supplier that is heavily dependent on coal with one station of 4 GW capacity intends to co-fire to 10 or 15% of that capacity with biomass. This biomass will be sourced globally in the form of anything transportable and burnable, such as straw, forest waste and rice husks. The carbon credit and renewables obligation system will force this company and others throughout the rich EU to spend large amounts of money on the world market, raising the price and hence pre-empting local market solutions in all situations where material can be bought and transported at competitive rates. The necessary world trade safeguards that prevent environmental standards in traded food will presumably also be developed with regard to energy. Indeed, the more aware environmental NGOs are already complaining that such safeguards with respect to origin and sustainability of supplies are not in place, yet they have lobbied for these supply options for over ten years.

Britain is heralded as a booming 'service' economy, particularly in banking, management and engineering consultancy, and this is promulgated as a path towards the advanced low-carbon economy. But Britain's financial centres are global in their operations. Large amounts of capital are being used for the expansion of industrial economies in India, China and South East Asia. These populations are pursuing essentially the same unreconstructed global development model. In China, for example, the massive trade surplus in dollars is currently financing the purchase of advanced electricity generating technology for an accelerated programme of coal-fired power stations. The Chinese economy already accounts for the decade-long surge in demand for oil, cement, metal ores and timber. It may now be faltering, but the current downturn is merely a crisis in confidence and not due to a lack of resources or an ability to pay for them.

These realities of global trade and investment are overlooked in the European arena of national targets and emission reductions. If carbon emission accounts had to reflect imports, then I doubt that many EU states would show any reduction. International corporations are already moving around the globe to take advantage of emission permits and will easily outmanoeuvre national or regional controls (and their added costs). As carbon trading moves to become international, there will be a premium on renewable sources, particularly of electricity. This will lead major players to target hydroelectric sources in hitherto remote and inaccessible places, many of high biodiversity and wilderness value.

In Britain, planning permission is being sought for very large wind turbine arrays in remote Scottish islands and highlands fabled for their beauty as well as for special areas of high conservation value. The electricity will either be taken by undersea cable or overhead power lines across recreational areas, or act as a focus for incoming high demand industries. In such cases, the 'international consensus' on the need for global action and the meeting of central policy targets have been weighed above local community and often wider needs for conserving landscape and biodiversity. The same arguments can be used across the globe to justify energy developments.

Wildlife groups have finally woken up to the implications. 'Wildlife Link' – an umbrella that includes the National Trust, Woodland Trust, the county wildlife trusts, RSPB and others – has issued a report on the threat to biodiversity from biofuels in agriculture, arguing that they are not against the development of renewable sources of power, but that safeguards must be in place. Greenpeace has begun a similar campaign to protect rainforests from palm-oil exploitation, just as Ugandan ministers are giving the go-ahead to convert long-protected forest reserves into sugar plantations for this purpose.

These wildlife groups have in the past naively taken up the cause of renewable energy supplies with no thought until now of the consequences. They have mounted no independent critical review of climate science and they have no effective critique of the global development model. What chance is there, given that even protected sites in the UK have suffered in the past decades of agricultural intensification *with* safeguards, that such pleas will deliver results? Once again, we are dealing with a major act of deception – even if, charitably speaking, it is a self-deception.

Consumption and migration

When the propensity for increased consumption is coupled with the need of developed economies to draw in immigrant labour and increase their housing and infrastructure (and also cater for the marital and family breakdowns consequent upon the level of emotional stress and dysfunctional community), then energy demand is set to rise – as most developed economies show and IPCC predicts to continue. Britain must build a further 3 million homes to add to its current 20 million, and although new build is planned to be 'zero carbon' by 2015, a substantial proportion of these citizens will be immigrants from low-carbon economies whose new found wealth will add to global demand.

Earthscan's recent *Atlas of Climate Change* provides a useful guide to the pattern of development. It lists, for example, how far countries have responded to the Kyoto Protocol. Only the UK (13%), France (5%), Luxembourg (16%), Germany (19%) and Sweden (5%) register a fall in emissions between 1990 and 2003. What is not shown is that a substantial proportion of those savings comes from substituting gas for coal and was an economic decision that would have happened anyway.

Interestingly, Denmark, a world leader in renewable technology, and without a heavy industrial base, records a 5% rise in emissions over the 1990 baseline. Austria, leader in wood-fired biomass, records a 13% rise, and Spain, another world leader in the purchase of renewables, records a 32% rise. The USA is up 20%, Canada 57%, Australia 5% and New Zealand 30%. Eastern European countries and Russia are down 30–70% following the collapse of their heavy industries. One has to look elsewhere for recent data on China (in 2004 it released 1000 million tonnes, up 67% since 1990 (*State of the World Report* by the Worldwatch Institute, 2006).

What are not shown in these figures are each state's balance sheets with regard to the embodied energy of imports and exports. In the case of the UK, for example, where the post-1990 period has seen a transfer of heavy industries and manufacturing to the Far East and the growth of the 'service' economy, it is relatively easy to lower the energy/GDP ration and the whole carbon equation looks positive, whereas if imports are included the equation turns negative.

With regard to mitigation of these rises, Earthscan's *Atlas* looks at carbon trading and financing the response. Up until 2004, 300 million tonnes of carbon (MtC) had been 'saved' on a project-based exchange scheme funded by the Global Environmental Facility; the UK, Holland and Japan purchased 50% of that, and other EU countries 32% (the USA purchased 4%). The 'allowance-based' market, which simply enables companies to purchase credits from countries that have no limit or have kept emissions below Kyoto targets, had reached 60 MtC by March 2005. The pace of growth in these carbon markets is accelerating, but they are still a tiny percentage of the investment devoted to supply technology (we shall return to a critique of carbon trading).

Thus mass movement of capital out of Europe can masquerade as emission reduction in what looks like a modern economy becoming less energy intensive. In fact, countries like China then accelerate their own emissions not just through the new industries but as a consequence of the vastly increased consumer demands that the new jobs and prosperity bring to the country.

The supply-side 'fix'

Within these global and regional scenarios, the only aspects of policy that survive are those related to supply-side options. This is partly because the consumer policies don't work and partly because the supply-side options prosper as overall energy demand rises. If there is new business and profit in the reduction of consumer demand it is not obvious to the corporate world, and this world is highly geared to economic opportunity and high returns in those industries that are currently taking full advantage of increasing demand. It should be noted that governments also see little merit for their exchequers in any form of reduced demand that affects tax receipts.

This leads me to another deception – the myth that renewable energy can reduce emissions. Certainly, any renewable source that is deployed will replace a fossil fuel source and thus less fossil fuel is used than would otherwise be the case. But this is not the same as effective emission reduction. If the renewable energy option exists within an economy experiencing expanding energy demand, as it invariably does if a globalized accounting index is adopted, then it will not and cannot hope to keep pace with that demand. This is because renewable energy is both constrained in terms of the resource available and in terms of the cost in relation to fossil fuels with the latter determining the degree to which the renewable source can penetrate the market.

Theoretically, as fossil fuel becomes more expensive, this should cause a shift towards renewables. But there are two other factors at play. The first is the massive reserves of 'tar sands' in North America which are held by some analysts to more than equal global oil reserves, and which become economic to mine at oil prices above $70 per barrel. The second issue concerns the health of economies hit by these high oil prices – they will not be able to afford to develop more expensive renewable sources (especially if these are constrained by environmental safeguards).

The recent $140 dollar spike in the oil price may not be a guide to future prices, but it demonstrated the global economy's vulnerability to a doubling of fossil fuel prices, the consequences of which are not yet fully apparent. This underlines my points about future projections based upon simplistic economic and non-real-world models. The cheapest renewable supply option, very large wind turbines on land, are currently about 200% more expensive than fossil fuels. Other smaller-scale options for household use vary between five (biomass) and ten times (solar PV) fossil fuel costs. Similar ratios apply to hydrogen cars, bioethanol and hybrid power vehicles. Any cost reductions in terms of production economies will be more than offset by higher demand and increased prices of the fuels. Non-fuel sources such as wind and solar cannot be expected to meet the near-term electrical demand from a socially un-restructured transport system.

For example, large wind turbines are cost effective (with subsidy) if sited in suitably windy areas – usually areas of natural beauty and recreational value. They also impact upon sensitive local rural communities. Even if these values were compromised, cost and other considerations limit wind to about 20% of electricity supply. Biofuels are similarly limited by land availability and cost, especially where appropriate environmental and biodiversity standards are applied. Many renewable supply scenarios have assumed a 20% biofuel component, and I think the system would struggle to find enough land for a quarter of that.

Supply side options are thus highly likely to be enacted over the next two decades but will hardly dent the problem of emission growth. They are also, as we have noted, even at low levels of penetration of the order of 5–20%, highly damaging to the environment. In that context, the deception is further illustrated by the Earthscan *Atlas*; it has a blindspot on this issue and maps only the consequences of the disease of global warming and not the supposed remedy. It has maps on climate-vulnerable biodiversity but not on the expected impact of turbines, tidal power or biofuels on that same wild habitat. This same blindspot has afflicted virtually all the campaigners for renewable energy supplies.

The environmental impact of arbitrary targets and timescales

The urgency to set targets has led to moves in the EU and in California (soon to be extended across the US) for a 5% biofuel component to oil and petrol in the transport sector. There has been recent pressure to extend that to 20% by 2020; and as we noted, there are already targets in place for renewable energy obligations as a component of electricity supply, also 20% by 2020, in the EU.

These supply-side initiatives have been activated before any standards have been set for the environmental provision of these fuels. Further, market pressures work against small-scale integrated development, driving up the size of biomass stations and turbine arrays. There will now be a rapid acceleration of production with a variety of consequent pressures:

- wheat, maize and soya crops from intensive production will be diverted (perhaps supported by subsidies) from food supplies;
- palm oil production in SE Asia and soya in South America will be accelerated with consequent loss of primary forest lands;
- new agricultural land may be taken from marginal land in Europe and the US, which is of high biodiversity and recreational value, and further areas of primary forest developed in the Amazon and savannah lands elsewhere, for soya, sugar and corn production;
- expansion of woodchip markets will lead to further pressure on hitherto low-grade 'unmanaged' woodland in Europe (especially cork-oak forests in Portugal and Spain) and secondary forest in the Baltic states and Russia; even primary forest may be threatened where it is not protected.

These pressures have environmental and human consequences that are far-reaching and under-researched. Safeguards are not in place, yet policies are being rapidly enacted.[6] In developed countries, rural communities face large-scale industrialization of the countryside with major implications for biodiversity and that part of the rural economy dependent upon recreation. In the UK the size of this rural recreation sector remained hidden until the outbreak of foot and mouth disease when measures to protect farming damaged a much greater part of the economy that had minimal representation in the decision-making process).[7]

Taking time for integrated development

There are slower and more integrated ways of responding to the crisis: firstly, to give time for prices, regulations and technological evolution to work on the demand side of the equation; secondly to creatively restore damaged ecosystems with a perspective on carbon sequestration, bio-diversity restoration, indigenous communities and local supplies of water and forest products; and thirdly to further dietary changes that will feed back positively to agricultural and forestry practice. Forest loss and other land-use changes are held to account globally for 25% of carbon emissions, and transport and food processing for 20% of energy use in agriculture; this area is an obvious part of a no-regrets strategy, and Britain, Europe and the USA could evolve a model for their own future development that would also act as a guide for developing nations.

The carbon emissions from soil and forest loss need to be halted not as part of an ineffective mitigation strategy but as an ecologically sound development goal in itself. Forest protection and ecosystem restoration conserve not just carbon, but water cycles, rural livelihoods and biodi-versity. In terms of emission reduction, about 25% could be saved if forest and soil protection was effective.

A further 25% of carbon emissions can then be reduced by enacting stricter regulation of vehicles, work-travel planning and transport, passive and active solar houses, efficiency of machines, together with the supply options of offshore wind and smaller-scale wildlife and community-friendly biofuel technologies. I am convinced that a 50% reduction in a fully worked scenario to 2050 would not require nuclear power, onshore wind turbines, large-scale industrial or tropical biofuels, or tidal barrages, and that fossil fuels could be maintained to supply the reduced demands of transport and large point-sources of power for industries.

This balancing of industrial society's carbon equation does not require either great urgency or excessive targets, at least not from a climate per-spective. These changes will contribute to the greater overriding concern of energy security and maintenance of social stability in the face of declining oil production.

Despite warnings to the contrary, in particular from the House of Lords Select Committee on Technology, the current dual approach of supply options and 'efficiency' which has characterized the past 20 years of policy looks set for at least another three years without change.[8] Yet there is no real evidence that it works. The problem is that financial savings from energy efficiency are reinvested in other areas of production and energy demand. Having a 'renewable' or 'green' component to supply may simply encourage further growth in demand by inducing a false sense of security, as in the classic case of the Spanish economy where heavy investment in renewables has coincided with a substantial rise in carbon emissions.

These issues are well rehearsed in professional circles but do not seem to advance to media discussion.

- Unless there is a culture of demand reduction, financial savings made through efficiency measures, whether in the domestic or industrial sector, will simply go towards other energy-consuming activities, thus increasing demand.
- High profile renewable energy projects such as wind turbines are likely to contribute to this demand rather than lower it, because consumers will feel more able to use more if they believe the sources are 'renewable'.
- The renewable component of supply cannot readily be increased beyond a token 5–10% across the fossil fuel spectrum without adding considerably to overall energy costs or taxation, and energy costs are a significant factor in inflation targets that are the mainstay of fiscal policy.

It is the demand-side that must be the focus of policy

There does seem to be an awakening as to the need for demand reduction. There is much media treatment of consumer 'carbon footprints', for example, and a growing number of businesses declaring their intentions to go 'carbon neutral'. There has been a tightening of house-building regulations and performance targets. Many firms are using 'carbon offsets' and investing in energy efficiency outside of their individual domain.

There is also much policy discussion on extending and improving the European 'carbon trading' agreements that currently relate to big energy users such as the electricity supply industry. But this 'pollution trading' has its critics. As Larry Lohmann commented in *New Scientist*, it carries the danger of 'why bother making expensive long-term structural changes if you can meet your targets by buying pollution rights from operations that can cut their carbon cheaply'.[9] He comments, 'it helps keep an oppressive, fossil-centred industrial model going at a time when society should be abandoning it'. Much as I sympathize with this conclusion, I realize that this is the reason why carbon trading has been developed. In the real world, the industrial model cannot be so readily 'abandoned'. It has to be restructured.

It remains to be seen how far the current exhortations and minor regulatory changes will impact upon the pursuit of a continually expanding economy. The economic crisis that has recently superimposed itself on these concerns alters the picture considerably. There are still major commitments to house-building and airport and road expansion; subsidizing such public works may well feature in the policies to restart the economy. There is no real discussion of restructuring that economy or reforming the global development model, yet if we are to address vulnerability to climate change, dwindling oil supplies, rising costs and food security, this model must be restructured. The greatest error would be to try to solve this problem with the same model and old modes of thinking.

Notes

1. It is now the end of January 2009 and the cycle has shown no more than the occasional flicker with NASA admitting they cannot predict what will happen and some solar scientists debating after whom a prolonged 'minimum' should be named. I would lobby for 'Landscheidt'.
2. In articles in *ECOS*, Vol. 28 (1), 104–106, and Vol. 29 (2), 81–88, I drew attention to Landscheidt's prediction in 2003 that temperatures would fall in 2007, and I warned that food would become a crisis issue.
3. Most of this information can be gleaned from the Wikipedia treatment of the Irish potato famine, and there is a historical review by R. Dudley Edwards and T. Desmond Williams (eds), *The Great Famine: Studies in Irish history 1845-52*, New York University Press (1957); paperback edition Lilliput Press (1994).
4. One of the most extreme recommendations comes from the influential writer George Monbiot in his *Heat*, where he argues that only an 80% reduction by 2030 will be adequate to mitigate the effects of global warming.
5. Contraction and Convergence are terms put forward by Aubrey Meyer of the Global Commons Institute proposing a movement toward equal per capita emission allowances for every planetary citizen, and it has gained widespread endorsement.
6. In 2008, the UK Government instigated a review by the UK Renewable Fuels Agency (known as the Gallagher Report) to examine the sustainability of the EU biofuels target and the UK's target of 3.5% biofuels in transport by 2010 rising to 5% by 2012. This first assessment found that serious attention needed to be given to issues of land-use change before carbon emission savings could be quantified. For example, if virgin forests or pasture lands were brought under cultivation, savings could be minimal due to soil carbon losses and emissions of nitrous oxide from fertilizer use. It recommended a slower development until such times as sustainability criteria had been agreed by the EU and that only marginal or unused land should be developed in order not to compromise food production. The UK Government has agreed to these principles. The report did

not detail biodiversity consequences nor disruption to indigenous peoples and rural communities.

7. I deal with this in a chapter on agricultural change in *Beyond Conservation*. The decision to close footpaths and effectively shut down the tourist section of the rural economy throughout upland Britain caused massive hardship. Whereas the agricultural sector qualified for financial aid, the small tourist businesses did not, and it was only realized after the epidemic that their economic contribution to rural life in those affected areas ran to near 20% compared to agriculture's 2–5%.

8. I review this issue in depth in 'No regrets energy options: choices in a changing climate' in *ECOS*, 21 (2), 2002.

9. See Larry Lohmann, *Carry on Polluting* and also his report published in the Dag Hammarskjöld Foundation's *Carbon Trading*: 'a critical conversation on climate change, privatization and power', quoted in *New Scientist*, 2 December 2006.

15

Vulnerability and Adaptation

The underlying message of the previous chapters is that dangerous climate change is an immediate problem that will inevitably get worse. It cannot be mitigated on any policy-relevant timescale. Whilst there are many prudent technological innovations to reduce carbon emissions, the main benefit relates to societies becoming less vulnerable to oil price shocks and adapting to an industrial future without oil. Given the longer-term nature of technological and financial cycles, industrial societies will necessarily adapt with rising costs driving efficiency and the use of renewable sources of energy. The need to intervene and attempt to regulate this process *on the basis of avoiding climate change* is not proven but there is still a strong case for intervention to ensure a stable and equitable transition.

Aside from energy costs and depleting oil and gas supplies, other vulnerabilities relate to the nature of a fluctuating climate at a time of rising population and resource demand. Ecological systems that are vulnerable to shifts in temperature and rainfall are intimately bound to and determined by economic and social factors. If these systems become stressed by rising prices and falling agricultural output, the stability of the dependent societies with respect to social order will become a more pressing factor than any aspects of technological change. The issues will range over health services, food supplies and the protection of the weak and vulnerable, as well as social stability related to unemployment and unrest, particularly in cities.

Despite the various organs of state devoted to maintaining this stability, a great deal depends upon the *financial* world, which has proven at times very fragile. It is not an easy matter to safeguard that financial stability; the consequences of error are at least as severe in the short term as climate change and, indeed, will be concurrent. (This paragraph was written six months before the current financial crisis surfaced. And it is all too evident that states are preoccupied with restarting economic growth and avoiding social disorder, with very little appetite for expensive climate control policies).

The priority of adaptation

Adaptation must take priority over mitigation. This will not be an easy transition to make. Current policies are 99% focused upon emission reduction, though there is an increasing amount of thought devoted to issues of flood protection, both from rivers and in relation to sea-level rise (particularly in the UK). Internationally there is much thought being devoted to water management with renewed interest in big dams (in Africa) and water transfer pipelines (in Australia). But little attention is given to food production, housing and transport, communities, health and biodiversity planning. Even less attention is given to the development model and in particular to the prospects following the 'oil peak' or to the burgeoning world population.

When scientific uncertainty is accepted, the lack of predictive power becomes a certainty, and a different kind of policy response evolves. Unfortunately, the global science and policy community is basing action upon very uncertain predictions and developing policies that can only have long-term and very uncertain effects. In this framework, almost all adaptation, as with mitigation, is geared to the predictions of climate models – whether flood defences from anticipated rainfall, coastal defences from sea-level rise, healthcare from rising temperature stress, recreational provisions, agricultural and water conservation strategies, and so on. But from what we have seen of the cyclic nature of climate change and the very real prospects of a global cooling episode, these policies may prove useless or even counter-productive. A no-regrets real-world strategy assumes no predictive power but focuses upon building an inherent resilience to climate changing in any direction.

A no-regrets policy would *include* reduction of demand for carbon fuels, not because using less would stabilize the climate in time but because their imminent high price and restricted availability will cause immense hardship. The less people need for their basic well-being the better off they will be. A warm, naturally heated and insulated home that costs less to maintain is the best insurance against global cooling *and* economic hardship. Other policy moves with regard to heat-stress, droughts, storm damage, food and water supplies and biodiversity require a great deal more study.

With regard to energy supplies, we should separate out those supply options that entail negative social and environmental impacts and put them on policy-hold for at least two decades, by which time physical as well as economic and political climates will have become clearer. This would apply to land-based wind turbines, estuarine barrages, biofuel programmes and the expansion of nuclear power. Instead, accelerated programmes of demand reduction should be prosecuted through housing design and public transport systems coupled to efficient planning and siting criteria.

Energy security through reduced demand

However, one of the key drivers of energy policy relates to energy security. There will always remain a vulnerability to political and market conditions relating to the geographical location of oil, gas and coal supplies. Reducing demand is the best insurance regarding price and availability in what is now a global market. In the longer term, renewable energy solutions have to be found — but they need to come *after* a large-scale transformation of the global development model. Such a transformation will require large expenditure, but this money needs to go to grass-roots programmes of adaptation and resilience. For example, safeguarding global drinking water, forestry and agricultural soils has been costed at less than $100 billion over a decade and at present much greater sums are being pumped into an ailing development model to stave off recession.

The problem is that no country has managed to grow economically without increasing its energy demand unless they have exported their demands elsewhere, as did Britain and western Europe to China. There are no obvious models for economic growth in an era of resource scarcity, high demand and high costs. The UK and some other nations managed to increase GDP and keep overall energy demand in check, but these industrial states moved into a mature phase where financial services (servicing the rest of the global economic boom) became a significant source of GDP. Sweden managed to reduce energy demand and carbon emissions but it has a small, highly skilled workforce with a strong technological heritage. The crucial areas of future global demand are not comparable — Indonesia, India, Brazil, China and Russia.

Critics of the carbon emissions policies based upon the IPCC approach and Kyoto Protocol continue to point out that it denies the potential for economic growth among the world's poor. This might be true were Kyoto and its successor capable of being implemented. The world's poor countries will be just as severely compromised by high oil prices. And only if they had buoyant growing economies could they afford alternative fuels, all of which are more expensive. The ability of the rich countries to transfer technology is limited by their own apparent needs to maintain their own economic advantage. In this respect, the critics are right – the only proven path to a modern industrial society has been through cheap fossil fuels, and the only proven method of reducing demand is through significantly increasing the price (except in times of recession).

I do not therefore propose to further examine scenarios of reduced demand. They were effectively rehearsed by the UK Royal Commission in 2001 and involved the usual methods of increased efficiency in the home, transport, energy production, etc. I assume that a 50% reduction in *current* energy demand in the UK is technically achievable. These figures will apply to many of the advanced economies, with the exception of those that have *already* invested heavily in energy efficiency, such as Sweden, Denmark, Germany and Japan, all of whom would struggle to find such a level of reduction. But the fundamental problem attached to reduction in demand is that it affects the economy of developed states, and hence brings all manner of social impacts from increased unemployment. This is immediately apparent when demand falls due to recessional (and largely psychological) factors which are cyclic, where demand only has to fall a few percentage points to impact upon the social fabric.

The economic price of reduced demand

Thus for an adaptation strategy that relies upon a reduction in the energy-hungry demands of goods and services there needs to be an adaptive response to the social impacts of unemployment. A significant number of jobs in the economy would not survive a low-energy demand restructuring. Some innovative social programme would be required to make this adjustment fair and hence politically feasible. It seems from the current financial crisis that a significant number of jobs would not survive a lower-demand future characterized by economic insecurity and a more frugal lifestyle. The irony is that governments would shy away from social engineering that entailed job losses, just as they did from housing standards that would raise house prices, only to find that these things happened anyway.

Any reduction in money flowing through energy-demand related sectors could be made up by expenditures in other sectors of lower energy demand – for example, in shifts to local organic food production. The problem relates to how this could be financed and furthered by governments during times of economic constraint and low tax take. This would also be the consequence of any engineered fall in demand for goods and services. It is a 'catch 22' situation, because only government-directed expenditure is likely to further adaptation and resilience to change. It is for this reason we have seen so little of that policy outside of the social democracies of continental Europe where, for example, Germany subsidized a programme of 100,000 solar roofs.

It would appear that the only feasible way forward is for a socially engineered development of resilience to both climate change *and* the economic repercussions of peak oil (as well as cyclical crises of confidence in financial markets). Such a major shift away from free markets would require a new paradigm of political philosophy, perhaps involving a return to public corporations (or some hybrid) for essential supplies such as water, food, housing, electricity and transport.

The concept of resilience

Resilience to climate change implies a robustness in the face of change. It implies resilience to change in whichever direction the climate changes. Natural ecosystems have evolved to be robust to such changes but if changes are rapid these can stress the system and cause population crashes and even extinctions. Human ecosystems have been long removed from the restrictions of the natural environment, and their robustness to change largely depends on human ingenuity and organization. Agriculture, for example, has evolved ever narrower crop types and manipulation of plant genetics, with heavy reliance upon artificial fertilizer and pesticides, and has emerged as a very simplified ecosystem compared to a natural system bounded by cycling of nutrients and predator-prey relations. Natural systems appear to have evolved diversity of species as a stabilizing force, whereas agriculture consists of monocultures, often with a narrow genetic base.

As a result, modern agricultural systems are extremely vulnerable to short-term collapse. Their narrow genetic base makes them particularly susceptible to diseases that may be transferred across a global market. Crop types may not adapt to rapid swings in rainfall and temperature. However, this artificial agricultural system can respond to rapidly changing *markets* and hence also to changing climates and choice of annual crops. And although more natural and organic practices are more robust to climate change, they supply more local and traditional markets.

Modern systems are, however, vulnerable to loss of soil fertility and long-term drought.

These issues of resilience are complex and there is no easy answer. However, some general principles are clear.

Local food

There is a growing movement towards local and organic food supplies, usually framed as action to reduce carbon footprints or to improve health, taste, rural employment and biodiversity. Arguably, drawing most of one's food from local sources adds resilience to the system but it becomes difficult to maintain this argument in a case where regional food supplies are compromised by poor harvests. In this case, importing food from areas that have not suffered such loss of production may be essential for survival. In extreme cases food may not be available from other regions if their own climate has undergone a shift, hence it is also prudent not to rely upon outside sources of food.

There is also the spectre of rich countries buying up dwindling food stocks by purchasing harvests from those countries in difficulties (or as we now see, *owning* that harvest in the first place). At present, free market rules apply where large food conglomerates have worldwide investment. In the general development model, food is regarded as a commodity and agriculture as an industry. This model is seen as successful in that it has successively increased production and more than kept pace with population expansion. However this model services the old structures of mega-cities and landless peasantry, which are extremely vulnerable to future changes.

Modernized agriculture has also come at a very steep price in loss of natural soil fertility, physical soil loss, damage to forests, wetlands and biodiversity. Modern production methods and global trade have also raised the incidence of serious disease both in animals and humans. A changing climate exacerbates these already serious areas of malfunction.

Diet itself is also an issue, not just in the aforementioned obesity epidemic but also in the ecological consequences of food choices. High levels of meat consumption mean that a large proportion of world grain stocks go to animal feed. Large-scale ranching in former rainforest areas degrades land even further.

Thus there are sound ecological and organizational reasons for focusing on local supplies of food that are fresh and unprocessed and on a lower-meat diet. If grown to organic standards then human labour and ingenuity replace oil-based chemicals, and farmland is more conducive to wildlife, soil contains higher levels of carbon, and plant species are selected that are more resilient to drought, pests and diseases. Local markets create further jobs and foster a sense of identity and community – something we shall come to consider as an essential prerequisite for robustness in the face of future changes.

At present, organic food is a lifestyle choice and regarded by the agricultural industry as a niche part of the market, which has nevertheless undergone rapid growth. Farmers are queuing up to participate in government funded conversion schemes. Twenty years ago, the industry laughed at attempts by ecologists to convince government that organic production had the merits that are now taken for granted. Indeed, my research group was the first to investigate the science and compare organic systems with conventional systems (**Arden-Clarke et al., 1990**). Our leading edge work was funded by the World Wildlife Fund when most government scientists and institutions thought it a fringe issue. We identified soil erosion, loss of fertility, susceptibility to disease, pollution by runoff, loss of biodiversity and rural unemployment as the key disadvantages of conventional production – all of which were ameliorated by organic systems of 'ecological' agriculture.

Though attitudes have changed, one thing has not – the industry still regards organic production as a niche market and customers pay a premium for the product. This policy avoids conceding the environmental merits. It is then generally wealthier and more concerned citizens who pay to protect the environment. Those who could afford the premium but do not pay then gain the advantage of spending less of their income on food and have more to dispose in other sectors. This is plainly unfair and can only be addressed by the phasing in of organic standards of production across the whole agricultural sector. Agriculture would then become more sustainable and arguably more resilient to climate change. There would also be a transfer of job opportunities to rural areas and an overall increase in employment. Of course, food prices would necessarily rise – but as with the housing debacle, this is likely to happen anyway. Organic production is generally about 20% more expensive, but this is also the level of oil-price and biofuel driven inflation that has shown itself in the last year and is likely to revisit.

There is general agreement within the ecological community that the current EU funded subsidies to agriculture should be maintained but paid only when 'cross-compliance' with environmental standards is enforced. At present, farmers can elect to receive extra subsidy in exchange for tougher standards with regard to the environment. In some 'less-favoured' areas of hill farming uptake is high, but over much of lowland agriculture only a small proportion are in the higher tier and environmentally meaningful 'stewardship' schemes. For the vast majority of lowland farms decisions are dominated by market forces and the operation of large corporate landowners. The supermarkets generate almost intolerable pressures on key agricultural sectors, and indeed have at times brought the embryonic organic sector close to bankruptcy.

The organic sector is in desperate need of government intervention. I do not believe it should be left to education and 'consumer' choice any more than such issues as fire safety standards or car insurance are optional in the housing or personal transport market. Just as the EU are pushing through ever stricter standards on vehicle efficiency, so EU-wide standards of ecological agriculture could be progressively phased in.

The one area where personal choice ought to remain sacrosanct is, of course, diet. But should everyone be at liberty to over-eat, become obese, threaten their own health, create pressure on global resources and strain a country's health services? Nobody is at liberty to pursue damaging (and illegal) drug habits or to abuse children, and obesity is an issue that falls into similar territory. The issue needs to be taken seriously and health professionals given the authority to intervene and recommend rehabilitation.

For the other main advantages of a better diet — the use of local food and lower meat consumption — education would be the main driving force. Meat and dairy products have been marketed on health grounds which are now known to be dubious, and education in the ecological merits of a vegetarian or semi-vegetarian diet should be pursued. The trends are moving in this direction, but there is still widespread ignorance. If meat were produced to organic standards, its high cost would lead to lower levels of consumption, but it would be essential to balance such a policy with education in schools with regard to balanced diets.

Diet and ecological agriculture constitute one of the most powerful forces for change on the planet. Government should devote considerable resources to monitoring food sources, production standards (at home and overseas), health issues related to diet, biodiversity and other environmental impacts. As with current certification schemes for forests (that are not yet binding) such approvals could be granted for sustainable food products. The World Trade Organization needs to be reformed such that states can enforce environmental standards for imported food (and bio-fuel).

The counter-argument to these policies would be that organic systems are less productive and the world is approaching serious food shortages. I don't believe the figures add up. There is evidence that organic systems can be as productive as conventional systems. But even if they were less so, then this shortfall would be made up by a percentage shift to less meat and dairy products in the diet. In the longer term the enhanced produc-tivity of conventional systems is likely to prove illusory as soils become exhausted, pests adapt to pesticides and diseases attendant on mass-production and global transport become more debilitating.

The current global development model leads to exactly the opposite trends in diet, consumption and means of production. This trend needs to be reversed if ecosystems, both natural and human, are to maintain their resilience.

Local employment and journeys to work

Every morning in my local town of 5000 households about 1500 leave the town for work in neighbouring towns or villages, and about 1500 come in from outside to work in the town. There are three major supermarkets, two of which are on the outskirts of town, and one new one is under construction. A new industrial estate is being constructed on a brown-field site and is anticipated to offer 1500 new jobs — though there is full employment in the town and also very full housing. A new housing estate is about to get planning permission and on land that is vulnerable to flooding. The new industrial estate was promised as meeting local needs and local people were consulted and asked what industries and facilities they would favour, but these wishes have been bypassed (they included a combined heat and power station fuelled by local sources of willow-coppice as well as leisure facilities and a visitor complex showcasing the estate as a leader in sustainable technologies and construction). Instead a mixed retail and small industries use with standard metal box construc-tion buildings is likely to manifest.

In this development no local council has the authority to specify the types of industry using the site (there is a strong likelihood these will be defence oriented) or the use of sustainable materials and energy. Equally, there is no provision for stipulating local job numbers and links to housing. Thus nearly a decade into the twenty-first century, and after at least two decades of lip-service to sustainability, and only a couple of years away from the end of the Kyoto treaty, virtually nothing has changed. And this town is at least typical and probably slightly greener than most.

Yet this is the one area in which great strides could be made in reduction of fuel dependency. People ought to be able to walk or cycle to work and no new industry should be developed without such provision. Further, the fostering of strong community requires not just consulting local people, but enacting policy that respects those values. In fact, regional government pulls in the opposite direction and this level of consciousness is prevalent throughout Britain and mirrors a general tendency on the part of bureaucracies in developed countries – they assimilate more control, they are closer to the business community than to any other, they align and support military objectives where almost all 'liberal' democracies are now involved in a coalition to fight foreign resources wars, either directly or by proxy, and with the cover of deception as to both the purpose and the legality.

In our neighbouring town, famous over nearly two centuries as a centre for quality shoe production, all production has ceased and been exported to cheaper operations overseas. In place, the industry has built a giant distribution centre – one of the biggest in Europe where the overseas production comes to be sorted and distributed throughput Britain by fleets of heavy lorries. The building is steel and has no solar panels. It has completely destroyed the sense of scale in the townscape and the surrounding views of the town. Interestingly, its design is that of a large aircraft hanger in naval warship hues of grey and pale blue (the townscape is one of subtle hues of red tiles and cream walls). Alternative designs and materials exist even for over-large distribution centres: hemcrete (hemp and lime) can effectively replace breeze block; wood composite beams can replace steel; roofs can be blended using 'sedum' (stonecrop) plants as 'turf' and also fitted with solar collectors. These alternatives are not prohibitively expensive; it is largely a matter of design and education, and at least one major distribution warehouse exists as a working model.

In my view there needs to be a national debate on the issues of adaptation to climate change, and it can act as a forum for awareness also of the global issues of population and peak oil. There has to be a major effort to raise consciousness sufficient to support new legislation and standards. These issues should take precedence over intrusive schemes for road-charging and surveillance, which are invasive and unpopular and contribute nothing to creating community and greater responsibility.

Water supplies

Britain is relatively well endowed with water supplies, yet its high population density and intensive industry and agriculture create regular water shortages in eastern counties, which are in a rain-shadow. Climate change has increased the propensity for summer drought. Winters have become warmer and wetter (thus far in the cycle), but rainfall can be erratic. At the time of writing the country is gripped by unprecedented levels of summer flooding caused by torrential rainfall, and climate pundits are claiming this was predicted as a consequence of global warming. Europe has also experienced intense floods and severe droughts in its various regions. Water supply as well as flood protection is now a central issue with regard to climate change. Further afield, the USA is gripped by drought in California and the south-west; and Australia, after almost a decade of drought, has experienced severe floods.

Water supply, whether rain-fed or from irrigation schemes, is vital to agricultural productivity in all regions of the world. And as human ecosystems reach their upper limits of population size and food production, these systems become vulnerable to both drought and periods of short and intense rainfall which can be destructive of crops and difficult to store and make use of. Rain-fed systems are becoming more important as groundwater supplies become stressed from overuse of fossil water and slow-replenishing aquifers for large-scale irrigation.

The conventional answer to water shortage is to build dams and reservoirs and to engineer mass transfer from one region to another. But behind this systemic response are the interests of corporate entities in a world where water supply has been privatized and made into a commodity. These systems foster corporate structure, finance and financial return. There is little such return in the fields of efficient use, recycling and conservation where the fixing of leaks and refurbishment of old supply systems (also of waste-water and sewage) are all seen as *costs*. Hence large-scale and more profitable technological fixes are an inevitable component of the global development model with its bias towards expanding economic activity and financial opportunity. The corporate solutions that have been developed in Europe and North America are now exported to Africa and South America, India and SE Asia, where the World

Bank is once again in favour of large dams (but this time under the guise of a response to climate change). These large-system responses also pave the way for the export of corporate agriculture to regions of water scarcity and cheap land, where irrigation becomes part of the development model. Such irrigation systems are notoriously vulnerable to climate shifts, silting dams and the salination of soils.

The appropriate response to climate-stressed water supplies is similar to that with regard to energy and food – use less and stay local. There are a myriad successful small-scale answers to water stress. But as the climate shifts in whatever direction, there are going to be some regions that will face severe difficulties, such as in the vulnerable Horn of Africa. However, before international development banks look to dams, irrigation and piped supply, which are in any case no guarantee of resilience because demand then expands and makes the whole resultant economy even more vulnerable, aid should look at the basics of subsistence agriculture, forest regeneration and protection, as well as smaller-scale means of adding an economic dimension to the lives of people in poor and remote areas. As with the supply of other utilities, whether public or private, such as centralized electricity and telephones, the rural poor will be expected to pay and hence become drawn further in to the global economy and the need for migration to cities – usually with the young men leaving their families on the smallholdings. This social pattern underlies the rapid spread of HIV in Africa.

Sustainable fuel supplies

There are few regions in any of the developed nations that have purely local supplies of fuel – whether for heating, cooking, transport or electricity generation. Indeed, the small city of Växjö in Sweden made news as I write this in becoming the first self-sufficient town in Europe. Its transport system runs on bioethanol and its electricity is supplied by a woodchip (sawmill waste) power station that also supplies waste heat to the town's housing. The city set a target of becoming fossil-fuel free over ten years ago when oil was $20 per barrel. Now oil-fired electricity would cost 30% more than the woodchip supply.

The Swedish region is blessed by extensive forests, and it is clear that as oil becomes scarce those regions with renewable sources will be at an advantage. But this advantage needs planning in advance. All renewable sources are diffuse and hence localized. They need a decentralized grid and, above all, decentralized and lower demand. It is not conceivable that major cities in the EU, even after a programme of 50% demand reduction, could be effectively supplied by woodchip powered heat and electricity or local biofuels and biogas.

And as we have seen from the figures for transport fuel, even a 5% supplement of fossil fuel with biofuel would stress land resources. Hydrogen fuels and engines are being developed, both with pure hydrogen and hydrocarbon gas with fuel cell technology, but the means of producing the hydrogen is currently limited by technology and the necessity of using some other cheap fuel source to power the hydrogen production.

The UK Royal Commission on Environmental Pollution (RCEP) concluded that in the low-demand sustainable development scenario to 2050 some fossil fuel use would be required for base-load electricity generation (coal with carbon capture) and load-following with gas, and that all liquid fuel such as oil would be reserved for transportation. This is a realistic scenario even though it assumes maximum efficiency and penetration of several currently very expensive technologies that act more as demand reducers than suppliers. It follows from my previous analysis than carbon capture is an unnecessary expense.

These demand-reducing technologies include solar water heating, solar photovoltaics and micro-generation using combined heat-and-power fuel-cell appliances. The latter technology was regarded as only five years away at the time of the RCEP in 2000 and is only now becoming available through the CERES product. The industries concerned have lobbied for a government-backed programme of 9 million homes fitted by 2030, supplying about 6 GW of decentralized power.

However, it is a rather standard pattern for commissions and reviews of energy, whether in relation to the environment or security issues, to focus upon technology. There are always long lists of what is available and the contribution that could be made. The problem is that social as well as economic barriers exist. In an *ECOS* review of the RCEP in 2000,[1] I wrote:

The main thrust of the report is on the technical solutions available to meet global environmental targets – there is little on the implications for Britain's local environment. Four scenarios are used as indicative of the kinds of solutions available: one with energy demand stabilized at current 1998 levels (no mean feat, considering demand is growing); two scenarios where it is reduced by 36% by 2050; and one with maximum demand reduction of 50%.

Under these four scenarios, a mix of technologies is proposed to replace fossil fuel supplies. Roughly 200 GW constitutes energy demand (all sources converted to watts), and in Scenario 1 this means 100 GW must be replaced with half coming from renewable sources and half either from fossil fuel with CO_2 capture or new nuclear stations. In Scenario 4 that reduces to 50 GW and no nuclear component.

In all scenarios, the main sources of supply, as percentages of the renewable component, are biomass (Short Rotation Coppice) at 10–20% (5–10 GWe), wind at 35–50% (17–25 GWe), and wave and tidal at 8–30% (4–15 GWe). The sheer scale of development required to meet these targets is indicated in the numbers of installations.

Biomass from short rotation coppice

The RCEP consider the maximum biomass resource of nearly 300–3000 stations using short rotation coppice varying in size from 1–10 MW in stations located close to towns to utilize CHP. This constitutes only 3% of the potential under Scenario 1, and 6% under Scenario 4, but would require, according to the RCEP, up to 800,000 ha of agricultural land (given as 15% of available productive land). One can readily see both the scale of the problem and the importance of reducing demand.

Onshore and offshore wind

The wind resource envisaged under the maximum demand scenario amounts to nearly 20,000 large offshore turbines (180 clusters of 100 each), 5000 large (50 clusters of 100 each) and 5000 medium onshore turbines (500 clusters of 10 each) – providing 17% of the required 100 GW under Scenario 1, and nearly 50% under the reduced demand Scenario 4. There is an assumed shift to greater use of electricity, for example as heat pumps for delivering low-grade heat.

The mix of technologies

The rest is made up of 15 million PV roof installations – but only under Scenario 1 where they constitute 10% of the 100 GW required; 7500 wave power units (3%), one tidal barrage (the Severn) at 2%, 500 tidal stream units at 0.25%, 4500 small scale hydro units at 0.3%, and small CHP plants for farm, forestry and municipal wastes at 7%. Even then, 46 base-load power stations are required which do not contribute CO_2, being either new nuclear, or using CO_2 capture technology, and 1000 back-up fossil fuel stations (40 MW each) and 120 (400 MW each) gas-fired plants for meeting peak demand with the latter two not using CO_2 capture but adding little to overall fossil fuel use (oil is reserved for transport).

It can thus be readily appreciated that without a large reduction in demand, there could be massive intrusion and change in the British countryside. Even with a 50% reduction, the figures are still daunting: the RCEP rely heavily upon wind − with 25 large onshore wind farms, 250 small, and 88 large offshore clusters − half of the maximal scenario, but nevertheless 5000 turbines on land (more than tenfold current numbers) and nearly 9000 at sea (*in fact these figures are underestimates considering the load-factors − the total requirement to meet 25 GW demand would be 100 GW of capacity or 50,000 turbines at 2 MW each*); tidal use remains the same; CHP biomass plants are reduced to 400 MW compared to 3000 MW. Oddly, the PV roofs are reduced to 0.75 million in Scenario 4 compared to 15 million in the high demand scenario. A significant component of Scenario 4 is 2.4 million domestic micro-CHP (2 kW) − contributing 2 GW in total.

Technical fixes and social choices

The RCEP devotes most of its thinking to technology and numbers, with a few recommendations as to institutional changes (such as a Sustainable Energy Agency) and policy initiatives to drive change, such as a focus upon heat distribution networks and CHP, and heat pumps as a better way of using electricity for heating. There are some rather unadventurous recommendations on taxes and subsidies. They do not deal with the environmental and social impacts of each scenario in any detail, nor is there any analysis of end-use, lifestyle, industrial ecology, and transportation issues that are linked to siting, planning, location of work and facilities.

One rather large hole exists with regard to housing issues and domestic demand − a sector that has performed badly over the last 15 years, where, as with transport, demand has *increased* by 30%, whereas the industrial sector has seen a 30% reduction in energy use *and* greater productivity. Much in line with government policy, the RCEP sees industry as the main candidate for change through taxation, which must be rather frustrating for that sector, when the consumer is still given little incentive to reduce demand (electricity and gas prices in 2002 were much less than 15 years ago). There is no analysis of the expected 20% increase in households over the next 20 years. This is far above normal population growth, and this factor drives a large part of the projected increase in energy demand.

If all new housing were built to Scandinavian standards of energy efficiency, making use of passive solar technology, heat pumps, micro-CHP and PV roofs – there could be a net supply rather than demand from the growth in that sector.

Costs, impacts and availability

The RCEP thus does not challenge the fundamental laissez-faire market-led philosophy of British energy policy. Those technologies that are nearest to 'market readiness' are the ones supported by subsidies, irrespective of their environmental costs – it having been assumed that we are going to need everything as soon as possible.

In practice, this benefits wind power (with its vociferous commercial interests), and drives controversial developments in the uplands, and doubtless soon, along wild and unspoiled coastlines. Recent government consultations of mechanisms for subsidizing renewables, largely relating to wind and biomass, offer a single tier structure of price support that will favour larger-scale development on 'least cost' sites – there is no incentive or compensation for developers to consider less damaging sites which put their costs up. Furthermore, technology that would significantly reduce centralized demand, and which is *technically* available, is considered not ready for the market simply because it costs too much compared to the cheap fuel options available and very largely driven by the (then) low cost of abundant gas. The government has just announced a lifting of the 'go-for-gas' moratorium on 6 GW of stations, all very large at 500–1000 GW each, and with no requirement for heat distribution networks. Such decisions by the DTI undermine renewable energy policy, at a time when government has surplus funds to direct development toward environmentally more benign technology.

Do we need to develop renewables now?

This question would appear nonsensical considering the pressures to meet international obligations by 2010, but the answer must be in doubt. All of the main options, such as wind, the Severn barrage (in all the RCEP scenarios), new small-scale hydro (especially development in National Parks), and even large-scale expansion of SRC, have serious implications for Britain's countryside. The least that is required is a careful evaluation of strategies on a regional level, with community participation, *and* a well thought out and fair mechanism of subsidies — none of which are in place. However, the main reason for not expanding these sources at this stage lies in the nature of the market and the technologies that are *being* developed.

Unless there is a major restructuring of energy policy, subsidies and fundamental rethinking of urban and rural planning, all of these damaging options will be exploited first, because they are the cheapest and nearest to market readiness. There *are* other options and these are mentioned in the report: for example, in Scenario 4 (but not 1), domestic micro-CHP (a gas boiler and generator the size of a normal domestic heating unit, possibly using fuel-cell technology), penetrates to 10% of the housing stock by 2050 — there is no detailed analysis because the technology is about 5 years away from commercial readiness.

If this 2 GW contribution were increased 8 x (80% penetration, a similar level to the 15 million PV roofs assumed in Scenario 1, which are available now but 'not commercially viable'), then 14 GW would be available to obviate the barrage and all of the onshore wind and a substantial proportion of the offshore wind too. If Scenario 4's 0.75 million PV roofs were replaced with Scenario 1's 15 million, then virtually no wind would be necessary to meet the RCEP's 2050 target.

I make the startling conclusion that *now* is not the time to expand renewable energy *supplies*. A precautionary, no-regrets strategy would put all available funding into reducing consumer demand, better housing regulations and design, urban and industrial planning, heat distribution, small-scale CHP, micro-CHP in homes, and PV roofing. The biomass sector has the potential for supplying local heating as well as power, and smaller-scale installations should be preferred and subsidized against the 'economies of scale' argument that would see greater environmental impact (lorries, infrastructure development such as bypasses), less community involvement and less local employment (2 large woodchip stations without CHP have been refused planning recently). Offshore wind and wave power in less-damaging locations could still supply a large component of electricity demand. These technologies could be subsidized in parallel with the demand reduction and decentralized embedded supplies in the domestic sector.

Other countries have seen the light. Germany and Japan have kick-started their solar industries with 100,000-roofs campaigns with subsidized PV tiling (several *hundred* million dollars compared to our £50 million for *all* sectors). The PV industry in these countries is undergoing rapid expansion. The major car companies are developing hydrogen fuel cells, and these can be readily adapted for domestic boiler use as micro-CHP units. A future for large numbers of small-scale decentralized technologically innovative installations can be mapped out to 2050 where the environmental impact and change in the countryside is minimized.

All of this argument still applies, though written seven years ago. The only key innovation that could be regarded as sustainable and sensible has been a government commitment to making the future housing stock 'zero carbon' by 2015, though how this will be achieved is not yet elaborated. Some progress has been made in tackling fuel poverty and there has been a relatively little-known small surcharge on electricity bills (£1 per year) that has been used to further energy efficiency and finance help with efficient low-cost heating to the most vulnerable members of the population.

But as I feared, market forces are left to dominate decision making and planning applications. And worse, government has outlined plans to limit any sort of public critique of technology choice and siting in order to overrule planning objections. This favours large-scale wind and biomass developments. Meanwhile, there are no demand-reduction strategies, and until the recent economic downturn consumer energy demand continued to expand.

The one thing that emerges from this analysis is the potential role of micro-generation and solar water/PV in the domestic sector. These technologies truly decentralize and act to reduce the demand from central large-scale point-sources such as 100–1000 MW(e) stations, which will still be required for many industrial processes, the commercial sector, lighting, etc. The deployment of these household technologies would be expensive, but that is the cost of sustainability (i.e. the protection of wildlife, recreational resources and rural communities). There are also incalculable financial and security benefits arising from reduced dependency on foreign supplies of gas.

There are roughly 20 million households in Britain and the average price of a house is £150,000 – about double what it was when I was writing seven years ago. At that time it was considered uneconomic and inflationary to require all new housing to be fitted with PV (at about

£10,000 or 15% of the then average price and readily factored in to a mortgage). Considering the pace of house-price inflation, that judgement now stands as totally misplaced and all new housing could have been fitted as standard – by now, over a million houses in total.

Micro-generation technology is now commercially available; although it uses gas, its enhanced efficiency cuts overall fuel use by half. A programme of solar-water and PV plus micro for all the housing stock over a 20-year period would entail about £15k per house (with cost reductions from large-scale production), totalling about £30 billion per year. This would be seen as a huge cost, but it is also *investment* – in jobs, energy security, lowered environmental impact (with costs saved also) and balancing of imports, as well as making people feel more secure in a time of rising costs and the potential for failure of supplies. If this programme were rolled out via the electricity utilities and each household capital costs absorbed in the bills over a ten-year period, coupled to a 50% government grant, it would add about £60 per month to the average fuel bill before reductions in use were factored in. If these reductions amounted to 50% at current prices the total extra cost would be about £30 per month or a 30% increase on average bills. These are high costs and would be regarded as unbearable with regard to inflation targets and the necessity to protect vulnerable people from fuel poverty. But £30 billion a year for ten years would also offset the environmental costs of nuclear and wind turbine build, reduce gas imports and enable the retirement of old centralized systems of fossil fuel generation. It would provide an *adaptive* network, create industrial jobs in a specialist and exportable technology, as well as provide widespread employment at community level for an army of fitters and electricians.

This is the kind of restructuring that is required to create a resilient housing sector. The fuel inflationary costs of 30% appear politically unacceptable but, as with previous experience, this level of inflation is likely to happen anyway and with most of that money flowing to suppliers external to the UK. We would also have the same situation with regard to house prices and the previously misplaced reluctance to add 10% or 15% to their price. The costs might also be amortized over a 25-year mortgage, with tax advantages given, and this could be a condition of all house purchases, thus ensuring retrofitting. This would be a less painful option, and could, with imagination, be extended to rental properties via policies adapted to that sector.

The resulting increases in costs should be seen as sensible investment.

Normal house price inflation fuelled a credit boom in luxury goods and travel, thus increasing overall energy demand. The additional costs of ensuring effective demand reduction in the housing sector will feed into sustainable industries, export potential, reduced balance of payments and dependency on foreign powers.

These would be bold moves but they would provide a remarkable degree of resilience to climate change, peak oil and future shocks. New houses would be super-efficient to Swedish standards of construction and hence of low demand and able to export power, and all older houses would be retrofitted with micro-CHP and where possible solar systems. The resulting reduction in end-use demand per household and additional supplies by those contributing to the grid could be of the order of 30% of primary energy use (as it would include the reduction in centralized electricity demand on the grid, which is associated with large transmission and production losses). These savings alone would amount to £7 billion per annum. This is a significant proportion of the required 50% reduction of emissions outlined by the RCEP. The rest would need to be achieved by the commercial, industrial, agricultural and transport sectors, and the RCEP identified feasible paths using existing technology that would achieve these reductions.

Adaptive housing design and construction

Having dealt briefly with the role of households as decentralized suppliers (and reducers of central demand), we can briefly look at housing design and construction. Current materials are energy-intensive and the house-building, mortgage and insurance sector notoriously conservative in its choices of design and materials. In practice, ecological houses sell quickly and at a premium. The main problem seems to be communication of best practice. Government has promised to bring in tougher regulations and this should include embodied energy, but it is unlikely to dictate use of local materials. In this case a well-developed system of financial incentives is required.

Greater use can be made of local materials such as cob walls, thatch, shingles, timber-frame and hemcrete blocks. In larger construction, composite timber beams can replace steel. Designs can be oriented to the sun to make maximum use of passive heat gain. High levels of insulation are already marked by new government regulations to be brought in progressively, though I cannot see the reason for any delay when these standards are best practice throughout Scandinavia. Further savings can be made with low voltage lighting and efficient cookers. In terms of physical resilience to climate change, Swedish housing copes with extremes of cold in winter and dry hot summers and should provide an adequate model for any weather we are likely to experience. Flood protection is a major issue, however, and this is a matter of the siting of future houses and more efficient protection for existing vulnerable housing stock. There has been a spate of building on floodplains and the summers of 2007 and 2008 have been unprecedented with regard to rapid onset of flood waters.

Practical defences against monsoon conditions are not easy but defence is made more difficult by the narrow channelling of rivers and the lack of expansive floodplains for holding up flood water. Runoff in the catchment is more rapid because of poor holding potential in the upstream river system, made worse by overgrazing and lack of riverine woodland. Greater attention needs to be paid to catchment management, reforestation, removal of grazing from upland river margins and redevelopment of wetlands.[2]

A large-scale programme of ecologically designed community housing in the hinterland of cities and towns focused upon local horticulture and smallholdings supplying the city markets, populated by retiring workers, the unemployable and those seeking a quieter and healthier lifestyle on a lower income, would also contribute to solving the problems of restructuring the industrial economy.

Whatever we do in the UK we are wealthy enough (today) to buy our way out of our vulnerabilities.[3] Other countries are less fortunate, and much of the technological issues we discuss here are not relevant to the world's poorest. Their basic requirements are already low and their vulnerability to climate change high. But an analysis of the situation in rural Africa, India, SE Asia and South America would likely show that if the economic stress that takes large numbers of, particularly, young males out of the rural economy into cities could be reversed these communities could be strengthened and become more resilient. Those cities need to study the Cuban example of an economy that had to manage without cheap fossil fuel and fertilizer and yet developed effective educational and health programmes and an efficient organic agriculture.

The most obvious development that runs counter to this programme is the Gulf States and Shanghai model, now adopted by India, of drawing large numbers of people into the global economy and high-rise cell-block living. This model demands roads, ports, dams, central power stations and industrialized agriculture and it is not sustainable; the western economies, however, have become reliant upon its financial returns. Curtailing this expansion is as important as any curtailment of forest losses and will have a bigger impact on future energy demands and carbon equations.

Cultural initiatives and community building

The most far-reaching and damaging planetary effect on human life has been, in my view, the forces that have destroyed community life, isolated and alienated people and led them into a drudgery of economic slavery. The creative human heart has lost ground in the face of mindless repetitive work. Vast numbers of people live in stressful conditions on the edge of cities and peripheral to the global economy. And within the more economically successful populations there are large discrepancies in opportunity and living conditions.[4] These factors coupled with the rise of international terrorism, organized crime and military control strategies are the greatest barriers to adaptive responses to climate change. The desire for better material circumstances drives energy consumption. The demand for oil and gas raises fears of resource security and generates policies of intervention. Fear, boredom, the desire for material security and comfort are all cultural issues, and they are attendant upon the current world development model that propels people from what may be a relatively simple and basically secure life on the land but, outside of the money economy, into what is often urban squalor.

These poorer populations of essentially disenfranchised people are the most vulnerable to climate change, not because of sea-level rise or heatwaves but because of the rising cost of fuel and food, low immunity, overcrowding and the prospect of disease.[5] This model is not going to change of its own accord, nor are there any strong political forces operating to change it. Therefore in the strategy I put forward I assume that little can be done to reverse the population shifts that have taken place, and that policies must be enacted that slow the move to cities and protect community life on the land.

When community life and creativity are vibrant, it follows that the desire for material distractions, including foreign travel, big cars, and huge new TV sets lessens. But for this there needs to be a reduction in the pressures that detract from community building. Much of the techno-logical responses to climate change, such as wind turbines, nuclear stations and biofuel farming, raises conflict and stress at community level. To their shame, this conflict is dismissed and denigrated by many so-called environmentalists through ignorance, albeit with the highest of motives of protecting vulnerable Bangladeshis from sea-level rise. As should be clear from the previous section, conflict is not an inevitable price to pay — there are responses that can be made that support local community.

Loss of community and vulnerable people are at one end of the scale. At the other are people who are quite successful, relatively well-off, well paid property owners with ample resources to move around and make adjustments — in effect, to buy their way out of the problems on the horizon. They are just as culturally afflicted but not at all conscious of what they have lost or what they may need when the going gets tough. They have been so far removed from real community that it does not occur to them that they have lost it. It is something that would only be noticed when serious social unrest emerged in their own neighbourhood. I am not sure that any work can be done with these populations until such time as their current world becomes unstable.

The currently rich and powerful upper 10% of the world's population — about 600 million people in Europe, North America, Australia and East Asia — have a choice to make in the face of climate change and the forces that are currently destabilizing human support systems: they can work hard to build a world community of trust or revert to control strategies — with walls, gated communities, immigrant controls, support for military domination of energy supplies, greater policing, surveillance, secluded second homes, savings, investments and insurances. It is perhaps to these people that James Lovelock in his *Revenge of Gaia* directs solutions that involve moving to nuclear-powered refuges in the Arctic (the only place cool enough as earth's natural homeostatic mechanisms become over-whelmed by carbon dioxide!).

The cultural issues of adaptation I outline here are not beyond the wit of government intervention — to enable planning systems, community building, effective consultation, openness and involvement at home and less militaristic solutions to international problems of resource security overseas. This is also a model that can be exported, if only by example, and one that will not contribute to the sense of exploitation and cultural alienation from the 'western' way.

The climate of fear

Finally, there is a cultural factor that is not often highlighted but is at the root of all the above highlighted problems – it is the prevalence of *fear* within the current development model. This fear operates consciously within the communities afflicted by crime and disorder, terrorism, disease and economic hardship. The last of these is perhaps rather basic to the economic system; few people feel secure and many decisions are made out of that fear. What is noteworthy is that virtually all climate campaigns blatantly add more fear to this pot.

I do not believe that any future strategy motivated by fear will work. Fear tends to manifest that which is feared, whereas positive and creative thought manifests positive action. A great deal of fear, however, is sub-conscious. Thus a consciously held 'dream' of an ideal world such as the American Dream of personal freedom and liberty fails so significantly to manifest because it is founded upon a basic insecurity in the American psyche that lies hidden in the subconscious. The most powerful military nation on earth appears to be afraid of its own shadow. At home the American Dream often descends into a nightmare reality with random killing, epidemic drug dependency, obesity, poverty (an estimated 20 million people in the USA live in harshly impoverished conditions) and mental ill health. Yet the dream is still projected onto global society as a perfect model for development.

In fact, American economic power was built using widespread tariffs and protectionism, cheap oil and coal, massive immigration, broken treaties with its indigenous peoples and genocidal land-grabs, followed by the steady purchase of foreign assets and control of commodity markets. Their diplomats then argue for a liberalization of markets, in Africa for example, and exposure of developing economies to competition with highly developed and sophisticated multinational corporations.[6] Other western nations collude with this dream both in its military and economic global projection.

Psychological motivations for foreign and economic policy are seldom factored into academic analysis. What drives some nations to act big on the world stage? What manner of endemic insecurities created within a national psyche get projected into foreign policy disguised, of course, as national economic interest? And if one wanted to delve into the machinations of policy formulation, how does creating and maintaining insecurities at home create an electorate supportive of such external policies?

In my view there is a clear link between the US system of 'social insecurity' and an electorate that supports programmes of hegemony and control justified by 'national security'. The actual policies pursued *increase* this insecurity, and thus the motivations must lie deeper within the structure and point to vested interests within the corporate world – what used to be more widely discussed as the 'military-industrial complex'.

This insecurity becomes a global contagion as other economies become dependent upon the USA as the engine of economic growth and their model for the corporatization of their economies. There is a fear that lack of growth will lead to social disorder, and all manner of inadvisable and unwise policies are then pursued because the majority of other nations fear for their economic security. They also then collude in the deceptive reasoning for these policies.

Real security

In this respect, I believe the policies we need to develop should foster a sense of well-being and basic security. It is no accident that global 'terror' takes Americans as its mortal enemy, nor that the CIA effectively created and nurtured the organization of al Quaeda as a counter to Soviet expansion in resource-rich areas. The Islamic rebellion is as much a rebellion of tribe against empire, of the disenfranchised 'son' against those behind the wall and in charge of the economies of the world. It is as old as the world that Mohammed inhabited when the Christian 'empire' began the appropriation of lands, and was represented later by the myth of the disenfranchised brother Ishmael in relation to Isaac. Both sides of the current conflict parade their mythic origins and this current shadow dance has the power to render every creative endeavour entirely fruitless. It is a dance that now deprives humanity of huge resources, personnel and intellect that would be better deployed in restoring the fragile ecology of human support systems.

It is not the planet we have to save. 'She' has seen much worse and recovered quite well, even if it took a few million years. Earth's natural propensity is to adapt to climate change. It is the place of humanity that is at risk and not, I think, in terms of survival but in terms of its spirit and integrity. Humanity is founded upon the principles of cooperation and compassion, not competitive economies and their resource wars. Human endeavour is more than a materialistic enterprise concerned with comfort and security. The current efforts to mitigate emissions, though not well-founded in science, nevertheless incorporate a sense not just of long-term self-interest but also of global equity and a deeper understanding of natural cycles and the vulnerability of ecosystems upon which everyone depends. Though the science may be uncertain, the issue has the capability of uniting nations in a common cause and it calls into question the limits to industrial growth and a development model that cannot serve a large proportion of the world population.

It is with this in mind that I make proposals for a no-regrets strategy that takes into account the scientific uncertainties and the deficiencies of the development model. It has the capability of strengthening the cooperative spirit and community of nations at a time when cynical forces of exploitation seek advantage from the general fear of ecological damage that has become so pervasive; and it works to set up a communication link between richer communities and poorer across the globe.

Notes

1. See my review of *Energy – The Changing Climate* (the 22nd Report of the Royal Commission on Environmental Pollution) in *ECOS*, 21(2), 79–83.
2. See, for example, *Valuing Ecosystem Services – the case of multifunctional wetlands*, ed. Turner et al., which provides case studies of positive cost benefit analysis of these solutions.
3. This appeared to be true at the time of writing this paragraph! The very rapid reversal of fortunes causes me great concern. Hundreds of billions have been not 'lost' but squandered in a consumer boom or exported to support 'growth' elsewhere, and without what may be an unacceptable tax regime most western governments will not now have the resources to restructure their economy and community towards greater resilience.
4. In a review in *New Scientist* of *Affluenza*, a book by Oliver James, Andrew Oswald, Professor of Economics at the University of Warwick, reports on his own work in the UK measuring psychological strain and finding increasing levels of unhappiness and evidence of declining mental health. He also points out that statistical evidence on causes and effects within a materialist culture are few (with little in Oliver James's book). Another recent survey reported by Jonathan Brown in the *Independent* (16 July 2007) relates happiness index to carbon footprints and this shows varying levels of contentment in European states with Scandinavian countries topping the rating for happiness as well as carbon frugality. So the issue has less to do with economics and more to do with how society and community is organized – though it has to be said that these are the wealthiest of European countries, with less inequality; and wealthy low-

population Scandinavian countries more readily increase their energy efficiency and GDP per capita-carbon-footprint.

5. There is a new report released by the UN Framework Convention on Climate Change on the ethical dimensions of climate change and reviewed by Fred Pearce in *New Scientist*, 11 November 2006, which highlights the plight of the world's poor and the human rights issues raised by the emission profiles of rich northern states. Africa is described as the continent least responsible for climate change and most at risk. Yet there is a continuing assumption that urgently addressing emissions (there is a call for 30% reductions by 2020 to follow the Kyoto Protocol) will somehow make a difference to Africa's future. There is no evidence for this and, as we note, Africa's future is already compromised — but by the nature of development and the naturally deteriorating climate. Most Africans aspire to the material goods of the western development model and it is the production of these goods at prices they can afford that causes the carbon imbalance they feel aggrieved by.

6. See for example Ha-Joon Chang's *Bad Samaritans: Rich Nations, Poor Policies and the Threat to the Developing World*, and Chang's article on protectionism in the *Independent*, 23 July 2007. He argues that more protectionism in developing countries — as practised in the developing phase of all the major developed countries — would enable faster economic growth and be in the self-interest of rich countries. He shows, for example, how sub-Saharan Africa is the only region of the developing world to experience negative growth in 1980–2004. But Chang is a Cambridge University economist and does not deal with the cultural dimension. There are other reasons to be protected from globalization! He also fails to consider that the vested interests of the rich countries are not wholly economic; there are subconscious psychological driving forces related more to the exercise of power, domination and even humiliation, as became evident during the occupation of Iraq by coalition forces.

16

Reflections from Anthropology

'We can't solve problems by using the same kind of thinking we used when we created them.'
Albert Einstein

Man-made global warming is exactly what it says on the label — a fabrication! It is an illusion borne of a particular way of looking at the world. It has become such a pervasive belief despite the uncertain science because that angle of view works very well for a lot of people. It is not a conspiracy. It is more a mass delusion coupled to massive collusion of interests. It follows that any attempt to 'fix' the climate serves merely to maintain the illusion and can come to no good. But how to break out of the delusive state and also make sure such a politically and economically damaging error does not occur again?

In the foregoing chapters I have limited myself very largely to the arena of science and politics. I have referenced my arguments to evidence. I have made a good case for a no-regrets strategy that would move society towards greater resilience in the face of natural changes and a rapidly growing vulnerability to those changes. It is possible that truth in the science of climate change will one day break through, although I have to admit I have never seen anything in all my experience like the monolithic hold that computer-generated realities now have on the institutions of science. Something deep and pervasive has happened to society's perceptions of the natural world.

Certainly many voices are now being raised and I am not alone in my view that this emperor has no clothes. As I edited this final chapter, Dr John Theon, a former senior scientist at NASA, who oversaw the department and initial work of James Hansen, one of the key instigators of global warming science, has now declared that he has no faith in the models. He is advising the US Senate that such simulations should form no basis for policy. Dr Theon, former chief of the Climate Processes Research Program, stated in a report to the Senate Environment and Public Works Committee on 15 January 2009: 'I add my name to those who disagree that global warming is man-made.' (The US Senate produced a Minority Report signed by 650 scientists questioning global warming theory in December 2008.)

The forces now arrayed to protect the orthodoxy are immense, however, and policies of emission reduction are now firmly entrenched and legally binding. So what kind of thinking can get us out of the mess? Even if we were to break through the fabrications and approach a real understanding of climate and humanity's vulnerability to its cycles, the kind of policy response I have argued for involves much of the same kind of thinking. It requires calculations and projections of technology, economics and strategy. As a long-term policy advisor I have watched how logical, rational and feasible policy recommendations have been ignored or received with pretences of action and very little then comes of it. The inability to act may simply be a matter of power and vested interests. Once a particular set of people and technology is favoured, it holds on to its position and quickly learns how to manipulate political processes. However, I believe this crisis has been created by a deeper level of disjunction than that of an error in computer models or the interplay of vested interests in politics, the media or environmental campaigns.

Climate as enemy

It is striking that a small group of men working behind computer screens created a virtual reality in which the future climate became the enemy of mankind. That original cabal was likely innocent of any underhand motivation and genuinely believed mankind faced a threat and that they would sound the alert and potentially stave off disaster. But sociologists will go a little bit further and look at the social environment that spawned the very concepts of the climate game, many of which we take entirely for granted. For example, the notion that humanity itself can be under threat or that the planet might need to be saved. These are very recent notions, at least from a societal perspective, and do not bear closer scientific scrutiny. My sense is that they have more in common with previous religious notions of an apocalypse but without the aspects of divine judgment. Instead, humanity judges itself and projects its own demise, with the ecological feedback system becoming the instrument of retribution.

Thus, humanity is culpable, but underlying this notion is the concept that humans now have the power to alter the planetary ecosystem and even to destabilize it. Such a sense of power is very recent and directly linked to the technologies of control available to an ever-increasing human population. The over-exploited earth readily becomes seen as 'fragile' and in need of care, and it is one step further to see humanity as a threat to itself simply because it threatens the ecosystems from which its material needs are drawn. The science of ecology has taught us about webs and interdependence, and a 'green' movement, of which I have been a part, has politicized this message.

There is an alternative view. In this, the earth is not fragile at all, and humanity's apparent dominion is an illusion created by a limited under-standing of the cycles of nature and an unwillingness to look dis-passionately at the rise and fall of former civilizations. In this view the earth is quite capable of cleansing itself, renewing all of its so-called ecosystems, and recreating whole dynasties of new creatures. And from such a natural process, humanity might emerge in lesser numbers but greater humility and a heightened state of awareness. This view is as valid scientifically as the orthodox scary climate story. But it would necessarily engender a wholly different attitude and collective response than that currently urged upon us in a twenty-first-century politics of ecology.

The view from anthropology

In what seems a long time back – in 1979 – I gave a parting seminar to the graduate school at Oxford's Institute of Social Anthropology entitled 'Steps towards a Political Ecology'. I warned that anthropologists needed to look at themselves in the same way they looked at apparently primitive tribes. I drew my examples from a study of the growing 'green' revolution in central Europe (**Taylor, 1977**). Ecologists were challenging the hege-mony of economics and political economy as theory. They were seeking a new theory, a political ecology, by which the revolution could be fur-thered. My question then was: why do we need social theory? Every sociologist from Adam Smith through Karl Marx to Emil Durkheim developed theory, much as an ancient Merlin might seek to enchant and advise the King. They were the modern equivalent of the seer, but now in competition with economists and scientists in a time out of favour with prophecy, though not foresight and planning. This I called the 'Merlin' position. It was, I believed, an unwritten supposition on the part of my sociological colleagues that this unnamed but pivotal position existed, and for that they strove in often ritual jostling, usually always losing out to the fancy-named French masters, where Claude Levy-Strauss was then the flavour in corridors of power.

It seems to have taken about 30 years for the political ecologists to get there, and oddly, without a particularly well-developed 'theory'. The new President of the USA, Barack Obama, has announced his advisors, and they are supposedly strongly 'green', including what the newspapers in Britain refer to as one of the world's most respected climate scientists, John Holdren of the Woods Hole Research Center. In the UK, the government has a Chief Scientific Adviser more likely to speak on environmental issues than pure science – recent incumbents have publicly derided opponents not only of climate policies but of the use of GMO crops, nuclear power and the gassing of badgers, all of which have significant value judgements beyond the remit of science. The UK Department of the Environment, Food and Rural Affairs now has a former official at IPCC as the head of its science team.

This causes me some reflection upon the days when environmentalist Max Nicholson wrote about ecologists as the new 'managers' of the planet, meaning also, of course, the economy. It used to be the ecologists' dream that such sanity would prevail and give them control. However, rather than following the party political road, the environmental activist as scientist sought out the Merlin position, and quite successfully it would seem. But again, that is a surface view. When I look more deeply at the characters involved in these influential positions I do not see 'environmentalists' – at least not in my old-school definition. For the most part I see physicists, chemists, computer specialists and lawyers, all wearing a green cloak of concern for the planet. They adhere to the dogma of 'global warming' or the carefully all-encompassing 'climate change' and are quite immune to any discussion. They have not studied the original science – indeed, few are real climate scientists. They have piggy-backed upon public concern. And to do this they have had to develop a very simplistic message and created a social defence reliant upon stigma and ridicule such that only an 'idiot' or a 'denier' would question them. They reside in an unquestionable 'authority' and consensus of all the experts, implying of course that anyone who does question is not among that particular expert elite.

Yet however successful this path in terms of reaching that key Merlin position, very little of any consequence comes of it. The real ecologist is emasculated in the process. And if that were now to change, such that 'green' policies were enacted, as seems likely if the USA now moves in that direction, then it is ironic that these policies would destroy more environmental value than they would create! A little reflection is therefore in order. Perhaps 'green' policies are now politically acceptable precisely because they are no longer green. They simply further the powerful interests of everything that has gone before. Wind turbines, barrages and nuclear power stations all require aerospace skills, concrete, high quality steel, plastics and engineers, and they are profitable within a subsidizing economy. In the Middle East, cash-rich nations now opt to build zero-carbon cities, but they are still building new cities dependent upon a global economy of migration and increased consumption of goods. Bio-fuels that are an inevitable part of a zero-carbon development require vast corporate empires of landowning agribusiness and these states are already purchasing land far from their own borders. Technologies of carbon capture still require all the skills and materials of the oil-pipeline industry. Carbon credits and carbon trading disproportionately benefit bankers and brokers, and costs are ultimately born by taxpayers and consumers. Nothing in the global warming bag of tricks actually offers anything to the human values of community, indigenous culture, food security and resilience in the face of climate change, or any redirection and restructuring of the faltering global development model.

On the surface, it does look like a conspiracy. A small group of scientists begin meeting in the mid-1980s, among them James Hansen of NASA, and develop the thesis that a doubling of carbon dioxide levels could lead to disaster. Levels are expected to double somewhere between 2050 and 2100. Within a few short years they managed to initiate a UN Working Group, and a Framework Climate Convention that is adopted at the Rio Summit in 1992. The Kyoto Protocol followed in 1997. All this at a time when environmental science, and perhaps science generally, had suffered a great loss of prestige. Humanity was recovering from the threat of nuclear holocaust, reactor meltdowns such as Chernobyl, dying seas, acid rain, lead in petrol and the loss of the ozone layer. Suddenly, scientists were putting themselves forward as world saviours.

Collusion not conspiracy

But we need not talk of conspiracies nor impugn the motives of earnest souls genuinely believing in a future climate chaos. We are dealing with a Zeitgeist that pulls together the interested parties and a collusion of interests that can propel an untruth forward to such an extent that people cannot afford not to believe it. There are other examples, such as the unquestioning belief in the existence of weapons of mass destruction in Iraq in advance of the invasion. On the climate issue we should therefore be equally aware of the potential for 'dodgy' dossiers to arise and underpin the political will. It would be naive in the extreme not to be aware of the scale on which modern deceptions can be practised, just as, in another field, the current financial meltdown reveals scams of hitherto unimaginable proportions, where for example one senior and respected banker can fraudulently 'lose' $50 billion.[1]

I think it would be foolish to assume that the climate-change edifice of science along with computer labs, committed politicians and environmentalists, media pundits and countless government functionaries, carbon accountants, brokers and the purveyors of new technology does not operate as a single-minded vested interest, that it does not have a group mind, an ideology and a purpose. Although, I am still foolish enough not to believe in a conscious purpose behind it all. Others talk of control, surveillance, taxation, a carbon economy and global governance, with all the newfangled and probably vastly more oldfangled mandarins such a new edifice would entail. It is a jobs-for-the-boys jamboree. But I still don't think it is that simple.

There are shades of religion at work, as if there is a need to believe, to have a mission, to proselytize and 'save' humanity from itself and even the planet from humanity. There is an underlying ideology of guilt coupled with a perverse form of hubris. In this hubris humanity is both powerful enough to destroy the planet's life-support system, and also to restore it to balance. This is mythic territory. The myth of climate change is very deeply entrenched in belief, not just in terms of scientific mechanisms but in the attendant notions of power.

In addition to this ideology of power, agency, guilt, reformation and restoration, another darker side has developed. This ideology seeks to restore not beauty or balance, not community, integrity or spirit, but dominance and control. The climate has got out of control through our own lack of foresight. The battle is now to regain that control, prevent chaos and reign in the runaway system. And the means are military. Indeed, all the language is of war. The advertisements speak of recruitment and 'doing your bit' and a 'carbon army'. The war on climate is more important that the 'war on terror' says a former UK government chief scientist. And the remedies all involve discipline, new forms of accounting and surveillance, the sacrifice of freedoms (and beauty) for the common good, as well as new technology and new forms of taxation.

Vulnerability and the strengthening of community

We could contrast this with the alternative view that climate change is natural, unavoidable and always dangerous. The truth is that humanity has become very vulnerable to climate change. And an even deeper truth is that we have been here before and that whole civilizations have perished because of this vulnerability. In the good times humanity prospers, and in the bad things can get very tough. In that truth, we are very small and nature is very powerful. Nature is also difficult to predict. Its cycles are not exact. There is an immense complexity of factors at work. These cycles interact with each other, at times amplifying, delaying, speeding up or cancelling out. The down-phase of the climate cycles always brings famine, competition for food and disease.

If this truth is accepted, rather than the illusion of control an entirely different response is engendered. Firstly, acceptance of the power of nature promotes humility rather than hubris. This acceptance faces us with a clear choice between cooperation and competition. The situation challenges humanity to become more humane. On a technological level, this means selecting technologies that promote cooperation and community, that strengthen the human spirit rather than undermine it. Conflict-generating technologies that in the illusory world appear to offer a solution to climate change – such as nuclear power, genetically modified crops, turbines and pylons across the landscape, vast acreage of biofuels, foreign ownership of productive croplands, and all the control and surveillance that these supposed remedies demand – have no place in the cooperative world of human community.

In the acceptance of vulnerability and the importance of community, a different technological response is born. It starts with the home and the neighbourhood and maintains scale and involvement. It fosters responsibility. It does not generate vast profits for corporate entities that have no connection with the community, but it can choose even the most modern and high technology solutions as long as they serve that community. For example, micro-generators, solar panels, underground heat exchangers, and passive solar construction make households net contributors to a decentralized grid. Here the high technology corporate world serves the communal interest. Public transport systems are expanded – but they still require engineers and good design. Small-scale biomass stations are fuelled from local materials. People walk and cycle to a workplace that is integrated into the neighbourhood's spatial planning. Cities require decades to reconstruct what are currently centres of very high energy demand.

Above all, there needs to be a cultural embrace of deeper human values that move away from passive entertainment, distractions and the mindless pursuit of material wealth. And in the latter, there needs to be an acceptance that in most developed countries there are millions enslaved by an economy from which they have little hope of escape – by that I mean, people who are not free, who cannot follow their own will for fear of losing the security that a job provides. This creates a dangerous pool, first of mindless consumerism, and then ultimately, when the natural cycle of abundance turns, of mindless violence.

This latter is the real threat to humanity and to the planet. And by the latter, I do not mean to imply life itself. The 'planet' is a symbol for the human world and its unacknowledged fragility. It is a projection of that fragility. As an ecologist well versed in geological and evolutionary history, I know that the planet is not endangered. I doubt also the survival of the human species is at stake. The earth's dynamic systems are far more powerful than even the most potent of nuclear technologies. There are also galactic forces at play that can change the face of the sun and shift the climate of the whole solar system into other phases that could last for millions of years. At times, these planetary and interplanetary forces have toppled whole dynasties of creatures, wiping the slate clean with skies of thunder, lightning and rain as strong as battery acid! And always evolution creates anew.

What is at stake is not the survival of a clever bipedal ape, but the future of a sensitive, creative and cooperative human being that cares for beauty. It is the age-old question of what profit there is in gaining the world and losing the soul. The apocalyptic vision of climate chaos is thus not only an illusion born of hubris and maintained by religious zealots; it is a deception on a grand scale and it serves the same powers and elites that the old stories served. It promises salvation and the defeat of an old enemy, but with more jobs and greater prosperity in a future green economy. That the green economy will be wearing a black shirt is not yet obvious, as such things seldom are at the outset.

But if this is the pit into which humanity has fallen, misled by science visionaries and abetted by visionless lawyers, accountants, bankers and publicists, how then to get out? This is where we have to think differently.

A leap in the dark

As I trust my analysis has shown, most of what is proposed by the climate conquistadores leads in the opposite direction to that which I have identified as the most rational response – the creation of *resilient systems*. The creation of such systems will not be furthered within the current paradigm, and there is no obvious sign of that paradigm shifting. I have talked about what *could* be done in the way of eco-sustainable communities – no new knowledge is needed. But this is not going to happen. Not unless the whole system collapses and the system grinds to a halt. I rather suspect that a collapse is coming and so, for my reflections, I want to look at the deeper lessons. I want to find a way towards the *kind* of new thinking that leads out of the mess, even a collapsed mess.

On the science alone, the imbalance of the superficial male mind would leap out at any gender-conscious sociologist. What is missing in the models is a science of cycles, irregular periods and spiral mathematics such as Fibonacci series and fractals. The limits of the linear electronic metaphor and the compartmentalized, structural way of thinking are hardly acknowledged – it is a form of hubris, and it purports to relegate anything outside its boundary to random 'chaos'. Few scientists would know that this latter word is derived from the Ancient Greek *Kaos*, a name for the ultimate force of the universe – symbolized in ancient wisdom as a female deity! Far from representing disorder, she symbolized creative power but through the portal of destruction.

A new way of thinking is unlikely to embrace ancient mythologies, though every anthropologist knows that a society lives by the nature of its creation myth. Nor can it just be a matter of introducing a mathematics of cycles or spirals. We have to become conscious of *why* the male mind amputated its right hemisphere. And the answer obviously lies deeper than the level of a more holistic mathematics. That other hemisphere works on a different wavelength – on another dimensional aspect of the mind that we loosely call intuition. Modern philosophy hardly has a language for the subtleties of knowledge available in this realm and, more particularly, any acknowledgement of the practical disciplines required. Intuitive knowledge requires a practised surrender of normal consciousness.[2]

Men *can* do this, of course, but only the great men – such as Einstein – made it a primary mode. In all the great scientific discoveries some kind of intuitive insight was involved. This is the realm of the 'feminine' mind in the sense that women have better access to the right hemisphere mode of thought. That does not mean that in this modern or even post-modern age they necessarily choose to use it! If a woman wants to get on in this world, she quickly learns which hemisphere brings more acknowledgement, power, prestige and prosperity. The male mind is structural, logical, calculating, measuring and motivated by the prospects of manipulating reality, whereas the natural feminine mind is intuitive, surrendered and creative – it also elevates feeling above thinking and works by trusting an initial state of emptiness and not-knowing. In this state of conscious awareness there evolves a natural and heartfelt love rather than fear and calculation. In this mode, technology is evaluated not in terms of external economic progress or illusory levels of security, but in terms of its impact upon community and relatedness. Most men have taught themselves not to balance their minds with the feminine mode because there are often severe penalties in doing so, especially in realms dominated by science.

The power of intuition ought to be more obvious and revered – after all, everything that is built or constructed must first have been passively dreamed or creatively imagined. But of course this is far from the case. No child of a modern educational system, at least past primary school, is given time to hone their dreaming abilities. Even the concept of training the intuition is foreign to modern education, whereas we can learn much from indigenous peoples who live closer to nature and move in tune with her cycles. From my own direct experience of tribal cultures this power of dreaming is rather stronger than our current male-minded culture cares to remember – if indeed, any cultural memory persists. An expanded intuitive mind can become aware far beyond the boundary of individual consciousness.

New thinking in an age of deception

In such a world of sharpened intuition, it is not so easy to be deceived. But any times when such an empowered condition existed are rather lost in distant history and we should at least be aware that the remnants of a powerfully educated feminine mind were systematically and ritually executed over the period of 500 years that saw the birth of scientific thinking. I am led to the conclusion that the systematic repression of the feminine mind is a necessary prerequisite for the imbalanced modern consumer mind and the dysfunctional society that seems incapable of right action, even when confronted with its own false but nevertheless *believed* analysis that it will eventually cause its own demise. Human society then essentially becomes, in its own mind, the mythic monster that eats its own children.

This rapacious and imbalanced consumerism is thus a consequence of a fear-based, conquest-oriented, military mind and carbon dioxide emissions are this monster's rather bad breath! I am thankful that this monster does not have the power to destroy the planet but not very hopeful that it can sensibly adapt to the natural cycles of the earth and learn to appreciate 'Nature', which, by the way, stems from the Latin *Natura* and means 'she who gives birth'. If colonial Latin had not replaced the Anglo-Saxon 'Birther' we might not have become so disconnected in our thought processes!

I could make many practical recommendations on the retrieval of a balanced mentality but this is not the place. Sadly, the modern mindset would resort to ridicule and seek to avoid the scientific criticism which is a central part of this book. Those recommendations will wait for a time when I can more fully explore the issues of how to create a more stable and resilient human world. I am not at all confident that such a task is achievable without first a period of hardship and decline.

Dogma and divine intervention

I spoke earlier of my impression that much of what moves the current dogma borders upon the religious. In actuality, of course, the *dogma* is irreligious in that it is science-based. In the scientist world-view there is however not only no divinity (other than that practised outside of science, which many scientists hold in some disdain), but no model for the interaction of consciousness and matter. Consciousness, like all species and forms, has arisen in the myth of purely accidental forces. Humans are simply more evolved forms of animal consciousness. Interestingly, within the realms of science, little interest is shown in the actual study of con-sciousness, at least with regard to *content*. For many decades, Carl Jung's psychology hardly penetrated the university campus. The possibility therefore that consciousness is *primary* – that it existed *before* matter – is not entertained. The idea that before the universe 'expanded', a purposeful consciousness existed is regarded as fanciful and *creationist*. This would of course lead back to the very concept of God and a reminder of the persecution that science faced over many centuries. Science cannot bring itself to embrace an alternative creation myth, however much it ties itself in knots whenever thought goes beyond the 'big bang' that gave birth to it all. Science as such is a negative image of the religious philosophy it has sought to replace. Where there was once divinity, there is simply a black hole. The possibility that black holes may wander around the galaxy consuming whole worlds whilst held by astrophysicists as a potential reality appears not to move them to models of higher levels of consciousness than that to which most scientists can attain.

Science can no more embrace the concept of a mythic destroyer than that of a creator. Ancient wisdom in other cultures, even quite advanced cultures such as in Hindu theology, usually have both mythic creators *and* destroyers. In supposedly grown-up, science-based cultures, the only creator is the human mind and that mind now takes upon itself the mantle of destructive power!

For those who might still believe in this burgeoning omnipotence, I urge some caution based on palaeostratigraphy (the science of sediments). There is still no accepted explanation for the sudden ending of ice ages, only for their slow onset. Nor is there an explanation for cycles of cosmic ray proxies that peak *throughout* the ice ages at 2500–5000 year intervals. The patterns form harmonic series and the phenomenon seems to create the 100,000-year main peak that ends the ice ages (see Fig. 2). I recommend a thorough reading of Richard Firestone's *Cycles of Cosmic Catastrophe* in which his team of NASA scientists survey the geological evidence for a major comet-like impact at the end of the last ice age and which also coincided with very high levels of cosmic isotopes recorded in the ice-cores. Virtually all large mammalian animals were eradicated from the northern hemisphere. It would appear that something rather regularly breaches the magnetic defences of the planet and it even looks like the sun itself lowers its shield in the face of this cosmic visitation.

This is not to imply a sense of millennialism or impending doom. Indeed, aversion to such cult phenomena is perhaps the main reason why consideration of extra-solar factors is not fashionable and anything 'cosmic' studiously avoided. However, this data is very real and very recent, and our understanding of the very longest cycles may be due for radical reappraisal.

On a more traditional ending note, here is my summary and conclusions:

- The proposition that the planet has warmed due to greenhouse gases of human agency and that humanity could in the near future destabilize the global ecosystem and the rich biodiversity of life is a *virtual reality* created by a small cabal of computer specialists.

- Those specialists had little real comprehension of ecology and in particular of past ecological environments. They were fundamentally mathematicians and physicists, chemists and computer technicians, and prone to all manner of ambitions to further their field of knowledge and make a play for saving the world from what they genuinely believed could be a future threat.

- Within a relatively short time, powerful global interests had allied to this cause, in particular those organizations in need of a new and truly global mission.

- The hypothetical threat from carbon dioxide was a perfect enemy. It lay at the heart of all the environmentally destructive tendencies occasioned by burgeoning economic growth; its global reach meant that no country was safe from its effects. Emissions reflected perfectly the global inequity of trade and wealth, pollution and the evils of laissez-faire capitalism.

- Thus, the war and the campaign began — alliances were sought, the media were activated and the debates politicized such that almost no party could gainsay the 'truth' of climate change.

- And when the UN's assembled scientists debated, disagreed and worked their caveats, a whole cadre of politicized drones reworked their wordings into something that 'policy makers' could act on — a series of 'targets' that related to emissions and percentages of renewable energy supply.

So successful has this operation become that the UK Parliament — on the day that I write this — just passed into law a target of 80% reduction of carbon emissions (or their equivalent in other gases) by the year 2050, thus effectively committing the nation to a decarbonized economy that no one but the deluded propagandist believes is achievable. And in the run-up to the US election, the Republicans, hitherto resistant to ideas of a low-carbon economy, hardly dared make it an election issue, and the new President has announced a Green Revolution. Within the European Union, every state is bound to a target of 30% reduction by 2020. Russia, which under Putin consistently refused to ratify the Kyoto Protocol though faced with minimal reductions, did so in 2004, against the advice of the Russian Academy of Science (since revised). Japan accepts the treaty without question. Only China and India, with their rapidly expanding fossil-fuel based economies, remain by common agreement outside of the Kyoto Protocol, though not outside the benefits of a Clean Development Mechanisms and Global Environment Fund that parallels the treaty.

In the academic worlds in Britain, western Europe, Japan, and India, there is barely a whisper of dissent, whereas in the USA, Canada, Australia, New Zealand and to a very small extent Scandinavia, there have been recent petitions numbering tens of thousands of professional scientists and meteorologists aimed at parliaments and the press. The major academies of science – in the USA, Russia and China – despite having expressed concerns in the past at the science-base for these policies are now toeing the line. And with regard to the press, there is a curious division between the right-wing supporters of less government control, taxation, regulation and restrictions on trade who have espoused the cause of the 'sceptics' and a left-leaning, liberal press who side with 'the environment' and the development of a new form of controlled economy.

This web of alliances and oppositions has created a global phenomenon that stretches from the lowliest jobbing builder working for government-backed home insulation schemes, through many levels of the Civil Service, the banks, the electricity companies, the turbine makers, the nuclear opportunists, the carbon credit brokers, the parliamentary lobbyists and last, but not least, the tweakers of the virtual-reality climate models who set this massive super-tanker of vested interest on its unstoppable course. It is almost unimaginable that it could be turned around.

And yet down on a street level the polls show that very few believe the scary climate story! In the USA, Canada and the UK, a majority do not trust the science, the campaigners or their governments. That majority believes it is a scam.

We are now in the midst of a global phenomenon that embraces currency, banking, all reaches of government and almost all branches of science, effectively all voluntary organizations concerned with environment and development, the press, radio and TV; and it is one which has developed its own language and ideology, including modes of defence and attack against any criticism. If, as I have come to conclude, the science base was flawed from the start and that a small, if deluded, cabal have used their position and all manner of collusion to foster this colossal erroneous child of our times, then this constitutes a mesmerism that surpasses that of any previous ideology or religion or even military alliance – and ironically, appears to do so by offering a global *unity* of good purpose.

Given such incredible virtue, who would not wish to support the movement — to sign up and do their bit? Well actually perhaps the great majority who, not believing, pay lip service, partly to avoid ridicule and partly out of some greater wisdom that all such follies must pass. And it is the unbelievers who are currently smiling as the coldest winter for decades sweeps across North America and western Europe, the PDO has turned, the Arctic records a 30% increase over last October's sea ice, and the North Atlantic remains as the last great heat store from the 1980–2000 period of illusion that I have detailed in the science.

Several oceanographers and solar specialists expect a global cooling of between 10 and 30 years, and one or two think we could descend into a new 'Little Ice Age' by 2030 — that is, a global cooling that would last another century or more. If this prognosis is correct, it will be interesting to see how this huge edifice seeks to justify itself and in particular how the institutions of science respond. It is a very large ass to cover. And of course if the modellers are right, and warming reasserts itself (although the real test would be that it would do so without a parallel rise in solar activity), they will deserve our gratitude and I will make a determined effort to eat my treasured Colombian fedora.

But on one level — a very sad level for someone of my sensibilities — the truth of this does not make much difference. Too much has been set in motion, in law, in targets and vested corporate interest. In the UK alone, 10,000 turbines and an extended electricity grid are scheduled to festoon the hills, wild places, rural countryside and coastal vista. Plans are being dusted off for tidal barrages and nuclear reactor stations. There is a developing acrimonious debate about whether holding off on new fossil fuel stations and the long delay times associated with barrages and nuclear development could plunge Britain into a major power crisis as old coal stations fall foul of new European legislation. And bio-fuel plants have hardly begun their assault on land cover, biodiversity and world food supplies (the latter causes me the greatest concern). The largest single coal-fuelled station in Europe, Drax in the UK, is about to co-fire with biomass, drawing from a global supply of forest and food residues, straw, rice husks and the like, thus using economic power to sequester renewable resources from those countries that are also being urged to develop their own renewable power base. No safeguards are in place for primary or secondary biofuels, whether for communities needing those resources, or for indigenous peoples or biodiversity, yet industrial giants are building plants that will come on stream within a few short years and begin to consume what is left of the world's current food surplus.

In China, India and the Middle East, nuclear power's prospects have never looked stronger, and even the USA and Britain look to a renaissance. Then there are the traditional fossil fuel interests whose profile in all of this, despite the protestations of collusion with sceptics and charges of behind-the-scenes conspiracy, are remarkably quiet. Perhaps there is little real threat to their immediate interests. Even with an inflated oil price, biofuels or hydrogen will remain more expensive for some time yet – time enough for oil companies to re-invest the profits. And for the constructors it matters little whether they build a hundred small turbines and electrical gear or one large central power station, just as long as they can build.

And thus, when I stand back and look at this army on the move, I can see how it works to everyone's interest to believe in the scary climate story. For many, the new story does not change much in the world. Not in the press, the research institutes, the corridors of government, the campaign groups or the industrial conglomerates. Indeed, the carbon story offers a convenient ideology of public works during an industrial downturn. Certainly in Britain the Prime Minister has announced that renewable and low-carbon energy projects will be brought forward to compensate for the looming recession.

This will, for the majority, be very much 'business as usual'. They will be taxed to pay for interventions in the market. And they will watch, for example, schemes such as Barclaycard pay China, with its trillion-dollar bank account, to put up wind turbines and build hydro-dams. They will know a scam when they see one. Not that I expect any kind of a revolution. There is no real critique of the world development model. No new thinking – at least not of the kind Einstein would wish for:

> A human being is part of the whole called by us universe, a part limited in time and space. We experience ourselves, our thoughts and feelings as something separate from the rest. A kind of optical delusion of consciousness. This delusion is a kind of prison for us, restricting us to our personal desires and to affection for a few persons nearest to us. Our task must be to free ourselves from the prison by widening our circle of compassion to embrace all living creatures and the whole of nature in its beauty. The true value of a human being is determined by the measure and the sense in which they have obtained liberation from

the self. We shall require a substantially new manner of thinking if humanity is to survive.[3]

Perhaps a major global crash will bring this forth. That is the real sadness. This whole movement *pretends* its cause – for the planet and the dispossessed, the Bangladeshi faced with a flooded home, the drought-stricken farmers, the Inuit and the polar bear, whichever icon can turn the heart and the subscriptions – but is blind to the fundamental realities of nature and human consciousness. It is to this blindness that I struggle to address any kind of analysis. It does not seem enough to follow Upton Sinclair because we are not dealing with a simple economic interest and refusal to entertain a truth. We are faced with missionary zeal.

Those peoples who live closest to nature and understand that power which gave them birth have seen all this before. They remain now the marginalized, yet they are the ones who can teach us what it means to be at one with the cycles of change. It involves a surrender of illusory power of the kind Einstein understood. Such a small step for most indigenous peoples, but somewhat of a giant leap for the rest of mankind.

Notes

1. The global financial crisis was itself built upon a wide-ranging deception in the selling of derivatives which did not have the value ascribed to them. For example, in Britain, many local authorities placed large sums on deposit in Icelandic banks offering high rates of return but without the assets to support them, and many have lost tens of millions leading to severe cuts in public services. The latest revelation concerns the Wall Street banker Bernard Madoff, who invented a pyramid selling technique that sucked in $50 billion dollars in a single scam (*Independent*, 17 December 2008).

2. Of course, the 'amputation' is cultural and the result of a long process of political repression of the feminine aspects of mind. Women practitioners of traditional healing, herbal lore and divination were branded as evil, then ritually executed on a grand scale. Few history books comment upon the devastating effect on woman's confidence in her unique abilities, or on the parallel evolution of scientific academies and societies emerging from their own refuges from such religious persecution in the realms of alchemy and Masonic magic during the seventeenth century.

3. Collected Quotes from Albert Einstein, Kevin Harris 1995.
   http://rescomp.stanford.edu/~cheshire/EinsteinQuotes.html

Glossary

Aerosol   A suspension of fine particles or droplets in a gaseous medium, e.g. of sulphate particles

Albedo   A measure of the reflectance of light from the surface of the planet

Anomaly   A deviation from the mean usually from one year to the next, in relation to a long-term average

Anthropogenic   Human generated, e.g. with regard to greenhouse gases

Anticyclone   Weather pattern associated with high pressure and downward flow of air at the centre of a spiral, generating clockwise winds in the northern hemisphere

AO   Arctic Oscillation – a 60–70-year cycle of sea-level pressure and surface temperature

CFC   Chlorofluorocarbons, man-made greenhouse gases containing chlorine and fluorine.

Conveyor   The global ocean circulation system from surface waters to depth in down-welling and, vice-versa, in up-welling zones

Cosmogenic   Generated by cosmic rays, e.g. isotopic changes to atmospheric gases caused by extra-solar (galactic) radiation

Cyclone   Weather pattern associated with low pressure and rising air at the centre of a spiral, generating anticlockwise winds

Decadal   Over a period of decades

Eemian   Previous interglacial period as represented by fossils in Europe

El Niño   See ENSO

ENSO   El Niño Southern Oscillation, a roughly four-year cycle of warm water in the eastern equatorial Pacific (La Niña, is the cold phase of the oscillation)

FoE   Friends of the Earth

GCM   General Circulation Model of the atmosphere or ocean or more recently 'coupled' ocean and atmospheric models in a computer simulation; also more specifically Global Climate Models with sea-ice and land-surface interactions

GCR   Galactic cosmic ray activity

GHG   Greenhouse gases, chiefly carbon dioxide, methane and nitrous oxide, as well as the halocarbons with chlorine, fluorine and bromine

GMO   Genetically modified organisms – usually referring to crop plants

GtC   Gigatonnes of carbon, giga = one thousand million

Gulf Stream   The surface current from the sub-tropical Atlantic to the North East Atlantic in Europe and north towards Scandinavia.

GW(e)   Gigawatts, giga = one thousand million, (e) is electrical output

Halocarbons   Hydrocarbon compounds containing chlorine, fluorine or bromine

Heliopause   The outer envelope of the solar system defined by the reaches of the magnetic field and the solar wind of electrons

HFC   Hydrofluorocarbons, a man-made greenhouse gas (usually refrigerants)

Holocene   A geological term for the current interglacial modern era (the last 10,000 years)

Insolation   The amount of light reaching the surface of the earth

IPCC   Intergovernmental Panel on Climate Change (set up by the UN)

IR   Infrared radiation wavelengths longer than the visible spectrum, the energy of which is also absorbed to produce heat. It is used by instruments to monitor clouds

Isotope   A variant of an element differing in mass (some are radioactive) but behaving the same chemically and allowing estimates of temperature when incorporated into ice or organic material. Some relate to cosmic ray activity

Jet stream   West-to-east flowing high-speed winds in the upper atmosphere that influence storm and cloud tracks at the surface

La Niña   See ENSO, of which it is the cold phase

LFO   Low-frequency oscillation

LIA   Little Ice Age, a period roughly from AD 1400 to 1800, with low temperatures, particularly in the northern hemisphere, but also recognized in the tropics, some parts of South America and New Zealand

LW   Long-wave radiation (infrared) from the heat of the planet's surface and clouds, associated with absorption by carbon dioxide and water vapour

Mean   A form of averaging used for temperature comparisons, particularly from year to year or month to month

Meridional   Longitudinal flow pattern in atmospheric or oceanic circulation. For example, in meridional 'overturning' the North Atlantic surface waters flow north and overturn to deep water off Iceland and then flow south

MtC   Megatonnes of carbon, mega = 1 million

MWP   Medieval Warm Period, approximately from AD 900 to 1300 and recognized throughout the northern hemisphere and parts of the southern

MW(e)   One million watts (electrical), with the suffix used to denote the power output of generating stations or turbines in electrical terms (after heat losses). Note: with wind turbines the capacity quoted is nominal and considerably reduced in terms of the amount of wind (typically of the order of 25% of rated capacity)

NAO   North Atlantic Oscillation – 60–70-year cycle

NAS   National Academy of Science in the USA

NASA   National Aeronautic and Space Administration in the USA

NGO   Non-governmental organization, generally referring to those bodies with access to inter-governmental meetings and covering a wide range of environmental and industrial lobby groups

NOAA   National Oceanic and Atmospheric Administration, the main research and monitoring body in the USA

Palaeoclimatology (also Paleo-)   The study of past climates, usually via sediments, tree-rings, stalagmites, fossil remains, isotopic ratios, etc.

Palaeoecology (also Paleo-)   The inference of past habitats from fossil remnants

PDO   Pacific Decadal Oscillation – a 30-year cycle

RF   Radiative forcing

RSPB   Royal Society for the Protection of Birds

SAT   Surface air temperature

Stratosphere   The upper reaches of the atmosphere out to 50 km

SW   Short-wave radiation in the visible light wavelength (the main warming energy from the sun)

Sunspot number   The number of dark spots on the sun and an indication of solar magnetic activity and the strength of the solar wind (plasma)

TAR   IPCC's 3rd Assessment Report

Teleconnections   Term used for the interaction between different ocean basins, for example in the transfer of heat, pressure systems and wind patterns

THC   Thermo-haline circulation – in the Atlantic especially, often referred to as the 'conveyor' system, a global circuit of up-welling and down-welling connected currents

TOA   Top of atmosphere

Troposphere   The lower reaches of the atmosphere (8 km thick at the poles and 16 km at the equator) and often used to measure global temperature change, e.g. at about 3000 metres

TSI   Total solar irradiance

UV   Ultraviolet radiation – wavelength approximately between the violet end of the visible spectrum and X-ray region of the spectrum

WTO   World Trade Organization

References

Akasofu, S. (2009) 'Recovery from the Little Ice Age', International Arctic Research Center, Fairbanks, Alaska.
http://People.iarc.uaf.edu/~sakasofu/pdf/recovery_little_ice_age.pdf.

Allen, M. et al. (2006), 'Observational constraints on climate sensitivity', in *Avoiding Dangerous Climate Change*, ed. Schellnhuber, Cambridge.

Archibald, D. (2006), 'Solar cycles 24 and 25 and predicted climate response', *Energy and Environment*, **17**, 1.

Arnold, N. and Neubert, T. (2002), 'The electric Earth: Cosmic influences on the atmosphere', *Astron. and Geophys.*, **43**, 6.9–6.12.

Bacon, S. (1998), 'Decadal variability in the outflow from the Nordic seas to the deep Atlantic Ocean', *Nature*, **394**, 871–984, doi:10.1038/29736.

Badalyan, O.G., Obridko, V.N. and Sykora, J. (2001), 'Brightness of the coronal green line and prediction for activity cycles 23 and 24', *Solar Physics*, **199**, 421.

Barnett, T.P., Pierce, D.W., Saravanan, R., Schneider, N., Dommenget, D. and Latif, M. (1999), 'Origins of the midlatitude Pacific decadal variability', *Geophys. Res. Lett.*, **26**, 1453–1456.

Barnett, T. et al. (2005), 'Penetration of human-induced warming into the world's oceans', *Science*, **309**, 284–287.

Barratt, J. (2005), 'Greenhouse molecules, their spectra and function in the atmosphere', *Energy and Environment*, **16**, 6, 1040–1046.

Beck, E.G. (2007), '180 years of atmospheric CO_2 gas analysis by chemical methods', *Energy and Environment*, **18**, 2.

Berger, A. and Loutre, M.F. (2000), 'CO_2 and astronomical forcing of the late Quaternary', in *European Space Agency Proc. 1st Solar and Space Weather Euroconference*.

Bianchi, G. and McCave, N. (1999), 'Holocene periodicity in North Atlantic climate and deep-ocean flow south of Iceland', *Nature*, **397**, February 1999.

Biondi, F.A., Gershunov, A. and Cayan, D. (2001), 'North Pacific Decadal Climate Variability Since AD 1661', *Journal of Climate*, **14** (1), 5–10.

Boberg, F., and Lundstedt, H. (2002), 'Solar wind variations related to fluctuations of the North Atlantic Oscillation', *Geophys. Res. Lett.*, **29** (15), doi:10.1029/2002GL014903.

Bochnicek, J., Bucha, V., Heijda, P. and Pycha, J. (1996), 'Relation between the northern hemisphere winter temperatures and geomagnetic or solar activity at different QBO phases', *J. Atmos. Terr. Phys.*, **58**, 883–897.

Bochnicek, J., Bucha, V., Heijda, P. and Pycha, J. (1999), 'Possible geomagnetic activity effects on weather', *Ann. Geophysicae*, **17**, 925–932.

Boehmer-Christiansen, S. (1990), 'Hidden persuaders v policy', Symposium on Energy, Environment and Climate (EEC 90), Stuttgart, Germany, 14–16 October 1990, *Energy Policy*, **19** (7), September 1991, pp. 695–696.

Boehmer-Christiansen, S. (1994), 'Global climate protection policy: the limits of scientific advice', Part 1, *Global Environmental Change*, 1994.

Bond, G. et al. (1997), 'A pervasive millennial-scale cycle in North Atlantic Holocene and glacial climates', *Science*, **278** (1257), 14 November 1997.

Bond, G. et al. (2001), 'Persistent solar influences on North Atlantic climate during the Holocene', *Science*, **294** (5549), 2130–2136.

Broecker, W. (1997), 'Thermohaline circulation, the Achilles' heel of our climate system: Will man-made CO_2 upset the current balance?', *Science*, **278**, 1582–1588.

Bryden, H.L. et al. (2005), 'Slowing of the Atlantic meridional overturning circulation at 25°N', *Nature*, **438**, 655–657.

Cai, M. (2005), 'Dynamical amplification of polar warming', *Geophys. Res. Lett.*, **32**, L22710, doi:10.1029/2005GL024481.

Calder, N. (1997), *The Manic Sun*, Pilkington, London.

Camp, C.D. and Tung, K.K. (2007), 'Surface warming by the solar cycle as revealed by the composite mean difference projection', *Geophys. Res. Lett.*, **34**, L14703, doi:10.1029/2007GL030207.

Camp, C.D. and Tung, K.K. (2007a), 'The influence of the solar cycle and QBO on the late winter stratospheric polar vortex', *J. Atmos. Sci.*, **64**, 1267–1283.

Camp, C.D. and Tung, K.K. (2007b), 'Stratospheric polar warming by ENSO in winter: A statistical study', *Geophys. Res. Lett.*, **34**, L04809, doi:10.1029/2006GL028521.

Clilverd, M., Clark, E., Ulich, T., Linthe, J. and Rishbeth, H. (2004), 'Reconstructing the long-term, aa index and forecasting solar activity', presented at COSPAR scientific assembly, Paris.

Compo, G. and Sardeshmukh, P.D. (2008), *Oceanic Influences on Recent Continental Warming*, Climate Diagnostics Center, Cooperative Institute for Research in Environmental Sciences, University of Colorado, and Physical Sciences Division, Earth System Research Laboratory, National Oceanic and Atmospheric Administration, Boulder, CO.

Countryside Agency (2002), *Renewable Energy in the Landscape of 2050*, Cheltenham.

Countryside Commission (1995), *Climate Change, Acidification and Ozone: Potential Impacts on the English Countryside*, CCP 458.

deMenocal, P. et al. (2000), 'Coherent high- and low-latitude climate variability during the Holocene warm period', *Science*, **288**, 2198–2202.

Dethlefsen, V., Jackson, T. and Taylor, P. (1993), 'The Precautionary Principle', in *Clean Production Strategies*, ed. Jackson, Stockholm Environment Institute.

Dickson, B. et al. (2002), 'Rapid freshening of the deep North Atlantic Ocean over the past four decades', *Nature*, **416**, 25 April 2002.

Dima, M., and Lohmann, G. (2007), 'A hemispheric mechanism for the Atlantic Multidecadal Oscillation', *Journal of Climate*, **20** (11).

Dima, M. and Lohmann, G. (2008), 'Conceptual model for millennial climate variability: a possible combined solar-thermohaline circulation origin for the ~1,500-year cycle', *Climate Dynamics*, **32** (2–3), 301–311.

Domingues, C.M. et al. (2008), 'Improved estimates of upper-ocean warming and multi-decadal sea-level rise changes', *Nature*, **453**, 19 June 2008, doi:10.1038/nature07080.

Douglas, D.H. and Christy, J. (in press), 'Limits on CO_2 Climate Forcing from Recent Temperature Data of Earth', *Energy and Environment*.

Firestone, R., West, A. and Warwick-Smith, S. (2006), *The Cycle of Cosmic Catastrophes*, Bear & Co., Rochester, Vermont.

Fleitmann, D. et al. (2003), 'Holocene forcing of the Indian monsoon recorded in a stalagmite from southern Oman', *Science*, **300** (5626), 1737–1739, doi:10.1126/science.1083130.

Foukal, P., Fröhlich, C., Spruit, H. and Wigley, T.M.L. (2006), 'Variations in solar luminosity and their effect on the earth's climate', *Nature*, 14 September 2006.

Friis-Christensen, E. and Lassen, K. (1991), 'The length of the solar cycle: an indicator of solar activity closely associated with climate', *Science*, **254** (5032), 698.

Fröhlich, C. (2006), 'Solar irradiance variability since 1978', *Space Science Reviews*, **125** (1–4), 53–65, doi:10.1007/s11214-006-9046-5.

Fröhlich, C. and Lean, J. (1998), 'The sun's total irradiance. Cycles, trends, and related climate change uncertainties since 1976', *Geophys. Res. Lett.*, **25**, 4377–4380.

Frölich, C. and Lean, J. (2004), 'Solar radiative output and its variability. Evidence and mechanisms', *Astronomy and Astrophysics Review*, **12**, 273–320.

Gallagher, E. (2008), *Review of the Indirect Effects of Biofuels*, Renewable Fuels Agency.

Gershunov, A. and Barnett, T.P. (1998), 'Interdecadal modulation of ENSO teleconnections', *Bull. Amer. Meteor. Soc.*, **79**, 2715–2725.

Gouretski, V.V. and Koltermann, K.P. (2007), 'How much is the ocean really warming?', *Geophysical Research Letters*, **34** L01610, doi:10.1029/2006 GL027834.

Gray, L.J., Haigh, J.D. and Harrison, R.G. (2005), *Review of the Influences of Solar Changes on the Earth's Climate*, Hadley Centre Technical Note No. 62, Met Office, Exeter, 82 pp.

Gray, S.T., Graumlich, L.J., Betancourt, J.L. and Pederson, G.T. (2004), 'A tree-ring based reconstruction of the Atlantic Multidecadal Oscillation since 1567 AD', *Geophys. Res. Lett.*, 31, L12205, doi:10.1029/2004GL019932.

Hanna, E. and Cappelen, J. (2003), 'Recent cooling in coastal southern Greenland and relation with the North Atlantic Oscillation', *Geophys. Res. Lett.*, 30, 1132, doi:10.1029/2002GL015797.

Hanna, E., Huybrechts, P., Janssens, I., Cappelen, J., Steffen, K. and Stephens, A. (2005), 'Runoff and mass balance of the Greenland ice sheet: 1958–2003', *J. Geophys. Res.*, 110, D13108, doi:10.1029/2004JD005641.

Hanna, E., McConnell, J., Das, S., Cappelen, J. and Stephens, A. (2006), 'Observed and modeled Greenland ice sheet snow accumulation, 1958–2003, and links with regional climate forcing', *J. Clim.*, 19, 344–358.

Hansen, B., Turrell, W. and Osterhus, S. (2001), 'Decreasing overflow from the Nordic seas into the Atlantic Ocean through the Faeroe Bank Channel since 1950', *Nature*, 411, 21 June 2001.

Hansen, J. et al. (2005), 'Earth's energy imbalance: Confirmation and implications', *Science*, 308, 1431–1435, doi:10.1126/science.1110252.

Hansen, J. et al. (2005), 'Efficacy of climate forcings', *J. Geophys. Res.*, 110, D18104, doi:10.1029/2005JD005776.

Harrison, R.G. (2002a), 'Radiolytic particle production in the atmosphere', *Atmos. Environ.*, 36, 169–160.

Harrison, R.G. (2002b), 'Twentieth century secular decrease in the atmospheric electric circuit', *Geophys. Res. Lett.*, 29, 1600–1603.

Harrison, R.G. (2004), 'The global atmospheric electrical circuit and climate', *Surv. Geophys.*, 25, 441–484.

Harrison, R.G. and Stephenson, D.B. (2006), 'Empirical evidence for a nonlinear effect of galactic cosmic rays on clouds', *Proc. Roy. Soc. London*, Ser. A, 462, 1221–1233.

Hátún, H. et al. (2005), 'Influence of the Atlantic subpolar gyre on the thermohaline circulation', *Science*, 309, 1841–1844.

Hegerl, G.C., Crowley, T.J., Hyde, W.T. and Frame, D.J. (2006), 'Climate sensitivity constrained by temperature reconstructions over the past seven centuries', *Nature*, 440, 1029–1032.

Hood, L.L. (2003), 'Thermal response of the tropical tropopause region to solar ultraviolet variations', *Geophys. Res. Lett.*, 30, 2215, doi:10.1029/2003GL018364.

Hoyt, D.V. and Schatten, K.H. (1997), *The Role of the Sun in Climate Change*, Oxford University Press.

IPCC (2007) *Climate Change 2007: The Physical Science Basis*, contribution of Working Group I to the Fourth Assessment Report of the Intergovernmental Panel on Climate Change, and the Synthesis Report, *Summary for Policymakers* (downloadable: *www.ipcc.ch*).

Jackson, T. and Taylor, P. (1992), 'The precautionary principle and the prevention of marine pollution', *Chemistry and Ecology*, 7 (1-4), 123-134.

Jacobowitz, H. et al. (2003), 'The Advanced Very High Resolution Radiometer Pathfinder Atmosphere (PATMOS) climate dataset: A resource for climate research', *Bull. Am. Meteorol. Soc.*, 84, 785-793.

Joyce, T. and Keigwin, L., 'Are We on the Brink of a New Little Ice Age?' – testimony to the US Commission on Ocean Policy, 25 September 2002, Woods Hole Oceanographic Institution, Mass., USA.

Karcher, M. et al. (2005), 'Arctic Ocean change heralds North Atlantic freshening', *Geophys. Res. Lett.*, 32, doi:1029/2005GL023861.

Keeling, C.D. and Whorf, T.P. (2000), 'The 1,800-year oceanic tidal cycle: A possible cause of rapid climate change', *Proc. Natl. Acad. Sci.*, 21 March 2000, 97 (8), 3814-3819, doi:10.1073/pnas.070047197.

Keenlyside, N.S., Latif, M., Jungklaus, J., Kornblueh, L. and Roeckner, E. (2008), 'Advanced decadal-scale climate prediction in the North Atlantic sector', *Nature*, 453 (May), doi:10.1038.

Kernthaler, S.C., Toumi, R. and Haigh, J.D. (1999), 'Some doubts concerning a link between cosmic ray fluxes and global cloudiness', *Geophys. Res. Lett.*, 26 (7), 863-866, doi:10.1029/1999GL900121.

Kerr, R. (2005), 'The Atlantic conveyor may have slowed, but don't panic yet', *Science*, 310, 1403-1404.

Kininmonth, W. (2004), *Climate Change: a natural hazard*, Multi-Science Publishing, Brentwood.

Knight, J.R. et al. (2005), 'A signature of persistent natural thermohaline circulation cycles in observed climate', *Geophys. Res. Lett.*, 32, doi:10.1029/2005GL024233.

Koutsoyiannis, D. Efstratiadis, A., Mamassis, N. and Christofides, A. (2008), 'On the credibility of climate predictions', *Hydrological Sciences/Journal des Sciences Hydrologiques*, 53 (4), 671-684.

Koutsoyiannis, D., Mamassis, N., Christofides, A., Efstratiadis, A. and Papalexiou, S.M. (2008a), 'Assessment of the reliability of climate predictions based on comparisons with historical time series', EGU General Assembly 2008, *Geophys. Res. Abstracts*, vol. 10, Vienna, 09074, European Geosciences Union (*www.itia.ntua.gr/en/docinfo/850/*).

Kristjánsson, J.E., and Kristiansen, J. (2000), 'Is there a cosmic ray signal in recent variations in global cloudiness and cloud radiative forcing?', *J. Geophys. Res.*, 105 (D9), 11851-11863.

Kristjánsson, J.E., Staple, A., Kristiansen, J. and Kaas, E. (2002), 'A new look at possible connections between solar activity, clouds and climate', *Geophys. Res. Lett.*, 29, doi:10.1029/2002GL015646.

Kristjánsson, J.E., et al. (2005), 'Response of the climate system to aerosol direct and indirect forcing: the role of cloud feedbacks', *J. Geophys. Res.*, **110**, D24206, doi:10.1029/2005JD006299.

Labitzke, K. (2001), 'The global signal of the 11-year sunspot cycle in the stratosphere. Differences between solar maxima and minima', *Meteorol. Zeitschrift*, **10**, 83–90.

Labitzke, K. (2004), 'On the signal of the 11-year sunspot cycle in the stratosphere and its modulation by the quasi-biennial oscillation', *J. Atmos. Solar Terr. Phys.*, **66**, 1151–1157.

Labitzke, K. et al. (2002), 'The global signal of the 11-year solar cycle in the stratosphere: Observations and models', *J. Atmos. Sol. Terr. Phys.*, **64**, 203–210.

Labohm, H., Rozendaal, S. and Thoenes, D. (2004), *Man-made Global Warming: unravelling a dogma*, Multi-Science Publishing, Brentwood.

Lamb, H.H. (1995), *Climate, History and the Modern World*, Methuen, London.

Landscheidt, T. (2003), 'New Little Ice Age instead of global warming?', *Energy and Environment*, **14** (2 & 3).

Latif, M. and Barnett, T.P. (1994), 'Causes of decadal climate variability over the North Pacific and North America', *Science*, **266**, 634–637.

Lean, J. (2000), 'Evolution of the sun's spectral irradiance since the Maunder Minimum', *Geophys. Res. Lett.*, **27**, 2425–2428.

Lean, J., and Rind, D. (1998), 'Climate forcing by changing solar radiation', *J. Climate*, **11**, 3069–3094.

Lean, J., Rottman, G., Harder, J. and Kopp, G. (2005), 'SORCE contribution to new understanding of global change and solar variability', *Solar Phys.*, **230**, 27–53.

Lefohn, A.S., Husar, J.D. and Husar, R.B. (1999), 'Estimating historical anthropogenic global sulfur emission patterns for the period 1850–1990', *Atmospheric Environment*, **33**, 3435–3444.

LeQuere, C. et al. (2007), 'Saturation of the southern ocean CO_2 sink due to recent climate change', *Science*, **316** (5832), 1735–1738.

Levitus, S.J., Antonov, I. and Boyer, T.P. (2005), 'Warming of the world ocean, 1955–2003', *Geophys. Res. Lett.*, **32**, L02604, doi:10.1029/2004GL021592.

Lindzen, R.S. (1991), 'Some uncertainties with respect to water vapor's role in climate sensitivity', *Proceedings of NASA Workshop on the Role of Water Vapor in Climate Processes*, October 29–November 1, 1990 in Easton, Maryland (D.O'C. Starr and H. Melfi, eds).

Lindzen, R.S., Hou, A.Y. and Farrell, B.F. (1982), 'The role of convective model choice in calculating the climate impact of doubling CO_2', *J. Atmos. Sci.*, **39**, 1189–1205.

Lindzen, R.S., Chou, M.-D. and Hou, A.Y. (2001), 'Does the earth have an adaptive infrared iris?', *Bull. Amer. Met. Soc.*, **82**, 417–432.

Lockwood, M., Stamper, R. and Wild, M.N. (1999), 'A doubling of the sun's coronal magnetic field during the past 100 years', *Nature*, **399**, 437–439.

Lockwood, M. and Foster, S. (2000), 'Long term variations in the magnetic fields of the sun and possible implications for terrestrial climate', *Proc. 1st Solar and Space Weather Euroconference*, 25–29 September 2000, Tenerif, Spain, European Space Agency, SP, **463**, 85–94.

Lockwood, M. and Fröhlich, C. (2007), 'Recent oppositely directed trends in solar climate forcings and the global mean surface temperature', *Proc. R. Soc. A.* doi:10.1098/rspa.2007.1880.

Lohmann, G. et al. (2004), 'Climate signature of solar irradiance variations: analysis of long-term instrumental, historical and proxy data', *Intl J. Climatology*, **24**, 1024–1056.

Lundstedt, H., Liszka, L. and Lundin, R. (2005), 'Solar activity explored with new wavelet methods', presented at EGU meeting in Nice 2004, in press *Annales Geophysicae*.

Lyman, J.M., Willis, J.K. and Johnson, G.C. (2006), 'Recent cooling of the upper ocean', *Geophys. Res. Lett.*, **33**, L18604, doi:10.1029/2006GL027033.

Macklin, M.G., Johnstone, E. and Lewin, J. (2005), 'Pervasive and long-term forcing of Holocene river instability and flooding in Great Britain by centennial-scale climate change', *The Holocene*, **15** (7), 937–943.

Mann, M.E. and Bradley, R. (1995), 'Global interdecadal and century scale climate oscillations during the past five centuries', *Nature*, **378**, 266–270.

Mann, M.E. and Park, J. (1996), 'Joint spatiotemporal modes of surface temperature and sea level pressure variability in the northern hemisphere during the last century', *J. Climate*, **9**, 2137–2162.

Mann, M.E., Bradley, R.S. and Hughes, M.K. (1999), 'Northern hemisphere temperatures during the past millennium: Inferences, uncertainties and limitations', *Geophys. Res. Lett.*, **26**, 759–762.

Mann, M.E. and Jones, P.D. (2003), 'Global surface temperatures over the past two millennia', *Geophys. Res. Lett.*, **30** (15), 1820, doi:10.1029/2003GL017814.

Maris, G., Popescu, M.D. and Besliu, D. (2004), 'Multi-wavelength investigations of solar activity', *IAU Symposium*, No. 233, Cambridge, Cambridge Press, pp. 127–128.

Markson, R. (1981), 'Modulation of the earth's electric field by cosmic radiation', *Nature*, **291**, 304–308.

Marsh, N. and Svensmark, H. (2000), 'Cosmic rays, clouds and climate', *Space Science Reviews*, **94**, 215–230.

Marsh, N. and Svensmark, H. (2003), 'Galactic cosmic ray and El Niño – Southern Oscillation trends in International Satellite Cloud Climatology Project D2 low-cloud properties', *J. Geophys. Res.*, **108**, 4195–4203.

Mason, C. (2003), *The 2030 Spike: the countdown to global catastrophe*, Earthscan, London & New York.

McCormack, J.P. (2003), 'The influence of the 11-year solar cycle on the quasi-biennial oscillation', *Geophys. Res. Lett.*, **30**, (22), 2162, doi:10.1029/2003GL018314.

McIntyre, S. and McKitrick, R. (2003), 'Corrections to the Mann et. al. (1998) proxy data base and northern hemisphere average temperature series', *Energy and Environment*, **14**, 3.

McIntyre, S. and McKitrick, R. (2005), 'The M&M critique of the MBH98 northern hemisphere climate index: update and implications', *Energy and Environment*, **16**, 1.

McIntyre, S. and McKitrick, R. (2005), 'Hockey sticks, principal components and spurious significance', *Geophys. Res. Lett.*, **32**, L03710, doi:10.1029/2004GL021750, 2005.

Meehl, G.A., Washington, W.M., Ammann, C.M., Arblaster, J.M., Wigley, T.M.L., Tebaldi, C. (2004), 'Combinations of natural and anthropogenic forcings in twentieth-century climate', *J. Clim.*, **17**, 3721–3727.

Meinshausen, M. (2006), 'What does a 2°C target mean for greenhouse gas concentrations? A brief analysis based on multi-gas emission pathways and several climate sensitivity uncertainty measurements', in *Avoiding Dangerous Climate Change*, ed. Schellnhuber, Cambridge.

Metz, B. and van Vuuren, D. (2006), 'How, and at what costs, can low level stabilization be achieved? An overview', in *Avoiding Dangerous Climate Change*, ed. Schellnhuber, Cambridge.

Minobe, S. (1997), 'A 50–70 year climatic oscillation over the North Pacific and North America', *Geophys. Res. Lett.*, **24**, 683–686.

Miskolczi, F. (2007), 'Greenhouse effect in semi-transparent planetary atmospheres', *Quarterly Journal of the Hungarian Meteorological Service*, **111**, (1), January–March 2007, 1–40.

Mitchel, J. (1861), *The Last Conquest of Ireland*, University College Dublin Press reprint, 2005, paperback.

Mitchell, J.F.B., Senior, C.A. and Ingram, W.J. (1989), 'CO_2 and climate: a missing feedback?', *Nature*, **341**, 132–134.

Moberg, A. et. al. (2004), 'Highly variable northern hemisphere temperatures reconstructed from low- and high-resolution proxy data', *Nature*, **433**, 613–617 (10 February 2005), doi:10.1038/nature03265.

Moberg, A. et al. (2005), 'Highly variable northern hemisphere temperatures reconstructed from low- and high-resolution proxy data', *Nature*, **433**, 613–617.

Monckton, C. (2008), 'Climate sensitivity reconsidered', *Forum for Physics and Society*, American Physical Society, July 2008.

Moss, R.H. (2005), 'The IPCC: policy relevant (not driven) scientific assessment: A comment on Sonja Boehmer-Christiansen's: "Global climate protection policy: the limits of scientific advice" ', *Global Environmental Change*, **5**, (3), 171–174.

Moy, C. et al. (2002), 'Variability of El Niño/Southern Oscillation activity at millennial time scales during the Holocene epoch', *Nature*, **420**, 162–164.

Muscheler, R. et al. (2005), 'What do our results mean for climate?', *Quat. Sci. Rev.*, **24**, 1849–1860.

Muscheler, R., Joos, F., Müller, S.A. and Snowball, I. (2005), 'Climate – How unusual is today's solar activity?', *Nature*, **436** (7050), E3–E4.

Muscheler, R., and Beer, J. (2006), 'Solar forced Dansgaard/Oeschger events?', *Geophys. Res. Lett.*, **33**, L20706, doi:10.1029/2006GL026779.

Muscheler, R. et al. (2007), 'Solar activity during the last 1000 yr inferred from radionuclide records', *Quat. Sci. Rev.*, **26**, 82–97), doi:10.1016/j.quascirev.2006.07.012.

National Academy of Sciences (2001), *Climate Change Science: an analysis of some key questions*, National Academy Press, USA.

National Academy of Sciences (2002), *Abrupt Climate Change: Inevitable Surprises*, National Research Council Committee on Abrupt Climate Change, National Academy Press, USA.

New Economics Foundation (2008), *Green New Deal: joined up policies to solve the triple crunch of credit crisis, climate change and high oil prices*, NEF, London.

Norris, J.R. (2000), 'What can cloud observations tell us about climate variability?', *Space Sci. Rev.*, **94**, 375–380.

Norris, J.R. (2005a), 'Multidecadal changes in near-global cloud cover and estimated cloud cover radiative forcing', *J. Geophys. Res.*, **110**, D08206, doi:10.1029/2004JD005600.

Norris, J.R. (2005b), 'Trends in upper-level cloud cover and atmospheric circulation over the Indo-Pacific region between 1952 and 1997', *J. Geophys. Res.*, **110**, D21110, doi:10.1029/2005JD006183.

Osborne, P.J. (1980), 'The late Devensian-Flandrian transition depicted by serial insect faunas from West Bromwich, Staffordshire, England', *Boreas*, **9**, 139–147.

Overland, J.E., Spillane, M.C., Percival, D.B., Wang, M. and Mofjeld, H.O. (2004), 'Seasonal and regional variation of pan-Arctic surface air temperature over the instrumental record', *J. Climate*, **17**, 3263–3282.

Overland, J.E. and Wang, M. (2005), 'The third Arctic climate pattern: 1930s and early 2000s', *Geophys. Res. Lett.*, **32**, L23808, doi:10.1029/2005GL024254.

Overland, J.E., Wang, M. and Salo, S. (2008a), 'The recent Arctic Warm Period', Tellus 60A, 589–597 (2008).

Overland, J.E., Walsh, J. and Wang, M. (2008b), Arctic Report Card 2008 – Atmosphere, NOAA, USA.

Pallé Bago, E. and Butler, C.J. (2000), 'The influence of cosmic rays on terrestrial clouds and global warming', *Astron. Geophysics*, **41**, 4.18–4.22.

Pallé Bago, E. and Butler, C.J. (2000), 'Sunshine, clouds and cosmic rays', *Proc. 1st*

Solar and Space Weather Euroconference, Tenerife, September 2000 ESA, SP-463, 147–152.

Pallé Bago, E. and Butler, C.J. (2001), 'Sunshine records from Ireland. Cloud factors and possible links to solar activity and cosmic rays', *International Journal of Climatology*, **21**, 709–729.

Pallé, E. et al. (2003), 'Earthshine and the earth's albedo: 2. Observations and simulations over 3 years', *J. Geophys. Res.*, **108** (D22), 4710, doi:10.1029/2003JD003611.

Pallé, E., Goode, P.R., Montañés-Rodriguez, P and Koonin, S.E. (2004), 'Changes in the earth's reflectance over the past two decades', *Science*, **304**, 1299–1301, doi:10.1126/science.1094070.

Pallé, E. et al. (2005), 'A multi-data comparison of shortwave climate forcing changes', *Geophys. Res. Lett.*, **32**, L21702, doi:10.1029/2005GL023847.

Pallé, E. (2006), 'Reply to comment by F.A-M. Bender on "A multi-data comparison of short-wave climate forcing changes"', *Geophys. Res. Lett.*, **33**, L15813, doi:10.1029/2006GL026101.

Pallé, E., Goode, P.R. and Montanes-Rodriguez, P. (2009), 'Inter-annual trends in earth's reflectance 1999–2007', *J. Geophys. Res.*, doi:10.1029/2008JDO10734 (in press).

Palmer, M.D., Haines, K., Tett, S.F.B. and Ansell, T.J. (2007), 'Isolating the signal of ocean global warming', *Geophys. Res. Lett.*, **34**, L23610, doi:10.1029/2007GL031712.

Perry, C.A. (2005), 'Midwestern streamflow, precipitation, and atmospheric vorticity influenced by Pacific sea-surface temperatures and total solar-irradiance variations', *Int. J. Clim.*, **26** (2), 207–218.

Perry, C.A. (2007), 'Evidence for a physical linkage between galactic cosmic rays and regional climate time series', *Advances in Space Research*, **40** (2007), 353–364.

Perry, C.A. and Hsu K.J. (2000), 'Geophysical, archaeological, and historical evidence support a solar-output model for climate change', *Proc. US National Academy of Sciences*, 10.1073/pnas.236423297.

Peterson, B.J. et al. (2002), 'Increasing river discharge to the Arctic Ocean', *Science*, **298**, 13 December 2002.

Pinker, R.T., Zhang, B. and Dutton, E.G. (2005), 'Do satellites detect trends in surface solar radiation?', *Science*, **308**, 850–854.

Pittock, B. (2005), *Climate Change*, Earthscan, London.

Polyakov, I. et al. (2002), 'Observationally based assessment of amplification of global warming', International Arctic Research Center, University of Alaska, Fairbanks, Alaska, *Geophys. Res. Lett.*, **29** (18), 1878, doi:10.1029/2001GL01111.

Polyakov, I. et al. (2003), 'Variability and trends of air temperature and pressure in the maritime Arctic, 1875–2000', *J. Climate*, **16** (12), 2067–2077.

Pritchard, H.D. and Vaughan, D.G. (2007), 'Widespread acceleration of tidewater glaciers on the Antarctic Peninsula', *J. Geophys. Res.*, **112**, F03S29, 1–10, doi:10.1029/2006JF000597.

Rahmstof, S. et al. (2007), 'Sea level rise', *Science*, **315** (5810), 368–370.

Richter-Menge, J. et al. (2006), *State of the Arctic Report*. NOAA OAR Special Report, NOAA/OAR/PMEL, Seattle, WA.

Roberts, N. (1998), *The Holocene: an environmental history*, Blackwell, Oxford.

Roesch, A. (2006), 'Evaluation of surface albedo and snow cover in AR4 coupled climate models', *J. Geophys. Res.*, **111**, D15111, doi:10.1029/2005JD006473.

Rosenzweig, C., Karoly, D., Vicarelli, M., Neofotis, P., Wu, Q., Casassa, G., Menzel, A., Root, T.L., Estrella, N., Seguin, B., Tryjanowski, P., Liu, C., Rawlins, S. and Imeson, A. (2008), 'Attributing physical and biological impacts to anthropogenic climate change', *Nature*, **453**, 353–357.

Rossow, W.B. and Dueñas, E.N. (2004), 'The International Satellite Cloud Climatology Project (ISCCP)', *Bull. Am. Meteorol. Soc.*, **85**, 167–172. *http://isccp.gis.nasa.gov*.

Rouillard, A.P., Lockwood, M. and Finch, I. (2007), 'Centennial changes in the solar wind speed and in the open solar flux', *J. Geophys. Res.*, **112** (5), A05103, doi:10.1029/2006JA012130, 2007.

Royal Commission on Environmental Pollution (2000), *Energy: the changing climate*, HMSO, London.

Royal Society (2005), *Ocean Acidification Due to Atmospheric CO_2*, RS, Policy Document 12/05, June 2005, *www.royalsoc.ac.uk*.

Salby, M. and Callaghan, P. (2004), 'Evidence of the solar cycle in the general circulation of the stratosphere', *J. Clim.*, **17**, 34–46.

Scafetta, N. and West, B. (2007), 'Phenomenological reconstructions of the solar signature in the northern hemisphere surface temperature records since 1600', *J. Geophys. Res.*, **112**, D24S03, doi:10.1029/2007JD008437,2007.

Schellnhuber, H.J. et al. (2006), *Avoiding Dangerous Climate Change*, Cambridge University Press.

Schweiger, A.J. (2004), 'Changes in seasonal cloud cover over the Arctic seas from satellite and surface observations', *Geophys. Res. Lett.*, **31**, L2207, doi:10.1029/2004GL020067,2004.

Shindell, D.T., Rind, D., Balachandran, N., Lean, J. and Lonergan, P. (1999), 'Solar cycle variability, ozone, and climate', *Science*, **284**, 305–308.

Shindell, D.T., et al. (2001), 'Solar forcing of regional climate change during the Maunder Minimum', *Science*, **294** (5549), 2149–2152.

Shukla, J., et al. (2006), 'Climate model fidelity and projections of climate change', *Geophys. Res. Lett.*, **33**, L07702, doi:10.1029/2005GL025579.

Soden, B.J. et al. (2005), 'The radiative signature of upper tropospheric moistening', *Science*, **310**, 841–844.

Solanki, S.K. (2002), 'Solar variability and climate change: is there a link?', *Astronomy and Geophysics*, **43**, 5.9–5.13.

Solanki, S. (2006), 'Solar variability of possible relevance for planetary climates', *Space Sci. Rev.*, 125 doi:10.1007/s11214-006-9044-7.

Solanki, S.K. and Krivova, N.A. (2003), 'Can solar variability explain global warming since 1970?', *J. Geophys. Res.*, **108**, 1200.

Solanki, S.K. et al. (2004), 'Unusual activity of the sun during recent decades compared to the previous 11,000 years', *Nature*, **431**, 1084–87.

Soon, W. (2006), 'Variable solar irradiance as a plausible agent for multidecadal variations in the Arctic-wide surface air temperature record of the past 130 years', *Geophys. Res. Lett.*, **32**. *http.//www.agu.org/pubs/crossref/2005/2005GL023429.shtml.*

Soon, W. and Baliunas, S. (2003), 'Climatic and environmental changes of the past 1000 years', *Climate Research*, **23**, 89–110.

Spencer, R.W. and Braswell, W.D. (2008), 'Potential biases in feedback diagnosis from observational data: a simple model demonstration', *J. Climate*, **21**, 5624–5628.

Stairs, K. and Taylor, P. (1992), 'Non-governmental organizations and the legal protection of the oceans', in *International Politics and the Environment*, ed. Hurrell & Kingsbury, Clarendon, Oxford.

Stamper, R., Lockwood, M.N.M. and Wild, M. (1999), 'A doubling of the sun's coronal magnetic field during the past 100 years', *Nature*, **399**, 437–438.

Stone, R.S. (1997), 'Variations in western Arctic temperatures in response to cloud-radiative and synoptic-scale influences', *J. Geophys. Res.*, **102** (D18), 21,769–21,776, 1997.

Stott, P.A. et al. (2006), 'Transient climate simulations with the HadGEMI climate model: Causes of past warming and future climate change', *J. Clim.*, **19**, 2763–2782.

Stroeve, J. et al. (2007), 'Arctic sea ice decline: faster than forecast?', *Geophys. Res. Abstracts*, **9**, 01362,2007

Svalgaard, L., Cliver, E.W. and Kamide, Y. (2005), 'Cycle 24: the smallest sunspot cycle in 100 years?', *Geophys. Res. Lett.*, **32**, doi:10.1029/2004GL021664.

Svensmark, H. (2007), 'Cosmoclimatology, a new theory emerges', *Astronomy and Geophysics*, **48**, 1.18–1.24.

Svensmark, H. and Friis-Christensen, E. (1997), 'Variation of cosmic ray flux and global cloud coverage – a missing link in solar climate relationships', *J. of Atmospheric and Solar-Terrestrial Physics*, **59**, (11), 1225–1232.

Svensmark, H., Marsh, N., Pepke Pederson, J.O., Enghoff, M. and Uggerhoj, U. (2006), 'Experimental evidence for the role of ions in particle nucleation under atmospheric conditions', *Proceedings of the Royal Society A*.

Svensmark, H. and Calder, N. (2007), *The Chilling Stars: a new theory of climate change*, Icon Books, Cambridge.

Taylor, P. (1977), 'Nuclear power in Central Europe', *The Ecologist*, 7, (6), 216–222.

Taylor, P. (1979), 'Nuclear Energy: how the odds are stacked against the opponents', *Nature*, 277, 594–595.

Taylor, P. (1980), *The Windscale Fire, October 1957*, PERG RR-7, Political Ecology Research Group, Oxford.

Taylor, P. (1982), *The Impact of Nuclear Waste Disposals to the Marine Environment*, PERG RR-8 Political Ecology Research Group, Oxford.

Taylor, P. (1985a), *The Disposal of Nuclear Waste to the Deep Ocean*, PERG RR-15, Political Ecology Research Group, Oxford.

Taylor, P. (1985b), 'Radionuclides in Cumbria: the international context', in *Pollution in Cumbria*, Institute of Terrestrial Ecology, Merlewood, HMSO.

Taylor, P. (1986), 'Risk decisions must involve the public', *Town and Country Planning*, 55, (3).

Taylor, P. (1987), 'The interpretation of monitoring results', in *Radiation and Health*, eds Southwood and Russell-Jones, Wiley, New York, pp. 19–45.

Taylor, P. (1988a), 'Environmental issues in nuclear risk assessment', *Nuclear Technology International*, 1988, pp. 219–223.

Taylor, P. (1988b), 'Land-use implications of radioactive contamination', *Land Use Policy*, 15 (1), 62–70.

Taylor, P. (1988c), 'Large consequence low probability accidents', in *Standing Conference on Health and Safety in the Nuclear Age*, CEC Radiation Protection Report, EUR 11608 EN.

Taylor, P. (1991), 'Environmental capacity and the limits of predictive science: the precautionary principle in the control of hazardous substances', in *Proc. Symp.: Joint International Symposium on Hazardous Waste Management*, Swedish Environmental Protection Agency and CEC, Stockholm.

Taylor, P. (1993), 'The state of the marine environment: A critique of the work and the role of the Joint Group of Experts on Scientific Aspects of Marine Pollution (GESAMP)', *Marine Pollution Bulletin*, 26, (3), 120–127.

Taylor, P. (1996), 'Climate change and air pollution: the implications for the work of the Countryside Commission', consultants report to Countryside Commission, Cheltenham.

Taylor, P. (2004), *Shiva's Rainbow*, Ethos, Oxford.

Taylor, P. (2005), *Beyond Conservation*, Earthscan, London.

Taylor, P. and Fraser, R. (2002), *Visualizing Renewable Energy in the Landscape of 2050*, project for the Countryside Agency, Cheltenham, available on Ethos website.

Taylor, P. and Jackson, T. (1991), 'The precautionary principle and the prevention of ocean pollution', paper to 1st International Ocean Pollution Symposium, Puerto Rico. Results also published in *Chemistry and Ecology* – see Jackson and Taylor, 1993.

Taylor, P., Jackson, T. and Dethlefsen, V. (1993), 'The Precautionary Principle', in *Clean Production Strategies*, ed. Jackson, Stockholm Environment Institute.

Theijl, P. and Lassen, K. (2000), 'Solar forcing of the northern hemisphere land air temperature', *Journal of Atmospheric Solar Terrestrial Physics*, 62, 1207–1213.

Tinsley, B.A. (2000), 'Influence of solar wind on the global electric circuit and inferred effects on cloud microphysics, temperature and dynamics in the troposphere', *Space Science Reviews*, 94, 231–258.

Tinsley, B.A., Rohrbaugh, R.P., Hei, M. and Beard, K.V. (2000), 'Effects of image charges on the scavenging of aerosol particles by cloud droplets, and on droplet charging and possible ice nucleation processes', *J. Atmos. Sci.*, 57, 2118–2134.

Tinsley, B.A. and Yu, F. (2004), *Atmospheric Ionization and Clouds as Links Between Solar Activity and Climate*, American Geophysical Union monograph, 141, 321–340.

Trenberth, K.E. and Hurrell, J.W. (1994), 'Decadal atmosphere – ocean variations in the Pacific', *Climate Dyn.*, 9, 303–319.

Udelhofen, P.M. and Cess, R.D. (2001), 'Cloud cover variations over the United States: An influence of cosmic rays or solar variability?', *Geophys. Res. Lett.*, 28 (13), 2617–2620.

Usoskin, I.G., Solanki, S.K., Schussler, M., Mursula, K. and Alanko, K. (2003), 'Millennium-scale sunspot number reconstruction: Evidence for an unusually active sun since the 1940s', *Phys. Rev. Lett.*, 91, doi:10.1103/PhysRevLett.91.211101.

Usoskin, I.G., Marsh, N., Kovaltsov, G.A., Mursula, K. and Gladysheva, O.G. (2004), 'Latitudinal dependence of low cloud amount on cosmic ray induced ionization', *Geophys. Res. Lett.*, 31, L16109, doi:10.1029/2004GL019507.

Usokin, I.G., Schuessler, M., Solanki, S.K. and Mursula, K. (2005), 'Solar activity, cosmic rays, and the earth's temperature: A millennium-scale comparison', *J. Geophys. Res.*, 110, A10102.

van Loon, H., Meehl, G.A. and Shea, D.J. (2007), 'Coupled air-sea response to solar forcing in the Pacific region during northern winter', *J. Geophys. Res.*, 112, D02108, doi:10.1029/2006JD007378.

Wagner, T. et al. (2005), 'Global trends (1996–2003) of the atmospheric H_2O column and cloud cover derived from GOME', *Geophys. Res. Abstracts*, 7, 08644, 2005 SRef-ID: 1607-7962/gra/EGU05-A-08644.

Wang, J., Rossow, W.B. and Zhang, Y.-C. (2000), 'Cloudsat vertical structure and its variations from a 20-year global rawinsonde data set', *J. Climate*, 13, 3041–3056.

Wang, J.-L., Gong, J.-C., Liu, S.-Q., Le, G.-M. and Sun, J.-L. (2002), 'The prediction of maximum amplitudes of solar cycles and the maximum amplitude of solar cycle 24', *Chin. J. Astron. Astrophys.*, 2, (6), 557–562.

Wang, Y. et al. (2005), 'The Holocene Asian monsoon: links to solar changes and North Atlantic climate', *Science*, 308, 854–856.

Wang, Y.M., Lean, J.L. and Sheeley, N.R. (2005), 'Modeling the sun's magnetic field and irradiance since 1713', *Astrophys. J.*, **625**, 522–538.

White, W.B., Lean, J., Cayan, D.R. and Dettinger, M.D. (1997), 'Response of global upper ocean temperature to changing solar irradiance', *J. Geophys. Res.*, **102**, 3255–3266.

White, W.B. et al. (2000), 'Global average upper ocean temperature response to changing solar irradiance: exciting the internal decadal mode', *Proc. 1st Solar and Space Weather Euroconference*, on the solar cycle and terrestrial climate, Tenerife, 23–30 September 2000, European Space Agency, SP-163.

White, W.B., Dettinger, M.D. and Cayan, D.R. (2003), 'Sources of global warming of the upper ocean on decadal period scales', *J. Geophys. Res.*, **108** (C8), doi:10.1029/2002JC001396.

Wielicki, B.A. et al. (2002), 'Evidence for large decadal variability in the tropical mean radiative energy budget', *Science*, **295**, 841.

Wielicki, B.A. et al. (2005), 'Changes in earth's albedo measured by satellite', *Science*, **308**, 825.

Wild, M. et al. (2005), 'From dimming to brightening: decadal changes in solar radiation at the earth's surface', *Science*, **308**, 847–850.

Willis, J.K. (2008), 'A new estimate of interannual variability in upper ocean heat content in light of recently discovered in situ data biases', *Geophysical Research Abstracts*, **10**, EGU2008-A-11464, 2008 SRef-ID: 1607-7962/gra/EGU2008-A-11464.

Willis, J.K., Roemmich, D. and Cornuelle B. (2004), 'Interannual variability in upper ocean heat content, temperature, and thermosteric expansion on global scales', *J. Geophys. Res.*, **109**, C12036, doi:10.1029/2003JC002260c.

Wong, T. et al. (2006), 'Reexamination of the observed decadal variability of earth radiation budget using altitude-corrected ERBE/ERBS nonscanner WFOV data', *J. Climate*, Vol. 19, 4028–4040.

WWF (2008), *Climate Change: faster, stronger, sooner. An overview of the climate science published since the UN IPCC 4th Report*, WWF, London.

Yalden, D. (1999), *The History of British Mammals*, Poyser, London.

Yu, F. (2002), 'Altitude variations of cosmic ray induced production of aerosols. Implications for global cloudiness and climate', *J. Geophys. Res.*, **107**, doi:10.1029/200IJ000248.

Zhang, D. and McPhaden, M.J. (2006), 'Decadal variability of the shallow Pacific meridional overturning circulation: Relation to tropical sea surface temperatures in observations and climate change models', *Ocean Modelling*, **15**, 250–273.

Zhang, Y.-C., Rossow, W.B., Lacis, A.A., Oinas, V. and Mishchenko, M.I. (2004), 'Calculation of radiative fluxes from the surface to top of atmosphere based on ISCCP and other global data sets: Refinements of the radiative transfer model and the input data', *J. Geophys. Res.*, **109**, 27.

Index